SURVEYING FOR ENGINEERS

OTHER TITLES FOR ENGINEERS

Civil Engineering Contract Administration and Control, 2nd Edition
I.H. Seeley

Civil Engineering Materials, 4th Edition
N. Jackson and R. Dhir

Civil Engineering Quantities, 5th Edition
I.H. Seeley

Fundamental Structural Analysis
W.J. Spencer

Practical Soil Mechanics
G. Barnes

Reinforced Concrete Design, 4th Edition
W.H. Mosley and J.H. Bungey

Structural Mechanics
J. Cain and R. Hulse

Understanding Hydraulics
L. Hamill

Understanding Structures
Derek Seward

Surveying for Engineers

J. UREN

Department of Civil Engineering
University of Leeds

W.F. PRICE

Department of Civil Engineering
University of Brighton

Third Edition

palgrave

First edition 1978
Reprinted five times
Second edition 1985
Reprinted nine times
Third edition 1994

Published by
PALGRAVE
Houndmills, Basingstoke, Hampshire RG21 6XS and
175 Fifth Avenue, New York, N. Y. 10010
Companies and representatives throughout the world

PALGRAVE is the new global academic imprint of
St. Martin's Press LLC Scholarly and Reference Division and
Palgrave Publishers Ltd (formerly Macmillan Press Ltd).

ISBN 0–333–57705–1

This book is printed on paper suitable for recycling and
made from fully managed and sustained forest sources.

A catalogue record for this book is available
from the British Library.

16 15 14 13 12 11 10
11 10 09 08 07 06 05 04 03

Printed and bound in Great Britain by
Antony Rowe Ltd, Chippenham and Eastbourne

Contents

Preface to the Third Edition

Since the publication of the second edition of *Surveying for Engineers*, much of the instrumentation now used by surveyors and engineers on site is electronic, especially theodolites, EDMs and total stations. Through the use of computers, electronic data acquisition and processing is now well established and there is a much greater emphasis on quality in surveying with the introduction of BS5750 and other similar standards.

When writing this third edition, we have attempted to reflect the changes occurring in current practice by introducing new chapters and revising others. This, together with the basic techniques of surveying, provide a book with an easy-to-read format that covers the equipment and methods essential for modern site surveying.

New chapters have been introduced dealing with measurements and errors, the global positioning system and deformation monitoring, all of which have gained prominence in engineering surveying in recent years.

Amongst other changes to the third edition are an extended introductory chapter, the chapters dealing with levelling and theodolites have been revised and the original chapter on distance measurement has been made into two separate chapters on taping and EDM/total stations. Methods of providing control on site are now dealt with in one chapter and the setting out chapter has been extended to include quality assurance.

Throughout the text, computerised surveying and associated software are discussed in some detail, especially those used for mapping and for applications in horizontal and vertical curves.

Although the book has been written with civil engineering students in mind, it is hoped that it will also be found useful by practising engineers as well as by any other students who undertake engineering surveying as a subsidiary subject.

The text covers engineering surveying up to the end of virtually all university first-year and most second-year degree courses in civil and environmental engineering, building, construction, engineering geology and other related disciplines and BTEC courses from level III at colleges of technology

J. UREN
W.F. PRICE

Acknowledgements

The authors wish to thank all those who have contributed in any way to the preparation of this book and, in particular, the following persons and organisations:

Sue Gardner, Department of Civil Engineering, University of Brighton, for providing Figures 1.1 and 1.2; the Ordnance Survey and, in particular, Mr Adrian Spence, for permission to publish figures 1.3, 1.13, 1.16 and 1.17; the Royal Institution of Chartered Surveyors for providing material on which sections 1.1, 1.2 and 1.3 are based and for permission to publish figure 9.1; Mr Stephen Booth and the Institution of Civil Engineering Surveyors for providing material on which sections 1.1 and 1.4 are based; Mr Stephen Kennedy and Pentak UK Ltd for providing figure 2.9; Ms Deborah Saunders, Mr Alan Murray, Dr Mike Grist and Leica UK Ltd for providing figures 1.4, 1.5, 2.5, 2.11, 2.12, 2.16, 2.17, 2.18, 2.24, 2.30, 3.3, 3.7, 3.8, 3.9, 3.10, 3.11, 3.14, 3.17, 3.18, 3.20, 3.21, 3.22, 3.32, 3.33, 5.10(a), 5.29, 5.36, 5.37, 7.18, 8.1, 8.6(a), 14.54, 14.61, 14.67, 14.68, 15.1, 15.2, 15.6 and 15.10; Mr Ralph Tiller, Mr Geoff Sinclair and Sokkia Ltd for providing figures 3.2, 3.5, 3.19, 3.23, 3.24, 3.26, 3.28, 5.7, 5.11, 5.12(b), 5.27, 5.40, 5.41, 5.42, 14.59 and 15.11; Ms Rosanne Pearce and Fisco Products Ltd, for providing figures 4.3, 4.4, 4.5, 4.11 and 4.12; The Building Research Establishment, in particular Dr Beryl Cook, for permission to publish figures 4.8, 4.9 and 15.3; Mr Brian McGuigan and Geotronics Ltd for providing figures 5.10(b), 5.14, 5.15, 5.22, 5.23, 5.24 and 5.34; Mr R.H. Wells, Survey Supplies Ltd, for providing figure 5.12(a); Mr Lawrence Smith and Carl Zeiss Ltd for providing figure 5.12(c); Hall and Watts Ltd for providing figures 5.13 and 5.33; Tellumat Ltd for providing figure 5.26; Husky Computers Ltd for providing figure 5.30; Psion UK PLC for providing figure 5.33; Mike Ewer of Chiltern Survey Equipment Ltd for providing figure 5.38; Ms Helen Knight and Trimble Navigation Europe Ltd for providing figure 8.6(b); Ms T. Hogg of the copy Centre, Headingley, Leeds, for help with section 9.3; Mr Ted Read and LM Technical Services Ltd for providing figure 9.17; Blue Moon Systems Ltd for providing figure 9.18; Mr Colin Beatty and Ashtech for providing figure 8.6(c); The Department of Transport for permission to publish tables 11.1, 12.1 and figures 12.4 and 12.5; Mr Katsumi Kaji and the Ushikata Mfg Co Ltd, Japan, for providing figure 13.11; Mr J.R. Dixon of the Civil Engineering Department, University of Leeds, for help with section 14.1; figures 14.1, 14.7, 14.8, 14.45 and 14.51 are reproduced from CIRIA/Butterworths-Heinemann book *Setting Out Procedures*; The National Swedish Institute for Building Research and, in particular, Dr John van den Berg, for permission to reproduce figures 14.6, 14.19, 14.21, 14.25, 14.39, 14.40, 14.14, 14.42, 14.44, 14.47, 14.49, 14.50, 14.55, 14.56 and 14.57; Mr Trevor Albinson and Pro-Set Profiles Ltd for providing figure 14.23; Mr Paul Finney and Zenith Surveys Ltd for providing figures 14.52 and 14.65; Mr Jim Pelham and AGL European Lasers Ltd for providing figures 14.62 and 14.64; Ms Nicky Newman and Spectra-Physics Ltd for providing figure 14.66; Mr Paul Kelly and John Kelly (Lasers) Ltd for providing figure 14.71; Mr Chris Boffey of the British Standards Institution for providing information on BS5750; Mr Steve Carey of Construction Instruments Ltd for providing laser diode information; extracts from BS5606: 1990 are reproduced with permission of the British Standards Institution – complete copies can be obtained by post from BSI Sales, Linford Wood, Milton Keynes, MK14 6LE.

Special thanks to Mr Roy Trembath, surveying technician in the Civil Engineering Department at the University of Leeds, for his considerable help with a number of the new photographs and illustrations used in the book.

Every effort has been made to trace all the copyright holders. If any have been inadvertently overlooked, the publishers will be pleased to make the necessary arrangements at the first opportunity.

1

Introduction

Surveying, to the majority of engineers, is the process of measuring lengths, height differences and angles on site either for the preparation of large-scale plans or in order that engineering works can be located in their correct positions on the ground. The correct term for this is *engineering surveying* and it falls under the general title of land surveying.

In the UK, professional matters relating to land and other types of surveying are primarily the responsibility of the *Royal Institution of Chartered Surveyors* and the *Society of Surveying Technicians*. However, the *Institution of Civil Engineering Surveyors* plays an important role on construction sites and elsewhere as it concerns itself solely with quantity surveying and land surveying.

These institutions and their aims are described briefly in the following section.

1.1 Surveying Institutions

Royal Institution of Chartered Surveyors (RICS)

The RICS is one of the largest professional bodies in the UK and it is regarded as a leader in matters relating to land, property and construction. The aims of the RICS are to provide its members (known as Chartered Surveyors) with support facilities, to promote and publicise to the general public the services offered by chartered surveyors and to regulate its members through rules of conduct to ensure the high standing of the profession.

Institution of Civil Engineering Surveyors (InstCES)

The aims of this institution are to promote the interests and advance the professional status of the quantity and land surveyors working in the civil

1

engineering industry. In the 1960s, many quantity surveyors involved with civil engineering projects felt that there was a distinct difference between their work and that of the quantity surveyor working in the building industry. Land surveyors, for their part, were aware at this time of their lack of appreciation by civil engineers, thus limiting the efficient application of surveying and dimensional control principles in the industry. For these reasons, land and quantity surveyors formed the Association of Surveyors in Civil Engineering which in 1980 became the Institution of Civil Engineering Surveyors.

The Society of Surveying Technicians (SST)

The Society of Surveying Technicians was formed under the auspices of the RICS in 1970 in order to establish a nationally recognised organisation for surveying technicians who were highly qualified specialists in a specific branch of surveying. The object of the society is to set a nationally accepted standard of technical ability which must be shown to have been reached before membership is granted to any applicant and to further the career development of all members. Although working in the closest cooperation with the RICS, the SST is a completely independent body governed by its own council.

1.2 What is Surveying?

According to the RICS, surveying is made up of seven specialisations, known as *sectors*, and these are shown in figure 1.1.

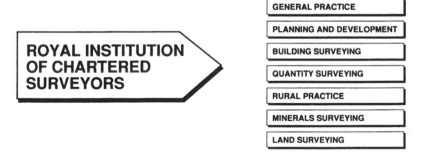

Figure 1.1 *Sectors of the RICS*

Almost half of all surveyors work in *general practice*, which is mostly concerned with valuation and investment. Valuation deals with issues such as how much a company's assets are worth – for stock flotation, acquisition or disposal, how much rent should be charged on commercial and industrial property and whether a home is good value. Valuation surveyors are experts in property markets, land and property values, valuation procedures and property law. Investment surveyors help investors to get the best possible return from property. They handle a selection of properties for purchase or sale by pension funds, insurance companies, charities and other major investors. In addition to these, some general practice surveyors specialise in housing policy advice, housing development and management and they work for local authorities, housing associations or charitable trusts.

Planning and development surveyors are involved in projects from their earliest planning stages right through to eventual completion. Projects vary enormously, but typically might involve instructions from a property developer to examine the viability of a proposed housing scheme. Planning and development surveyors would analyse the existing level of residential accommodation against regional population forecasts and they would prepare valuations to help gauge profitability. Additionally, they would be involved in preparing planning applications and negotiating with local authority planners to obtain planning permissions.

Over the last decade, the trend towards large-scale urban refurbishment has led to a doubling of chartered *building surveyors*. Their work involves advising on the construction, maintenance, repair and refurbishment of all types of residential and commercial property and may include specialisation in other types such as historic or listed buildings. All buildings must conform to a wide range of regulations and technical standards, so building surveyors need to be familiar with these to ensure that quality and safety levels are achieved. The analysis of building defects is an important part of a building surveyor's discipline. Wherever new buildings are being contemplated or existing buildings repaired, building surveyors play a major role in advising on design and maintenance.

Whenever any building project is proposed it is vital to know in advance the costs involved – the costs of preparing the site, construction, labour, materials and plant costs, fitting out costs, professional fees, taxes and other charges as well as the likely running and maintenance costs for the new building. The *quantity surveyor* is trained to evaluate these costs and to advise on alternative proposals. Once the decision has been taken to build a project, the quantity surveyor advises the client on appropriate contract arrangements as well as the legal contract and conditions under which the building will be constructed. Acting on behalf of the client, they advise the architect and engineer on the cost implications of different construction methods, alternative choices of materials and they ensure that each element of a project agrees with the cost plan allowance and that the overall project remains within budget. Quantity surveyors also assess the implications of changes in design, site conditions and working arrangements and are able to give the client accurate budget and time estimates.

At present, the pressures on the countryside are enormous. Developments in agriculture have inadvertently put the livelihoods of many farmers at risk, new forests are needed to meet timber shortages, urban centres are spilling over into rural areas, people want more access to the countryside for recreation and tourism and there is a widespread concern for the rural environment. Surveyors in *rural practice* advise landowners, farmers and others with interests in the countryside. They are responsible for the management of country estates and farms, the planning and execution of development schemes for agriculture, forestation, recreation and the valuation and sale of property and livestock.

Minerals surveyors plan the development and future of mineral workings. They work with local authorities and the owners of land on planning applications and appeals, mining law and working rights, mining subsidence and damage, the environmental effects of mines and the rehabilitation of derelict land. A minerals surveyor also manages and values mineral estates and surveys mineral workings in open cast or deep underground mines.

This book is concerned solely with the next sector of surveying which is known as *land surveying*. Traditionally, the land surveyor has been trained to measure land and its physical features accurately and to record these in the form of a map or plan. Such information is used by commerce and industry for planning new buildings and by local authorities in managing roads, housing estates and other facilities. In fact, anyone who uses a map to find a way round town or countryside is using information gathered by land surveyors. Nowadays, a large part of the land surveyor's work is to undertake positioning and monitoring for construction works, and it is these aspects of land surveying that are emphasised throughout the book.

1.3 Land Surveying

As with surveying in general, land surveying can be broken down into several subsections as shown in figure 1.2. However, it must be stressed that there is a considerable overlap between these sections, particularly as regards the basic methods and instruments used. That part of land surveying which is

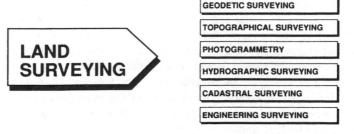

Figure 1.2 *Components of land surveying*

Figure 1.3 *Satellite position fixing at Shakespeare Cliff for the Channel Tunnel (© Crown copyright)*

relevant to civil engineering and construction is engineering surveying and, as stated in the previous section, is covered in some detail in this book. Before introducing engineering surveying, the other specialisations of land surveying are described briefly.

Geodetic surveys cover such large areas that the curved shape of the Earth has to be taken into account. These surveys involve advanced mathematical theory and require precise measurements to be made to provide a framework of accurately located points. These points can be used to map entire continents, they can be used to measure the size and shape of the Earth or they can be used to carry out scientific studies such as the determination of the Earth's magnetic field and detection of continental drift. Position fixing by satellite (see figure 1.3) and sophisticated computers and software are a feature of modern geodetic surveys.

Topographical surveys establish the position and shape of natural and man-made features over a given area, usually for the purpose of producing a map of an area or for establishing a geographic information system. Such surveys are usually classified according to the scale of the final map or terrain model formed. Small-scale surveys cover large areas such as an entire continent, country or county, and may range in scale from 1:1 000 000 to 1:50 000 like the familiar Ordnance Survey Landranger maps of Great Britain. Medium-scale maps range in scale from about 1:10 000 to 1:1000 and may cover the area of a small town. Large-scale maps show details and present information which is not often available from a map purchased in a shop and are therefore usually commissioned for a specific purpose. These maps range in scale from 1:500 up to 1:50 or larger and are often provided to meet the needs of architects, civil engineers, or government departments.

Figure 1.4 *Photogrammetric stereoplotter (courtesy Leica UK Ltd)*

Topographic surveys at most scales may be undertaken by *photogrammetry* using photographs taken with special cameras mounted in an aircraft. Viewed in pairs, the photographs produce three-dimensional images of ground features from which maps or numerical data can be produced, usually with the aid of stereo plotting machines and computers (see figure 1.4). Close-range photogrammetry uses photographs taken with cameras on the ground and is used in many applications.

Hydrographic surveyors gather information in the marine environment and their traditional role for centuries has been to map the coastlines and sea bed in order to produce navigational charts. More recently, much of their work has been for offshore oil exploration and production. Hydrographic surveys are also used in the design, construction and maintenance of harbours, inland water routes, river and sea defences, in control of pollution and in scientific studies of the ocean.

Cadastral surveys are those which establish and record the boundaries and ownership of land and property. In the United Kingdom, cadastral surveys are carried out by Her Majesty's Land Registry, a government department, and are based on the topographical detail appearing on Ordnance Survey maps. Cadastral work is mainly limited to overseas countries where National Land Registry Systems are under development.

1.4 Engineering Surveying

The term *engineering surveying* is a general expression for any survey work carried out in connection with the construction of civil engineering and building projects. Engineers and surveyors involved in site surveying are responsible for all aspects of dimensional control on such schemes. The main purposes of engineering surveying are:

At the concept and design stage to provide large-scale topographical surveys and other measurements upon which projects are designed. Since this data forms the basis for an entire project, the reliability of the design depends to a great extent on the precision and thoroughness with which the original site survey is carried out. In most cases, the initial survey will be in digital form and computerised equipment will be used to collect and process data.

At the construction stage to provide the precise control from which it is possible to position the works and, most importantly, to ensure that engineering projects are built in their correct relative and absolute positions (this is known as *setting out*). In addition to these, data for the measurement of the works is also collected to enable volumes of material to be estimated during construction. Occasionally, as-built records of the project are surveyed as construction proceeds.

At the post construction stage to monitor for structural movement on major retaining structures such as dams.

Engineering surveys are usually based on horizontal and vertical control frameworks which consist of fixed points called control stations. *Horizontal control*, as its title suggests, defines points on an arbitrary two-dimensional horizontal plane which covers the area of interest. *Vertical control*, although usually treated separately from horizontal control as far as fieldwork and calculations are concerned, is the third dimension added to the chosen horizontal datum.

Horizontal and vertical control are established by measuring angles, distances or a combination of both of these in well-established techniques such as traversing, triangulation, intersection, resection and levelling. However, increasing use is being made of artificial satellites for position-fixing and it is expected that these will eventually become as commonplace as terrestrial methods.

On site, a wide variety of equipment is used for establishing control and for setting out. This includes theodolites for measuring angles, levels for measuring vertical distances (heights), tapes and electronic instruments for measuring distances. Typical examples of this equipment are shown in figure 1.5. Computers are also used extensively in engineering surveying for applications such as network analysis, automated data processing for plan production and the computation of setting out data and quantities.

In order to ensure that reliable measurements are taken for engineering surveys, equipment and techniques of sufficient precision should be used both before and during construction. However, it is not always necessary to

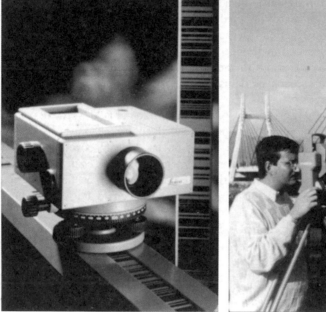

Figure 1.5 *Surveying instruments (courtesy Leica UK Ltd)*

use the highest possible precision; some projects may only require angles and distances to be measured to 1' and 0.1 m, whereas others may require precisions of 1" and 0.001 m. It is important that the engineer realises this and chooses equipment and techniques accordingly. To help with this choice, the precisions of the various equipment and techniques are emphasised throughout the book.

1.5 Coordinate Systems

In the previous section it was stated that engineering surveys are all based on horizontal and vertical control stations, the positions of which can be fixed by a variety of survey methods. For all applications of engineering surveying, the horizontal positions of control points are defined using rectangular coordinates irrespective of the method used to survey them. An understanding of the use of coordinates in all aspects of modern engineering surveying is vital and an introduction is given here.

Rectangular Coordinates

The coordinate system adopted for most survey purposes is a plane, rectangular system using two axes at right angles to one another as in Cartesian geometry. One is termed the north (N) axis and the other the east (E) axis. The scale along both axes is *always* the same. With reference to figure 1.6, any particular point P has an easting (E) and a northing (N) coordinate, always quoted in the order easting, northing unless otherwise stated. The position of each control station in a network in relation to all the others is specified in terms of these E and N coordinates. Bearings are related to the north axis of the coordinate system.

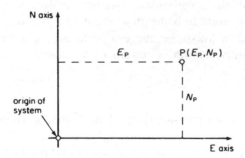

Figure 1.6 *Rectangular coordinate system*

For *all* types of survey and engineering works, the origin is taken at the extreme south and west of the area so that all coordinates are *positive*. If, at some stage in a survey, negative coordinates arise, the origin should be moved such that all coordinates will again be positive.

North Directions

The specified reference or north direction on which a coordinate system is based may be true north, magnetic north, some entirely arbitrary direction assigned as north, or grid north.

True north

The accurate determination of this direction is undertaken only for special construction projects and it is not normally used in engineering surveys. However, an approximate value can be scaled from Ordnance Survey (OS) maps.

Magnetic north

This is determined by a freely suspended magnetic needle and can be measured with a prismatic compass.

The direction of magnetic north varies with time and the annual change or drift in magnetic north is known as *secular variation*, the behaviour of which can only be predicted from observations of previous secular variations. The daily or *diurnal variation* in magnetic north causes it to swing through a range of values according to the time of day. In addition to these, *irregular variations* caused by magnetic storms and other disturbances can cause magnetic north to become unstable and to vary considerably.

Allowing for all these variations, it is not possible to determine the direction of magnetic north to better than about ±15'. Consequently, magnetic north is only used in low-order surveys or to give a general indication of north when an arbitrary north is chosen for a survey.

Arbitrary north

Arbitrary north is most commonly used to define bearings in engineering surveys.

Any convenient direction is usually chosen to represent north even though it is not, in general, a true or magnetic north direction.

Grid north

This north direction is based on the National Grid, which is discussed later in this section.

Whole-circle Bearings

To establish the direction of a line between two points on the ground, its bearing has to be determined. The whole-circle bearing (WCB) of a line is measured in a clockwise direction in the range 0° to 360° from a specified reference or north direction. Examples of whole-circle bearings are given in figure 1.7.

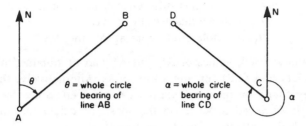

Figure 1.7   *Whole-circle bearings*

Polar Coordinates

Another coordinate system used frequently in surveying is the polar coordinate system shown in figure 1.8. When using this, the position of a point H is located with reference to a point G by polar coordinates D and θ where D is the horizontal distance between G and H and θ the whole-circle bearing of line GH. Polar coordinates define the relative position of one point with respect to another and, in surveying, are not normally used as absolute coordinates such as eastings and northings on a rectangular coordinate system.

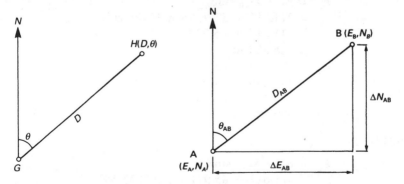

Figure 1.8   *Polar coordinate system*   Figure 1.9   *Calculation of rectangular coordinates*

Calculation of Rectangular Coordinates

Figure 1.9 shows the plan position of two points A and B. If the coordinates of A (E_A, N_A) are known, the coordinates of B (E_B, N_B) are obtained from A as follows

$$E_B = E_A + \Delta E_{AB} = E_A + D_{AB} \sin\theta_{AB} \tag{1.1}$$

$$N_B = N_A + \Delta N_{AB} = N_A + D_{AB} \cos\theta_{AB} \tag{1.2}$$

where

$$\Delta E_{AB} = \text{eastings difference from A to B}$$
$$\Delta N_{AB} = \text{northings difference from A to B}$$
$$D_{AB} = \text{horizontal length of AB}$$
$$\theta_{AB} = \text{whole-circle bearing of line AB.}$$

If a calculator is used, values of ΔE and ΔN can be obtained directly from equations (1.1) and (1.2) for any value of θ. Alternatively, the polar/rectangular key found on most calculators can be used. Since the method by which this is achieved depends on the make of calculator, the handbook supplied with the calculator should be consulted.

Worked example: calculation of rectangular coordinates

Question
The coordinates of point A are 311.617 m E, 447.245 m N. Calculate the coordinates of point B where $\theta_{AB} = 37°11'20''$ and $D_{AC} = 57.916$ m and point C where $\theta_{AC} = 205°33'55''$ and $D_{AB} = 85.071$ m.

Solution
With reference to figure 1.9 and equations (1.1) and (1.2)

$$\begin{aligned}
E_B &= E_A + D_{AB} \sin\theta_{AB} \\
&= 311.617 + 57.916 \sin 37°11'20'' \\
&= 311.617 + 35.007 \\
&= \mathbf{346.624 \ m} \\
N_B &= N_A + D_{AB} \cos\theta_{AB} \\
&= 447.245 + 57.916 \cos 37°11'20'' \\
&= 447.245 + 46.139 \\
&= \mathbf{493.384 \ m}
\end{aligned}$$

Similarly

$$\begin{aligned}
E_C &= E_A + D_{AC} \sin\theta_{AC} \\
&= 311.617 + 85.071 \sin 205°33'55'' \\
&= 311.617 - 36.711 \\
&= \mathbf{274.906 \ m} \\
N_C &= N_A + D_{AC} \cos\theta_{AC} \\
&= 447.245 + 85.071 \cos 205°33'55'' \\
&= 447.245 - 76.742 \\
&= \mathbf{370.503 \ m}
\end{aligned}$$

Calculation of Polar Coordinates

In the previous section it was shown that, knowing the whole-circle bearing and length of a line, the coordinates of one end of the line could be computed if the coordinates of the other end were known.

For the reverse case, where the coordinates of two points are known, it is possible to compute the whole-circle bearing and the horizontal distance of the line between the two points. This is known as an *inverse calculation* and is commonly used in engineering surveying when setting out works by polar coordinates. The WCBs and horizontal distances can be calculated by one of two methods; either by considering the *quadrant* in which the line falls or by using the *rectangular/polar conversion key* found on most calculators. If this type of calculation is undertaken using a computer which does not have a rectangular/polar facility, any programs written must be based on the quadrants method.

Referring to figure 1.10, suppose the coordinates of stations A and B are known to be (E_A, N_A) and (E_B, N_B) and that the whole-circle bearing of line AB (θ_{AB}) and the horizontal length of AB (D_{AB}) are to be calculated. The procedure for calculating these polar coordinates using quadrants is as follows.

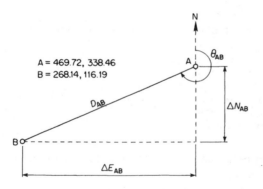

Figure 1.10 *Inverse calculation*

(1) A sketch showing the relative positions of the two stations should *always* be drawn in order to determine in which *quadrant* the line falls. This is most important as the greatest source of error in this type of calculation is wrong identification of quadrant. For whole circle bearings the quadrants are shown in figure 1.11.

(2) θ_{AB} is given by (in figure 1.10)

$$\theta_{AB} = \tan^{-1}(\Delta E_{AB}/\Delta N_{AB}) + 180°$$

$$= \tan^{-1}[(E_B - E_A)/(N_B - N_A)] + 180°$$

$$= \tan^{-1}\left[\frac{268.14 - 469.72}{116.19 - 338.46}\right] + 180°$$

$$= \tan^{-1}\left[\frac{-201.58}{-222.27}\right] + 180°$$

$$= \tan^{-1}(0.906\ 915) + 180°$$

$$= 42°12'19'' + 180°$$

Hence

$$\boldsymbol{\theta_{AB} = 222°12'19''}$$

It must be realised that, *in general*, the final value of θ_{AB} will depend on the quadrant of the line and a set of rules, based on the quadrant in which

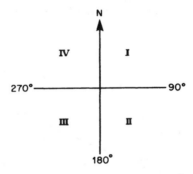

Figure 1.11 *Quadrants*

the line falls, can be proposed to determine the whole circle bearing. These rules are shown in table 1.1.

(3) Having now found θ_{AB}, D_{AB} is given by

$$D_{AB} = (\Delta E_{AB}/\sin\theta_{AB}) = (\Delta N_{AB}/\cos\theta_{AB})$$

For figure 1.10

$$D_{AB} = \frac{-201.58}{\sin 222°12'19''} = \frac{-201.58}{-0.671\ 789} = \boldsymbol{300.06\ m}$$

$$= \frac{-222.27}{\cos 222°12'19''} = \frac{-222.27}{-0.740\ 743} = \boldsymbol{300.06\ m}\ \text{(check)}$$

When evaluating D, *both* of the above should be calculated as a check against gross error. In the case where *small* differences occur between the

TABLE 1.1

Quadrant	I	II/III	IV
Formula	$\theta = \tan^{-1}(\Delta E/\Delta N)$	$\theta = \tan^{-1}(\Delta E/\Delta N) + 180°$	$\theta = \tan^{-1}(\Delta E/\Delta N) + 360°$

Note: ($\Delta E/\Delta N$) must be calculated allowing for their signs.
For a line XY, $\Delta E_{xy} = E_y - E_x$ and $\Delta N_{xy} = N_y - N_x$.

two results, the correct answer is given by the trigonometric function which is the slower changing. For example, if $\theta = 5°$, D found from ($\Delta N/\cos \theta$) gives the more accurate answer since the cosine function is changing less rapidly than the sine function at this angle value.

Alternatively, D may be given by $D = (\Delta E^2 + \Delta N^2)^{1/2}$. If an electronic calculator is being used then this method will give a satisfactory answer but provides no check on the result. For this reason, the method involving the trigonometrical functions is preferred.

If a calculator is available which is fitted with a rectangular/polar key, values of D and θ can be obtained directly. When using this function, the coordinate values must be entered into the calculator in the correct sequence otherwise the wrong bearing will be obtained. In all cases, if θ is displayed as having a negative value, 360° must be added to give the correct whole-circle bearing. This is due to the fact that calculators display θ either between 0° and +180° or between 0° and −180°.

The National Grid

In Great Britain, a rectangular coordinate system covering the whole of the country has been established by the Ordnance Survey (OS) as shown in figure 1.12. (The OS is described in more detail in section 1.9.)

When specifying position on the National Grid, eastings and northings are used and a point A could have National Grid coordinates of 473 867.74 m E, 264 213.45 m N. It is important to note that as with other coordinate systems, eastings are always quoted first on the National Grid. This coordinate system has a number of applications in engineering surveying and these are discussed in later sections of the book.

1.6 Scale

All engineering plans and drawings are produced at particular scales, for example, 1:500, 1:200, 1;100, and so on in 5,2 or 1 ratios as used in the

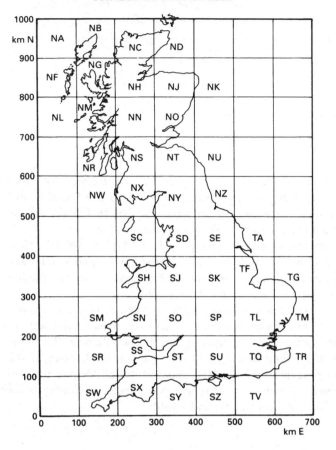

Figure 1.12   *National Grid of Great Britain*

metric system. The scale value indicates the ratio of horizontal and/or vertical plan distances to horizontal and/or vertical ground distances that was used when the drawing was produced, for example, a horizontal plan having a scale of 1:50 indicates that for a line AB

$$\frac{\text{horizontal plan length AB}}{\text{horizontal ground length AB}} = \frac{1}{50}$$

and, if line AB as measured on the plan = 18.2 mm, then

$$\text{horizontal ground length AB} = 18.2 \times 50 = 910 \text{ mm}$$

The term 'large-scale' indicates a small ratio, for example, 1:10, 1:20, whereas the term 'small-scale' indicates a large ratio, for example, 1:50 000.

On engineering drawings, scales are usually chosen to be as large as possible to enable features to be drawn as they actually appear on the ground. If too small a scale is chosen then it may not be physically possible to draw true representations of features and in such cases conventional symbols are used; this is a technique commonly adopted by the Ordnance Survey.

It must be stressed that the scale value of any engineering drawing or plan must always be indicated on the drawing itself. Without this it is incomplete and it is impossible to scale dimensions from the plan with complete confidence.

1.7 Units

Wherever possible throughout the text, Système International (SI) units are used although other widely accepted units are introduced as necessary. Those units which are most commonly used in engineering surveying are as follows.

Units of length

$$
\begin{aligned}
&\text{millimetre (mm), metre (m), kilometre (km)}\\
&1 \quad \text{mm} = 10^{-3}\ \text{m} = 10^{-6}\ \text{km}\\
&10^3\ \text{mm} = 1 \quad \text{m} = 10^{-3}\ \text{km}\\
&10^6\ \text{mm} = 10^3\ \text{m} = 1 \quad \text{km}
\end{aligned}
$$

Units of area

$$\text{square metre (m}^2)$$

Although not in the SI system, the hectare (ha) is often used to denote area where 1 ha = 100 m $\times$ 100 m = 10 000 m^2.

Unit of volume

$$\text{cubic metre (m}^3)$$

Units of angle

The SI unit of angle is the radian (rad). However, most surveying instruments measure in degrees (°), minutes (′) and seconds (″) and some European countries use the gon (^{g}), formerly the grad, as a unit of angle. The relationship between these systems is as follows

$$1\ \text{circumference} = 2\pi\ \text{rad} = 360° = 400^g$$

and, taking π to be 3.141 592 654 gives

$$90° = 1.570\ 796\ 327\ \text{rad} = 100^\text{g}$$
$$1° = 0.017\ 453\ 293\ \text{rad} = 1.111\ 111\ 111^\text{g}$$
$$1' = 0.000\ 290\ 888\ \text{rad} = 0.018\ 518\ 519^\text{g}$$
$$1'' = 0.000\ 004\ 848\ \text{rad} = 0.000\ 308\ 642^\text{g}$$

$$1\ \text{rad} = 57.295\ 779\ 513° = 63.661\ 977\ 236^\text{g}$$
$$0.01\ \text{rad} = 34.377\ 467\ 708' = 0.636\ 619\ 772^\text{g}$$
$$0.0001\ \text{rad} = 20.626\ 480\ 625'' = 0.006\ 366\ 198^\text{g}$$

$$100^\text{g} = 90° = 1.570\ 796\ 327\ \text{rad}$$
$$1^\text{g} = 0.9° = 0.015\ 707\ 963\ \text{rad}$$
$$0.1^\text{g} = 5.4' = 0.001\ 570\ 796\ \text{rad}$$
$$0.01^\text{g} = 32.4'' = 0.000\ 157\ 080\ \text{rad}$$

A useful approximate relationship which can be used to convert small angles from seconds of arc to their equivalent radian values is

$$\theta\ \text{rad} = \frac{\theta''}{206\ 265} = \theta'' \sin 1'' = \frac{\theta''}{\text{cosec}\ 1''}$$

1.8 Surveying Computations

Today, engineers and surveyors use pocket calculators, microcomputers and sometimes main-frame computers for performing calculations and for data processing. Although a large amount of data is produced for the site engineer and surveyor by microcomputers or even larger computer systems, the ability to be able to calculate by hand using a pocket calculator is essential on site. A good example of the dual nature of surveying computations often occurs when setting out the centre line of a road curve where it is likely that a computer will produce a print-out of centre line coordinates defining the position of the curve and where the engineer or surveyor has to convert these, by hand, into bearings and distances that can be set out from the most convenient control stations. The important point to note here is that the computer has not replaced the pocket calculator for day-to-day calculations on construction sites: it has simply made the task of processing large amounts of data quicker and easier and has made possible complicated calculations that were considered almost impossible in the past.

Since the pocket calculator is used extensively on construction and other sites, anyone thinking of purchasing a calculator should study table 1.2 before doing so in order to ensure that the one chosen has those features which will be most useful for surveying.

Like fieldwork, computations should be carefully planned and carried out in a systematic manner and all field data should be properly prepared before calculations start. Where possible, standardised tables or forms should be used to simplify calculations. If the result of a calculation has not been

TABLE 1.2

Pocket Calculator Functions for Engineering Surveying

Function or facility	Notes
Display	Should be at least 8 digit, preferably 10.
Arithmetic	Basic functions required.
Trigonometrical	$\sin$, $\sin^{-1}$, $\cos$, $\cos^{-1}$, $\tan$, $\tan^{-1}$ essential.
Degrees, rad, gon (grad)	Facility for using trigonometrical functions in degree, rad and gon (grad) modes useful.
Decimal degrees	Conversion between deg, min, sec and decimal degrees needed.
Polar/Rectangular	Conversion between polar (bearing and distance) and rectangular form (ΔE and ΔN) simplifies coordinate calculations greatly.
Programmable	Useful (but not essential) for most calculations provided that program storage is available for repeat calculations. Some models use plug-in modules to extend programming capability.
Printer connection	Hard-copy facility avoids transposition errors.
General purpose	$1/x$, x^3, $\sqrt{x}$, y^x occur frequently in engineering surveying.
Logarithms	$\log x$, 10^x, $\ln x$, e^x sometimes used.
Floating point	Essential when dealing with large or small numbers.
Rechargeable batteries	Preferable with AC current use.
Storage registers (memory)	Useful in complicated problems.
Pre-programmed constants	π required.
Statistical functions	$\bar{x}$ and s not essential but sometimes convenient.

checked, it is considered unreliable and for this reason, frequent checks should be applied to every calculation procedure, however simple. Examples of the use of tables and checks in surveying calculations are shown throughout this book.

Significant Figures

The way in which numbers are written in surveying is important, whether these numbers represent measurements or are derived from calculations. As far as measurements are concerned, an indication of the precision achieved by a measurement is represented by the number of significant figures recorded. For example, a distance may be recorded as 15.342 m implying a precision of 1 mm in the measurement whereas the same distance recorded as 15.34 m implies a precision of 10 mm. A number such as 15.342 contains five significant figures, 15.34 has four and the difference between these implies quite a difference in the equipment used to take the measurement. The position of the decimal place in a number does not indicate significant figures, for example, 0.000 652 1 has four significant figures and 0.098 has two significant figures.

Determining significant figures for quantities or numbers that are derived from observations is not as easy as for the observations themselves because calculations are involved. Care must be taken to ensure that the correct number of significant figures is carried through a calculation and this is especially important when using a pocket calculator (and computer) since these are capable of displaying many digits, some of which may not be significant. In general, it must be realised that any quantity calculated cannot be quoted to a higher precision than that of the data supplied or that of any field observations.

Various rules exist for determining significance for numbers resulting from a calculation. In general, it is the least precise component in the calculations which determines the precision of the final result.

For addition and subtraction, an answer can only be quoted such that the number of figures shown after the decimal place does not exceed those of the number (or numbers) with the least significant decimal place.

For multiplication and division, an answer can only be quoted with the same number of significant figures as the least significant number used in the calculation.

The following example demonstrates these principles.

Worked example: significant figures

Question
(a) Calculate the sum of 23.568, 1103.2, 0.3451 and 0.51.
(b) Calculate the difference between 45.451 and 38.9.
(c) Multiply 23.65 by 87.322.
(d) Divide 112 by 22.699.

Solution
(a) Listing each number gives

$$
\begin{array}{r}
23.568 \\
1103.2 \\
0.3451 \\
\underline{0.51} \\
1127.6231 = 1127.6
\end{array}
$$

The result is quoted with five significant figures to agree with the number with the least significant decimal place, 1103.2, despite the fact that 0.3451 and 0.51 have fewer significant figures.
(b) The difference between 45.451 and 38.9 is

$$45.451 - 38.9 = 6.6$$

since 38.9 has the least significant decimal place.
(c) Multiplying gives

$$23.65 \times 87.322 = 2065.1653 = 2065$$

since the least significant number, 23.65, has four significant figures
(d) Dividing gives

$$112/22.699 = 4.934\ 138\ 068 = 4.93$$

since 112 has three significant figures.

In some cases, results may be quoted to more significant figures than the above rules suggest. For example, the arithmetic mean of 14.56, 14.63, 14.59, 14.62 and 14.58 is

$$\frac{14.56 + 14.63 + 14.59 + 14.62 + 14.58}{5} = \frac{72.98}{5} = 14.596$$

since the number 5 is an exact number and retaining an extra significant figure is justified for a mean value which is more reliable than a single value (see chapter 6). Many circumstances occur in surveying in which numbers can be treated as exact and these have to be carefully defined.

1.9 The Ordnance Survey

The Ordnance Survey (OS) is the principal surveying and mapping organ-isation in Great Britain. Its work includes geodetic surveys and associated scientific studies, topographical surveys and the production of maps of Great Britain at various scales.

Ordnance Survey Maps

The range of map production from the OS is extremely wide and maps are available from the small-scale Routeplanner map, which is revised every year and contains the whole of Great Britain on one sheet at a scale of 1:625 000 to Superplan products, some of which are available at 1:200 scale. As far as engineering surveying is concerned, the OS maps of particular interest are those at the basic scales of 1:1250, 1:2500 and 1:10 000 and Superplan products.

Large-scale mapping

1:1250 scale
This is the largest scale of map data held by the OS and maps are available for cities and other significant urban areas throughout Great Britain. Street names, house numbers or names are shown, as well as parliamentary bound-aries. Height information and some survey control points are also shown. Each map represents an area of 500 m by 500 m on the ground and has

National Grid lines overprinted at 100 m intervals. There are over 57 000 maps in this series, an example of which is shown in figure 1.13*a*.

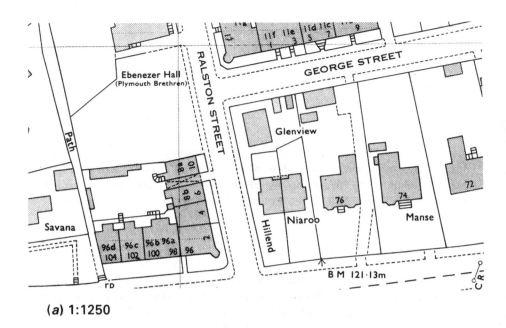

(*a*) 1:1250

(*b*) 1:2500

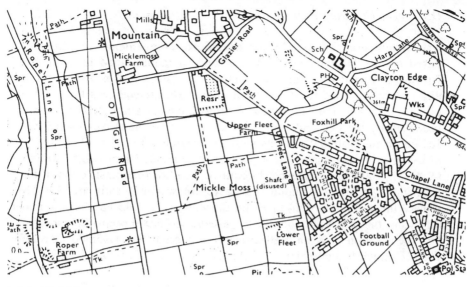

(*c*) **1:10 000**

Figure 1.13   *Examples of large-scale OS mapping (© Crown copyright)*

1:2500 scale

Maps in this series cover all those parts of Britain other than the significant urban areas covered at the larger 1:1250 scale and mountain and moorland areas. They provide detailed information about small towns, villages and the countryside for the same sort of purposes that 1:1250 maps provide for the larger built up areas. As well as buildings, roads, railways, canals, lakes, lochs, rivers and antiquity sites, almost all permanent tracks, walls, fences, hedges, ponds, watercourses and a great many other features are shown. Height information is provided by means of bench marks and spot heights. Administrative and parliamentary boundaries are also shown. Normally, each map covers an area of 2 km east to west and 1 km north to south, although some have a 1 km by 1 km format. National Grid lines are shown at 100 m intervals. An example of a 1:2500 map can be seen in figure 1.13*b*.

1:10 000 scale

These maps cover the whole of Britain, and an example is shown in figure 1.13*c*. They are the largest scale of OS mapping to cover mountain and moorland areas and to show contours. Some maps are at 1:10 560 scale with contours at 25 feet vertical interval, but they are being replaced by 1:10 000 scale maps with contours at 10 m interval in mountainous areas and 5 m interval elsewhere. The National Grid is shown at 1000 m intervals.

OS Large-scale Map References

The system used by the OS for providing large-scale map references is based on the National Grid coordinate system described in section 1.5. However, instead of using coordinates directly, a combination of letters and numbers is used to define sheet edges. This reference system is based on the 1:25 000 first series maps (now known as the Pathfinder maps) produced by the OS, and begins by allocating two letters to each 100 km square on the National Grid as shown in figure 1.12.

Using grid coordinates for a point A of 473 867.74 m E, 267 213.45 m N, the 1:25 000 map reference is obtained as follows. By inspection of figure 1.12, the two letters defining the square in which A falls are SP. These replace the 4 in 473 867.74 m E and the 2 in 267 213.45 m N. The numbers following these in each coordinate give the 1:25 000 map reference as SP 76. The position of A on a 10 km square defining the 1:25 000 map reference is shown in figure 1.14.

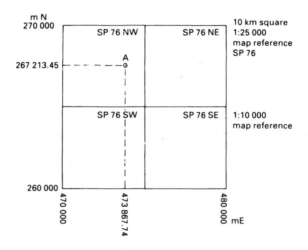

Figure 1.14   *1:25 000 and 1:10 000 National Grid map references*

The 1:10 000 map reference for A is obtained by adding two further letters to SP 76 to define a 5 km square within that used for the 1:25 000 map reference. In the example, A has a 1:10 000 map reference of SP 76 NW.

The 1:2500 map reference is obtained by adding the next digit in each National Grid coordinate to the letter prefixes: this defines a 1 km square. For example, A has a 1:2500 map reference of SP 73 67 (see figure 1.15) and subdividing into a 500 m square gives a map reference of SP 73 67 SE for A at 1:1250 scale.

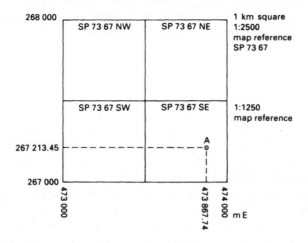

Figure 1.15   *1:2500 and 1:1250 National Grid map references*

Digital Mapping

The OS has been making digital maps since 1970 and at first, digital technology was used simply as a means of producing traditional printed maps on paper. In recent years however, digital data has found a market of its own.

A digital map may be defined as a map in computer-readable form. Once converted into this format, digital map data is suitable for use in such facilities as geographic information sytems and computer-aided design systems, both of which are used in engineering surveying. Once map data has been digitised and stored in a computer it can be displayed on a computer screen, merged with other graphical data or plotted on paper and film. The data may be displayed or plotted at any scale and individual features may be distinctively coloured or omitted so that maps are made to meet user requirements. The OS provides a digital update service which enables the digital map user to be supplied with the very latest data. As soon as OS continuous revision surveyors have completed their work, the new detail is incorporated into the database. Updated digital map files are available to users at regular intervals.

Conversion of large-scale maps into digital form is a major task for the OS. At present, all 57 400 of the 1:1250 scale urban maps are available as digital data. Progress converting the rural 1:2500 scale maps continues and a start has been made on digitising the basic 1:10 000 series.

In addition to large-scale mapping, the OS offers a number of different versions of digital map data including such diverse products as a computerised set of 1:50 000 Landranger maps and digital height information, available to special order only, at a scale of 1:10 000.

Superplan

The latest product available from the OS is known as Superplan. Maps in this format are obtained directly from the digital map data derived by the OS from the 1:1250 and 1:2500 basic map series. The underlying principle of Superplan is that a map can be produced to suit any requirement rather than be restricted to those in a published series.

To obtain a Superplan product, an OS Network Superplan Agent is contacted. At present, fifteen National Agents throughout Great Britain have Superplan plotting systems on site and can produce Superplan plots to order. Regional Agents and Sub-Agents offer advice on how to obtain Superplan products and can order the relevant maps from National Agents.

The Superplan product range is as follows. The *Standard (NG) Format* is similar to the existing 1:1250 and 1:2500 map series but has twelve plot options giving a range of map information and presentation outputs. However, the *Pre-Defined (Site Centred) Format* allows customised plots to any combination of plot area, size, scale and map/layout specification which the Superplan plotting system can accept, as shown in figure 1.16.

The OS is engaged in maintaining its large-scale surveys at 1:1250, 1:2500 and 1:10 000 scales, and change is being continuously surveyed. These changes are transferred to microfilm from the digital database at frequent intervals. The *Superplan Instant Printout* is a paper copy of the latest available microfilm of an area and provides the most up-to-date information showing change as soon as possible after it occurs. An Instant Printout is, however, restricted to National Grid sheetlines and cannot be customised.

Other OS Services

In addition to its wide range of maps and digital information, the OS provides many other services, some of which are of interest to engineering surveyors.

To produce its maps, the OS has a framework of horizontal and vertical control points known as triangulation stations (see section 7.17) and bench marks (see section 2.2) which are located all over Britain. These points are frequently used in engineering work and survey information is available as follows.

Triangulation stations: There are over 6000 of these throughout Great Britain and they are either the familiar triangulation pillar (see figure 1.17), generally located on hilltops, or they are intersected stations such as church spires, chimneys, masts and so on. Station descriptions are available which

Superplan...

A4 Extract - 1   Site Centred

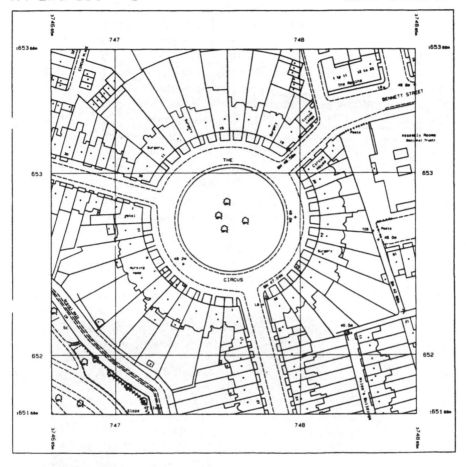

A Superplan Standard Symbols leaflet
is available on request from
Ordnance Survey Agents.

Scale 1: 1250

Figure 1.16   *Superplan site-centred plot (© Crown copyright)*

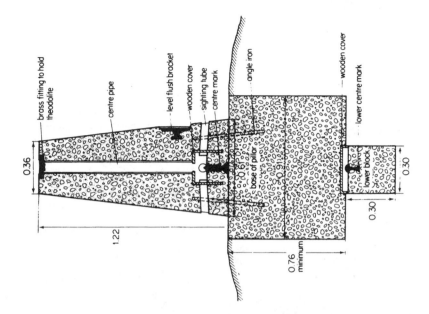

Figure 1.17 OS triangulation pillar (©Crown copyright)

include the name, nature of station, National Grid coordinates, elevation and many other details.

Minor control points: Detailed descriptions of these points which are established in urban areas to control large-scale mapping are available. These points are located on permanent features such as manhole covers or buildings and information about them is only held by OS Agents.

Levelling: Height information in Great Britain is available from the OS in the form of bench mark lists. Each list provides information on bench marks in a one kilometre square and carries the same National Grid reference number as the corresponding 1:2500 scale map. The locations and heights of bench marks are also marked on 1:1250 and 1:2500 scale maps.

1.10 Aims and Limitations of this Book

Engineers use land surveying simply as one of the means by which they can undertake their work and there is a definite limit to the surveying knowledge required by them, beyond which the surveying becomes of interest rather than importance. Specialist land surveyors are usually called in to deal with any unusual problems.

The main aim of this book, therefore, is to provide a thorough grounding in the basic land surveying techniques required in engineering. Its originality is not so much in the topics it contains, but more in the emphasis placed on each topic and the depth to which each is covered.

Although it is likely that engineers will come into contact with other branches of land surveying such as geodesy and photogrammetry at some stage in their careers, these are specialist subjects which require considerably more space to cover to sufficient depth than that available here. Consequently, they are not included and space is instead given to a thorough discussion of the equipment and techniques used in general site work and plan production. Further information on land surveying can be found in some of the references listed in the following section.

The text is limited mainly to *plane surveying*, that is, the effect of the curvature of the Earth is ignored. This is a valid limitation since the effect of the curvature of the Earth is negligible for areas up to 200 km^2 and it is unlikely that many engineering projects or construction sites will exceed this. The only time that curvature is considered in the text is in the section dealing with trigonometrical heighting where long sight lengths may be used.

Modern equipment is discussed at all times except where the more traditional equipment is ideal for illustrating a particular technique, and the use of electronic calculators and computers is discussed wherever applicable.

A note of caution must be introduced at this point. Although the methods involved in engineering surveying can be studied in textbooks, such is the practical nature of the subject that no amount of reading will turn a student into a competent engineering surveyor. Only by undertaking some practical surveying, under site conditions, and learning how to combine the techniques and equipment as discussed in this text will the student eventually become proficient and produce satisfactory results.

Further Reading

Russel C. Brinker and R. Minnick, *The Surveying Handbook* (Van Nostrand Reinhold, 1988).

C.D. Burnside, *Mapping from Aerial Photographs*, 2nd Edition (Collins, London, 1985).

A.E. Ingham, *Hydrography for the Surveyor and Engineer*, 2nd Edition (Crosby Lockwood Staples, London, 1984).

Institution of Civil Engineering Surveyors – Yearbook 1990/91 (InstCES, 1990).

T.J.M. Kennie and G. Petrie, *Engineering Surveying Technology* (Blackie, Glasgow and London, 1990).

Making Land, Property and Construction Work (RICS, Cambridge, 1990).

Ordnance Survey, *Digital Map Data Catalogue 1993* (Ordnance Survey, Southampton, 1993).

Ordnance Survey, *Information Leaflets, 30, 31, 33* and *48* (Ordnance Survey, Southampton, 1988).

Ordnance Survey, *Large Scale Mapping, Special Products and Services Catalogue 1993* (Ordnance Survey, Southampton, 1993).

P. Vanicek and E.J. Krakiwsky, *Geodesy: The Concepts*, 2nd Edition (Elsevier, Amsterdam, 1986).

P.R. Wolf, *Elements of Photogrammetry*, 2nd Edition (McGraw-Hill, Tokyo, 1983).

The following are a useful source of information for those interested in engineering and other aspects of surveying.

Civil Engineering Surveyor, journal published by the InstCES, ten issues per annum.

The Surveying Technician, journal published by the SST, five issues per annum.

Surveying World, journal published by GITC in the Netherlands and distributed to all chartered surveyors in the UK by the RICS, six issues per annum.

2

Levelling

Levelling is the name given to the process of measuring the difference in elevation between two or more points. In engineering surveying, levelling has many applications and is used at all stages in construction projects from the initial site survey through to the final setting out. In practice, it is possible to measure heights to better than a few millimetres when levelling and this precision is more than adequate for height measurement on the majority of civil engineering projects.

Specialised equipment is required to undertake levelling and traditionally this has been an optical level with its tripod and staff. However, levelling is now also carried out using digital and laser levels. This chapter deals with conventional and digital levelling: laser levels and their applications in setting out are discussed in chapter 14.

2.1 Level and Horizontal Lines

The terms level line and horizontal line are used frequently in levelling and need to be carefully defined.

A *level line* (or surface) is defined as a line along which all points are of the same height. Because the Earth is curved, level lines are also curved as shown in figure 2.1. Consequently, when using an optical or digital level to determine height differences, measurements should be taken from curved level lines.

A *horizontal line* is one which is normal to the direction of gravity (the vertical) at a particular point such as P in figure 2.2. Horizontal lines (or surfaces) are, therefore, tangential to level lines (or surfaces) at individual points (see figure 2.3).

When an optical or digital level is set up correctly, it defines a horizontal line for measurement of height differences and this would seem to con-

31

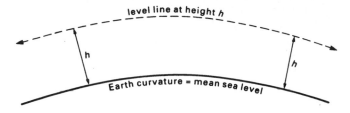

Figure 2.1 *Level line*

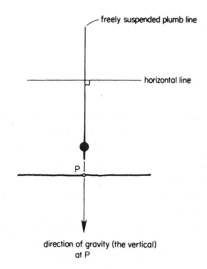

Figure 2.2 *Horizontal line*

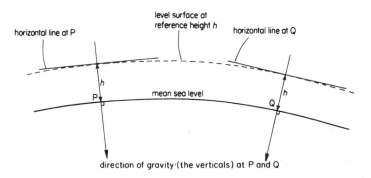

Figure 2.3 *Level and horizontal lines*

tradict the need to measure along (curved) level lines. However, for most survey work, the difference between a horizontal line and a level line (called curvature) is small enough to be ignored and it can be assumed that level and horizontal lines are the same. This is discussed further in section 2.17.

2.2 Datums and Bench Marks

For all surveys, a level line is chosen to which the elevation of all points is related and is known as a *datum* or *datum surface*. This can be any surface but the most commonly used datum is *mean sea level* and, for Great Britain, this is the mean sea level as measured at Newlyn in Cornwall. Since the Ordnance Survey (OS) of Great Britain use this datum, it is called the *Ordnance Datum* and any heights referred to Ordnance Datum are said to be *Above Ordnance Datum* (AOD). All heights marked on OS maps and plans will be AOD.

On many construction and civil engineering sites, mean sea level is not often used as a datum for levelling. Instead, a permanent feature of some sort is chosen on which to base all work and this is given an arbitrary height to suit site conditions. Whatever the chosen datum, the height of a point relative to a datum is said to be its *reduced level*.

Bench marks are permanent reference marks or points, the reduced levels of which have been accurately determined by levelling.

Ordnance bench marks (OBMs) are those which have been established by the Ordnance Survey throughout Great Britain and are based on the Ordnance Datum. The most common type are permanently marked on buildings and walls by a cut in vertical brickwork or masonry, an arrow or crowsfoot mark indicating the bench mark. On horizontal surfaces, OBMs consist of a rivet or bolt, the position of the RL being shown in figure 2.4 for both types.

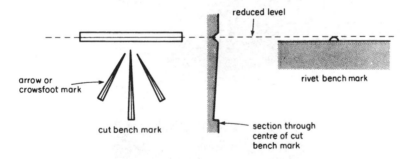

Figure 2.4 *OS bench marks*

All Ordnance Survey bench marks have been in place for some time and may have been affected by local subsidence or physical disturbance since the date they were verified. To guard against this passing unnoticed, it is always advisable to include at least two OBMs in levelling schemes where Ordnance datum is being used.

Temporary or *transferred bench marks* (TBMs) are marks set up on stable points near construction sites to which all levelling operations on that particular site will be referred. These are often used when there is no OBM close to the site. The height of a TBM may be assumed at some convenient value, usually 100.00 m, or may be accurately established by levelling from the nearest OBM. Various suggestions for the construction of TBMs are given in chapter 14.

2.3 Automatic Levels

The general features of the automatic level are shown in figure 2.5. These instruments establish horizontal lines of sight at each point where they are set up and consist of a telescope with a compensator. The *telescope* provides an accurate line of sight and enables the level to be used over distances suitable for surveying purposes. The *compensator*, built into the telescope, ensures that the line of sight viewed through the telescope is horizontal even if the optical axis of the telescope itself is not horizontal.

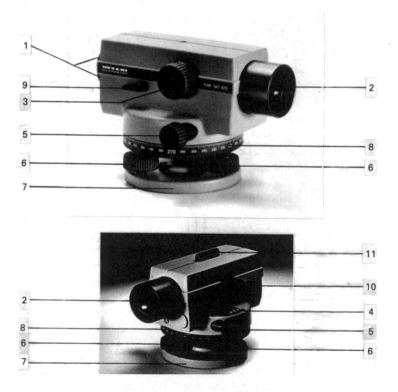

Figure 2.5   *Automatic level: 1. eyepiece; 2. objective; 3. focusing screw; 4. circular bubble; 5. tangent screw (slow motion screw); 6. footscrew; 7. baseplate; 8. horizontal circle; 9. compensator test lever; 10. mirror for bubble; 11. sight (courtesy Leica UK Ltd)*

2.4 The Surveying Telescope

Since the type of telescope used in levels is also used in theodolites (see chapter 3), the method of construction is considered in detail.

The surveying telescope is *internally focusing* as shown in figure 2.6. Incorporated in the design of the telescope are special cross lines which, when the telescope is adjusted correctly, are seen clearly in the field of view. These lines provide a reference against which measurements can be taken. This part of the telescope is called the *diaphragm* (or *graticule*) and consists of a circle of plane glass upon which a series of lines is etched, the more common patterns being shown in figure 2.7. Conventionally, the vertical and horizontal lines are called the *cross hairs*.

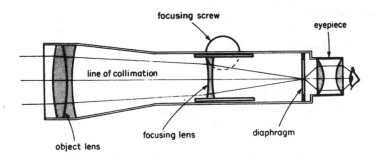

Figure 2.6   *Internal focusing telescope shown correctly adjusted*

Figure 2.7   *Diaphragm patterns*

The object lens, focusing lens, diaphragm and eyepiece are all mounted on the same optical axis and the imaginary line passing through the centre of the cross hairs and the optical centre of the object lens is called the *line of collimation* or the *line of sight*. When using the level, all readings are taken using this line. The diaphragm is held in the telescope by means of four adjusting screws so that the position of the line of collimation within the telescope can be moved (see section 2.11).

The action of the telescope is as follows. Light rays from a distant point pass through the object lens and are brought to focus in the plane of the diaphragm by axial movement of the concave lens. This is achieved by mounting the concave lens on a tube within the telescope, this tube being connected, via a rack and pinion, to a focusing screw attached to the side of the telescope. The eyepiece, a combination of lenses, has a fixed focal point that lies outside the lens combination and, by moving the eyepiece, this point can be made to coincide with the plane of the diaphragm. Since the image of the object has already been focused at the diaphragm, an observer will see in the field of view of the telescope the distant point focused against the cross hairs marked on the diaphragm. Furthermore, the optical arrangement is such that the object viewed through the eyepiece is magnified.

A problem often encountered with outdoor optical instruments is water and dust penetration. In order to provide protection from these, the telescope and compensator compartment of some levels are sealed and filled with dry nitrogen gas. This is known as *nitrogen purging* and since the gas is pressurised, water and dust are repelled and use of a dry gas prevents fogging of the objective.

2.5 Parallax

It must be realised that for the surveying telescope to operate correctly the image of a distant point or object must fall exactly in the plane of the diaphragm and the eyepiece must be adjusted so that its focal point is also in the plane of the diaphragm. Failure to achieve either of these settings results in a condition called *parallax* and this is a major cause of error in both levelling and theodolite work. Parallax can be detected by moving the eye to different parts of the eyepiece when viewing a distant object; if different parts of the object appear against the cross hairs then the telescope has not been properly focused and parallax is present, as seen in figure 2.8.

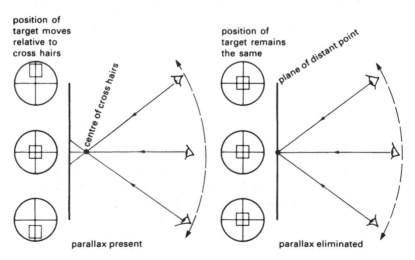

Figure 2.8 *Parallax*

It is impossible to take accurate readings under these circumstances since the line of sight alters for different positions of the eye. Parallax *must* be removed *before* any readings are taken when using *any* optical instrument with an adjustable eyepiece.

To remove parallax, the eyepiece is first adjusted while viewing a light background, for example, the sky or a booking sheet, until the cross hairs appear in sharp focus. The distant point at which readings are required is now sighted and brought into focus and is viewed while moving the eye. If the object and cross hairs do not move relative to each other then parallax has been eliminated; if there is apparent movement then the procedure should be repeated.

2.6 The Compensator

In an automatic level, the function of the compensator is to deviate the horizontal ray of light at the optical centre of the object lens through the centre of the cross hairs. This ensures that the line of sight (or collimation) viewed through the telescope is horizontal even if the telescope is tilted.

Whatever type of automatic level is used it must be levelled within approximately 15–30' of the vertical to allow the compensator to work. This is usually achieved by using a three footscrew arrangement in conjunction with a small circular level (sometimes called a pond bubble) which is mounted somewhere on the level.

Figure 2.9 shows a compensator and the position it is usually mounted in the telescope and the action of the compensator is shown in figure 2.10, which has been exaggerated for clarity. The main component of the compensator is a prism which is assumed to be freely suspended within the

Figure 2.9 *Compensator (courtesy Pentax UK Ltd.)*

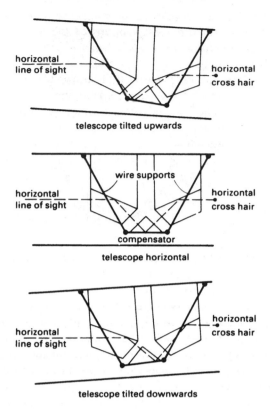

Figure 2.10 *Action of compensator*

telescope tube when the instrument has been levelled and which takes up a position according to the angle of tilt of the telescope. Provided the tilt is within the working range of the compensator, the prism moves to a position to compensate for this, and a horizontal line of sight (collimation) is always observed at the cross hairs.

The wires used to suspend the prism are made of a special alloy to ensure stability and flexibility under rapidly changing atmospheric conditions, vibration and shock. The compensator is also screened against magnetic fields and it uses some form of damping, otherwise the compensator, being light in weight, would tend to oscillate for long periods when the telescope is moved or affected by wind and other vibrations.

2.7 Use of the Automatic Level

The first part of the levelling process is to set the tripod in position for the initial readings, ensuring its top is levelled by eye after the tripod legs have been pushed firmly into the ground. Following this, the level is attached to the tripod using the clamp provided and the circular bubble is centralised using the three footscrews.

When an automatic level has been roughly levelled, the compensator automatically moves to a position to establish a horizontal line of sight. Therefore, no further levelling is required after the initial levelling.

As with all types of level, parallax must be removed before any readings are taken.

In addition to the levelling procedure and parallax removal, a test should be made to see if the compensator is functioning before readings commence. One of the levelling footscrews should be moved slightly off level and, if the reading to a levelling staff remains constant, the compensator is working. If the reading changes, it may be necessary to gently tap the telescope to free the compensator. On some automatic levels this procedure is not necessary since a button is attached to the level which is pressed when the staff has been sighted (see figure 2.11). If the compensator is working, the horizontal hair is seen to move and then return immediately to the horizontal line of sight. Some levels incorporate a warning device that gives a visual indication to an observer, in the field of view of the telescope, when the instrument is not level.

A disadvantage with automatic levels is that either a strong wind blowing on the instrument or machinery operating nearby will cause the compensator to oscillate, resulting in vibrating images. To overcome this the mean of several readings should be taken or a tilting level could be used instead of an automatic level. This problem is often encountered on construction sites, particularly roadworks where the site is sometimes narrow.

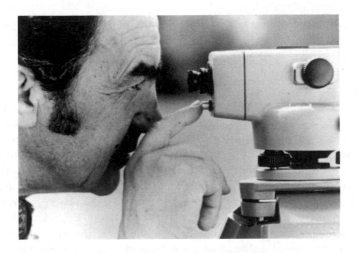

Figure 2.11 *Compensator check (courtesy Leica UK Ltd)*

2.8 The Tilting Level

Figure 2.12 shows a photograph of a tilting level. On this instrument the telescope is not rigidly attached to the base of the level and can be tilted a small amount in the vertical plane about a pivot placed below the telescope. Hence the name *tilting level*. The amount of tilt is controlled by the tilting screw which is usually directly underneath the telescope eyepiece.

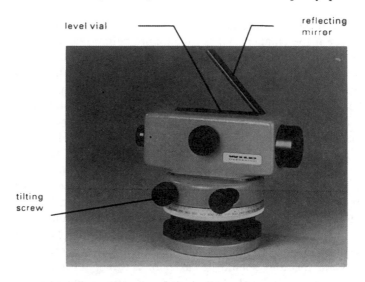

Figure 2.12   *Tilting level (courtesy Leica UK Ltd)*

Unlike an automatic level, a tilting level will have a level vial fixed to its telescope to enable horizontal line of sight to be set. A *level vial* (figure 2.13) is a barrel-shaped glass tube, sealed at both ends, that is partially filled with purified synthetic alcohol. This is used because it is non-freezing, quick acting and maintains a stable length for normal temperature variations. The remaining space in the tube is an air bubble and, marked on the glass vial, is a series of graduations that are used to locate the relative position of the bubble within the vial. The imaginary tangent to the surface of the vial at the centre of these graduations is known as the axis of the level vial. When the bubble is centred in its run, that is, when it takes up a position in the tube with its ends an equal number of graduations (or divisions) either side of the centre of the vial, the axis should be horizontal, as shown in figure 2.13.

By attaching a level vial to a telescope such that the axis of the vial is parallel to the line of collimation, a horizontal line of sight may be set. This is achieved, with a tilting level, by adjusting the inclination of the

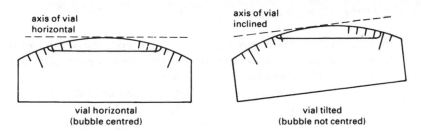

Figure 2.13   *Level vial*

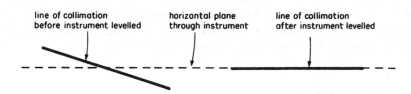

Figure 2.14   *Principle of the tilting level*

telescope with the tilting screw until the vial's bubble lies in the middle of its graduations. This is known as levelling the instrument and the principle is shown in figure 2.14.

A feature of many tilting levels is the coincidence bubble reader in which the bubble is centred by bringing both ends together, as shown in figure 2.15. This is achieved using a prism to view both ends of the bubble simultaneously and in most instruments, a magnified image of the bubble ends is seen enabling a very accurate setting of the bubble to be achieved.

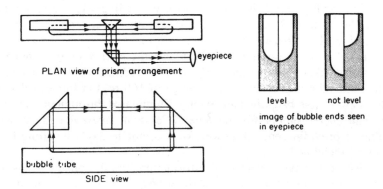

Figure 2.15   *Coincidence bubble reader*

2.9 Use of the Tilting Level

As with an automatic level, having set up the tripod, the tilting level is attached to it and the footscrews are used to centralise the circular bubble. This ensures that the instrument is almost level.

Parallax is now removed and the telescope rotated until it is pointing in the direction in which the first reading is required. The tilting screw is now turned until the main bubble is brought to the centre of its run or coincidence is obtained. This ensures that the optical axis of the telescope or, more precisely, the line of collimation is exactly horizontal in the direction in which the reading is to be taken. When the telescope is rotated to other directions, the main bubble will change its position for each setting of the telescope since the standing axis is not exactly vertical. Therefore, *the main bubble must be relevelled before every reading is taken.*

2.10 The Digital Level

The well known Swiss Company, Leica, introduced the Wild NA2000 in 1990 as the first level to measure, calculate and record electronically. Further developments of the NA2000 have led to the production of the NA2002 and NA3000.

Shown in figure 2.16, the Wild digital level uses electronic image-processing techniques and interrogates a specially made bar-coded staff in order to obtain readings. In operation, it is set up in the same way as an optical level by attaching it to a tripod and centralising a circular bubble using footscrews: this enables the compensator to set the line of sight horizontal. When levelling, the bar-coded staff is sighted, the focus is adjusted and the measuring key is pressed. There is no need to read the bar-coded staff as the display will show the staff reading four seconds after the measuring button has been pressed. In addition to staff readings, it is also possible to display the horizontal distance to the staff with a precision of 10 mm. The reverse side of the bar-coded staff has a normal 'E' type face (see section 2.12) and if it is not possible to take electronic staff readings, optical readings can be taken in the same way as with ordinary levels. In good conditions, the Wild digital level has a range of 100 m, but this can deteriorate if the staff is not brightly and evenly illuminated throughout its scanned area. The power supply for the level is a small internal battery which is capable of providing enough power for a complete day's levelling before it has to be recharged.

Because it generates electronic information, the digital level has a great advantage over conventional levels since observations can be automatically stored in a plug-in recording module supplied as an integral part of the instrument. This removes two of the worst sources of error from levelling:

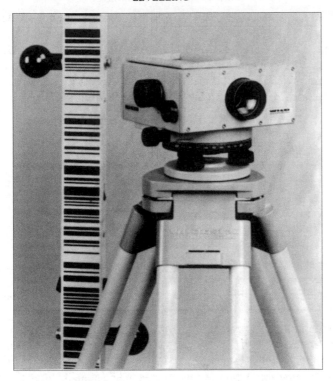

Figure 2.16   *Wild digital level (courtesy Leica UK Ltd)*

reading the staff incorrectly and writing the wrong value for a reading in a fieldbook. The Wild digital level also has a number of resident programs built into it, including one for the calculation of heights, and data generated by this program can also be stored in the recording module together with staff readings. This removes another source of error from levelling: the possibility of making mistakes in calculations. In order to be able to run programs and code readings a control panel, which is essentially a special keypad, is fitted to the front of the digital level. All data stored on the recording module can be transferred to a computer using a reader. The module is removed from the level and is inserted into the reader which is a device not unlike a disc drive (see figure 2.17): the contents are then sent by the reader to whatever computer is interfaced with it. The user can then pro—cess the data using any software but a typical print direct from the reader is shown in figure 2.18.

The digital level can be used for any type of levelling in the same way as an optical level but has the advantage of being able to measure and record electronically. The specialised application of the Wild digital level in sec-tioning is described in section 2.21.

Figure 2.17   *Wild recording module reader (courtesy Leica UK Ltd)*

2.11 Permanent Adjustment of the Level

In the preceding sections, the way in which automatic, tilting and digital levels are set up have been described. In each case, it is necessary to level the instruments such that the line of collimation, as viewed through the eyepiece, is horizontal. Adjustments of this nature are called *temporary adjustments* since these are carried out for every instrument position and, in some cases, for every pointing of the telescope.

So far, for every level discussed, the assumption has been made that once the temporary adjustments have been completed, the observed line of collimation is exactly horizontal. This, however, will only occur in a perfectly adjusted level, a case seldom met in practice. Hence, some checking method is required to ensure that the level is correctly adjusted. This is known as a *permanent adjustment* and should be undertaken at regular intervals during the working life of the equipment, for example, once a week, depending on its usage.

L E I C A (U K) Ltd.

WILD NA2000 FIELDBOOK PRINTOUT

File Name I:\SAMPLE\NA2000.DAT

Point Number	Back Sight	Fore Sight	Int. Sight	Reduced Level	Distance	S/N
140502	0.0000	0.0000	0.0000	95.7760	0.000	
140502	0.1623	0.0000	0.0000	95.7760	32.230	
1	0.0000	0.1560	0.0000	95.8395	0.000	
1	0.0853	0.0000	0.0000	95.8395	84.230	
2	0.0000	0.1772	0.0000	94.9208	56.100	
2	0.0236	0.0000	0.0000	94.9208	66.730	
3	0.0000	0.2057	0.0000	93.0994	37.440	
3	0.0478	0.0000	0.0000	93.0994	47.390	
4	0.0000	0.2292	0.0000	91.2847	47.070	
4	0.0026	0.0000	0.0000	91.2847	66.250	
5	0.0000	0.2195	0.0000	89.1154	47.880	
5	0.0879	0.0000	0.0000	89.1154	47.720	
6	0.0000	0.0000	0.1590	88.4044	19.430	
7	0.0000	0.1636	0.0000	88.3588	40.730	
7	0.0655	0.0000	0.0000	88.3588	82.830	
8	0.0000	0.1490	0.0000	87.5240	64.230	
8	0.0442	0.0000	0.0000	87.5240	10.000	
9	0.0000	0.2122	0.0000	85.8436	20.040	
9	0.0458	0.0000	0.0000	85.8436	77.850	
10	0.0000	0.1259	0.0000	85.0432	24.800	
10	0.2625	0.0000	0.0000	85.0432	47.140	
11	0.0000	0.0235	0.0000	87.4328	33.550	
11	0.2584	0.0000	0.0000	87.4328	23.040	
12	0.0000	0.0743	0.0000	89.2742	23.790	
12	0.1469	0.0000	0.0000	89.2742	13.740	
13	0.0000	0.0687	0.0000	90.0559	11.620	
13	0.0125	0.0000	0.0000	90.0559	64.620	
14	0.0000	0.2539	0.0000	87.6412	18.180	
14	−0.0006	0.0000	0.0000	87.6412	36.660	
14	0.0047	0.0000	0.0000	87.6412	37.160	
15	0.0000	0.2234	0.0000	85.4545	0.000	
15	0.1600	0.0000	0.0000	85.4545	79.830	
16	0.0000	0.1497	0.0000	85.5578	61.950	
16	0.2354	0.0000	0.0000	85.5578	68.880	
17	0.0000	0.0664	0.0000	87.2482	70.020	
17	0.2368	0.0000	0.0000	87.2482	33.600	
18	0.0000	0.1976	0.0000	87.6408	38.450	
Code : 00000900						
10039002	00000000	00000000				
149901	0.0000	0.0000	0.0000	87.6480	0.000	
149901	0.0514	0.0000	0.0000	87.6480	35.760	
1	0.0000	0.2631	0.0000	85.5306	27.410	
1	0.1139	0.0000	0.0000	85.5306	10.390	
2	0.0000	0.1825	0.0000	84.8446	9.330	
2	0.1530	0.0000	0.0000	84.8446	39.690	
3	0.0000	0.1837	0.0000	84.5375	38.790	
3	0.1973	0.0000	0.0000	84.5375	68.920	
4	0.0000	0.0231	0.0000	86.2799	21.930	
4	0.2037	0.0000	0.0000	86.2799	21.240	
5	0.0000	0.0671	0.0000	87.6463	31.420	
Code : 00000900						
00160290						
20	0.0000	0.0000	0.1524	50.5551	4.710	
21	0.0000	0.0000	0.1524	50.5552	4.710	

Figure 2.18 *Example print-out for NA2002 (courtesy Leica UK Ltd)*

Automatic and Digital Level Adjustment

The only permanent adjustment check necessary for an automatic and digital level is to ensure that the compensator and diaphragm are set such that horizontal readings are taken when the circular bubble is centralised.

If horizontal readings are not being taken then a *collimation error* is present in the level.

The usual method of testing and adjusting a level is to carry out a *two-peg test* which is carried out as follows, with reference to figures 2.19*a* and *b*.

(1) On fairly level ground, hammer in two pegs A and B a maximum of 60 m apart. Let this distance be L metres.
(2) Set up the level exactly midway between the pegs at point C and level carefully.
(3) Place a levelling staff (see section 2.12) at each peg in turn and obtain readings S_1 and S_2, as in figure 2.19*a*.

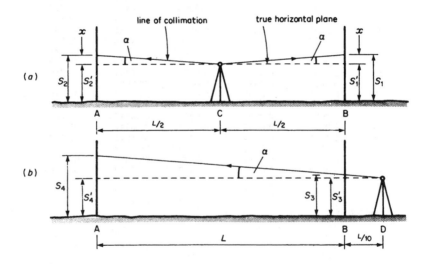

Figure 2.19 *Two-peg test*

Since AC = CB the error, x, in the readings S_1 and S_2, will be the same. This error is due to the collimation error, the effect of which is to incline the line of collimation by angle α. This gives

$$S_1 - S_2 \;=\; (S_1' + x) - (S_2' + x) = S_1' - S_2' \qquad (2.1)$$

$$=\; \textit{true difference in height between A and B}$$

In figure 2.19 the assumption has been made that the line of collimation, as set by the compensator and diaphragm, lies above the true horizontal plane. Even if this is not the case it does not affect the calculation procedure since the sign of the collimation error is obtained in the calculation as shown in the example at the end of this section.

(4) Move the level so that it is, preferably, $L/10$ m from peg B at D (see figure 2.19b) and take readings S_3 at B and S_4 at A. Compute the *apparent difference in height* between A and B from $(S_3 - S_4)$.
If the instrument is in adjustment $(S_1 - S_2) = (S_3 - S_4)$.
If there is any difference between the apparent and true values, this has occurred in a distance of L metres and hence

$$\text{Collimation error } (e) = (S_1 - S_2) - (S_3 - S_4) \text{ m}$$

$$\text{per } L \text{ metres} \tag{2.2}$$

If the error is found to be less than ±3 mm per 60 m the level is not adjusted. Instead, any readings taken must be observed over equal or short lengths so that the collimation error cancels out or is negligible.

(5) To adjust the instrument at point D, the correct reading that should be obtained at A, S_4', is computed from

$$S_4' = S_4 - [\text{Collimation error} \times \text{Sighting distance}]$$

A check on this reading is obtained by computing S_3' and by comparing $(S_3' - S_4')$ with the true difference in height (see the worked example in this section).

(6) With the level still at D and having deduced the correct reading S_4', the adjustment can be made by one of two methods.
For most instruments the cross hairs are moved using the diaphragm adjusting screws until the reading S_4' is obtained.
In some levels, however, it is necessary that the compensator itself is adjusted. Since this is a delicate operation, the level should be returned to the manufacturer for adjustment under laboratory conditions.
Some automatic levels have, in addition to a movable diaphragm, a special adjusting screw for the compensator. When adjusting such an instrument, the compensator screw should never be touched as its setting is precisely carried out by the manufacturer.

(7) The test should be repeated to ensure that the adjustment has been successful.

Tilting Level Adjustment

The only permanent adjustment check necessary for a tilting level is to ensure that the line of collimation is parallel to the axis of the level vial so that when the bubble is centred the line of sight is horizontal. Consequently, the two-peg test must be carried out as for the automatic level.

Having deduced the correct reading S_4', and with the level still at D while observing the staff at A, the tilting screw is adjusted until a reading of S_4' is obtained. However, this causes the main bubble to move from the centre of its run, so it is brought back to the centre by adjusting the level vial. As before, the test should be repeated at this stage to ensure that the test has been carried out correctly.

Worked Example: Two-peg Test

Question

The readings obtained from a two-peg test carried out on an automatic level with a single level staff set up alternately at two pegs A and B placed 50 m apart were as follows:

(1) With the level midway between A and B

> staff reading at A = 1.283 m
> staff reading at B = 0.860 m

(2) With the level positioned 5 m from peg B on the line AB produced

> staff reading at A = 1.612 m
> staff reading at B = 1.219 m

Calculate

(1) The collimation error of the level per 50 m of sight
(2) The reading that should have been observed on the staff at A from the level in position 5 m from B.

Solution

(1) Referring to figures 2.19a and b

$$S_1 = 0.860 \text{ m} \quad S_2 = 1.283 \text{ m} \quad S_3 = 1.219 \text{ m} \quad S_4 = 1.612 \text{ m}$$
$$\text{collimation error, } e = (0.860 - 1.283) - (1.219 - 1.612)$$
$$= \textbf{-0.030 m per 50 m}$$

(2) For the instrument in position 5 m from peg B, the reading that should have been obtained on the staff when held at A is

$$S_4' = 1.612 - \left[-\frac{0.030}{50} \right] 55 = 1.645 \text{ m}$$

This is checked by computing $(S_3' - S_4')$ and by comparing with $(S_1 - S_2)$ as follows

$$S_3' = 1.219 - \left[-\frac{0.030}{50}\right] 5 = 1.222 \text{ m}$$

Hence

$$(S_3' - S_4') = 1.222 - 1.645 = -0.423 = (S_1 - S_2)$$

$$\text{(checks)}$$

When solving problems of this nature it is important that the lettering sequence given in figure 2.19 for S_1 to S_4 is adhered to. If it is not, incorrect answers will be obtained.

2.12 The Levelling Staff

The *levelling staff* enables distances to be measured vertically above or below points on which it is held relative to a line of collimation. Many types of staff are in current use and these can have lengths of between 2 and 4 m although 4 m is the normal length for a staff. The staff can be rigid, telescopic, hinged, folding or socketed in as many as four sections for ease of carrying and is usually made of metal. The staff markings can take various forms but the 'E'-type staff face conforming to European DIN Standards, as shown in figure 2.20, is the most common. The staff can be read directly to 0.01 m, with estimation to 0.001 m.

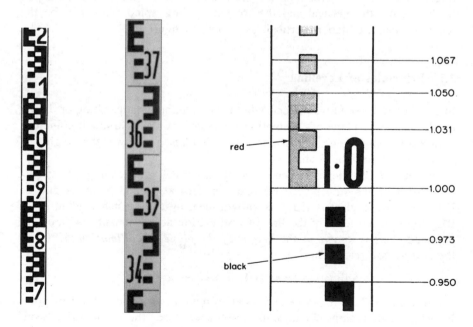

Figure 2.20 *Levelling staffs with example readings*

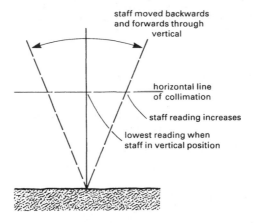

Figure 2.21   *Vertical reading on staff*

Since the staff is used to measure a vertical distance it must be held vertically and some staves are fitted with periscope-type handles and a circular bubble to assist in this operation. If no permanent bubble is fitted, a detachable circular bubble may be used. This device is mounted on a metal angle bracket and is either fixed to or is held against the staff when levelling. If no bubble is available, the staff should be slowly swung back and forth through the vertical and the lowest reading noted. This will be the reading when the staff is vertical, as shown in figure 2.21.

2.13 Principles of Levelling

In a correctly levelled instrument, whether it is an automatic, tilting or digital level, the line or plane of collimation generated by the instrument coincides with a horizontal plane. If the height of this plane is known, the heights of ground points can be found.

In figure 2.22, a level has been set up at point I_1 and readings R_1 and R_2 recorded with the staff placed vertically in turn at ground points A and B. If the reduced level of A (RL_A) is known then, by adding staff reading R_1 to RL_A, the reduced level of the line of collimation at instrument position I_1 is obtained. This is known as the *height of the plane of collimation* (HPC) or the *collimation level*. Thus

$$\text{collimation level at } I_1 = RL_A + R_1 \qquad (2.3)$$

From figure 2.22 it can be seen that to obtain the reduced level of point B (RL_B), staff reading R_2 must be subtracted from the collimation level, hence

$$RL_B = \text{collimation level} - R_2 = (RL_A + R_1) - R_2$$
$$= RL_A + (R_1 - R_2) \tag{2.4}$$

Since the direction of levelling is from A to B, the reading on A, R_1, is known as a *back sight* (BS) and that on B, R_2, a *fore sight* (FS).

From the above expression for RL_B and considering figure 2.22, the height difference between A and B is given by, in both magnitude and sign, $(R_1 - R_2)$. Furthermore, since R_1 is greater than R_2 and hence $(R_1 - R_2)$ is positive, the base of the staff must have risen from A to B and the expression $(R_1 - R_2)$ is known as a *rise*.

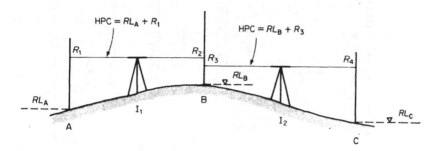

Figure 2.22   *Principles of levelling*

Referring to figure 2.22, assume the level is now moved to a new position I_2 in order that the reduced level of C may be found. Reading R_3 is first taken with the staff still at point B but with its face turned towards I_2. This will be the back sight at position I_2, the fore sight R_4 being taken with the staff at C. At point B, both a FS and a BS have been recorded consecutively, each from a different instrument position. A point such as B is called a *change point* (CP).

From the staff readings taken at I_2, the reduced level of C (RL_C) is calculated from

$$RL_C = RL_B + (R_3 - R_4)$$

The height difference between B and C is given both in magnitude and sign by $(R_3 - R_4)$. In this case, since R_3 is smaller than R_4, $(R_3 - R_4)$ is negative. The base of the staff must, therefore, have fallen from B to C and the expression $(R_3 - R_4)$ is known as a *fall*.

In practice, a BS is the first reading taken after the instrument has been set up and is always to a point of known or calculated reduced level. Conversely, a FS is the last reading taken before the instrument is moved. Any readings taken between the BS and FS from the same instrument position are known as *intermediate sights* (IS).

2.14 Field Procedure

A more complicated levelling sequence is shown in cross-section in figure 2.23a, in which an engineer has levelled between two TBMs to find the reduced levels of points A to E. The readings could have been taken with any type of level and figure 2.23b shows the levelling in plan view. The field procedure is as follows.

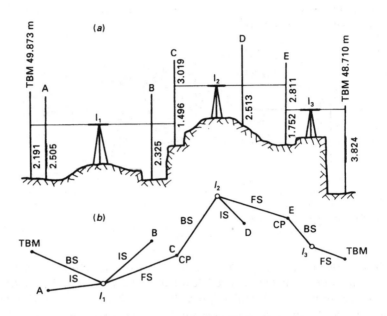

Figure 2.23 *Levelling sequence*

(1) The level is set up at some convenient position I_1 and a BS of 2.191 m taken to the first TBM, the foot of the staff being held on the TBM and the staff held vertically.

(2) The staff is moved to points A and B in turn and readings taken. These are intermediate sights of 2.505 m and 2.325 m respectively.

(3) A change point must be used in order to reach D owing to the nature of the ground. Therefore, a change point is chosen at C and the staff is moved to C and a reading of 1.496 m taken. This is a FS.

(4) While the staff remains at C, the instrument is moved to another position, I_2. A reading is taken from the new position to the staff at C. This is a BS of 3.019 m.

(5) The staff is moved to D and E in turn and readings taken of 2.513 m (IS) and 2.811 m (FS) respectively, E being another CP.

(6) Finally, the level is moved to I_3, a BS of 1.752 m taken to E and a FS of 3.824 m taken to the final TBM.

(7) The final staff position is at a point of known RL. This is most important as all levelling fieldwork must start and finish at points of known reduced level, otherwise it is not possible to detect misclosures in the levelling (see section 2.15).

2.15 Booking and Reduced Level Calculations

The booking and reduction of the readings discussed in section 2.14 can be done by one of two methods.

The Rise and Fall Method

The readings are shown booked by the rise and fall method in table 2.1. These are normally recorded in a level book containing all the relevant

TABLE 2.1
Rise and Fall Method
(all values in metres)

BS	IS	FS	Rise	Fall	Initial RL	Adj	Adj RL	Remarks
2.191					49.873	—	49.873	TBM 49.873
	2.505			0.314	49.559	+0.002	49.561	A
	2.325		0.180		49.739	+0.002	49.741	B
3.019		1.496	0.829		50.568	+0.002	50.570	C (CP)
	2.513		0.506		51.074	+0.004	51.078	D
1.752		2.811		0.298	50.776	+0.004	50.780	E (CP)
		3.824		2.072	48.704	+0.006	48.710	TBM 48.710
6.962		8.131	1.515	2.684	48.704			
8.131			2.684		49.873			
−1.169			−1.169		−1.169			

columns. Each line of the level book corresponds to a staff position and this is confirmed by the entries in the *Remarks* column. The calculation proceeds in the following manner, in which the reduced level of a point is related to that of a previous point.

(1) From the first TBM to A there is a fall (see figure 2.23). A BS of 2.191 m has been recorded at the TBM and an IS of 2.505 m at A. The

resulting height difference is given by (2.191 − 2.505) = −0.314 m. The negative sign indicates the fall and is entered against point A. This fall is subtracted from the RL of the TBM to obtain the initial reduced level of A as 49.559 m.

(2) The procedure is repeated and the height difference from A to B is given by (2.505 − 2.325) = +0.180 m. The positive sign indicates a rise and this is entered opposite B. The RL of B is (RL$_A$ + 0.180) = 49.739 m.

(3) This calculation is repeated until the initial reduced level of the final TBM is calculated, at which point a comparison can be made with the known value (see (6) below).

(4) When calculating the rises or falls the figures in the FS or IS columns must be subtracted from the figures in the line immediately above, either in the same column or one column to the left. At a CP, the FS is subtracted from the IS or BS in the line above and the BS on the same line as the FS is then used to continue the calculation with the next IS or FS in the line below.

(5) When the *Initial* RL column of the table has been completed, a check on the arithmetic involved is possible and must always be applied. This check is

$$\Sigma(\text{BS}) - \Sigma(\text{FS}) = \Sigma(\text{RISES}) - \Sigma(\text{FALLS})$$

$$= \text{LAST RL} - \text{FIRST RL} \qquad (2.5a)$$

It is normal to enter these summations at the foot of each relevant column in the levelling table (see table 2.1). Obviously, agreement must be obtained for all three parts of the check and it is stressed that this only provides a check on the *Initial RL* calculations and does not provide an indication of the accuracy of the readings.

(6) In table 2.1, the difference between the calculated and known values of the RL of the final TBM is −0.006 m. This is known as the *misclosure* and gives an indication of the accuracy of the levelling. If the misclosure is outside the *allowable misclosure* (see section 2.16) then the levelling must be repeated. If the misclosure is within the allowable value then it is distributed throughout the reduced levels. The usual method of correction is to apply an equal, but cumulative, amount of the misclosure to each instrument position, the sign of the adjustment being opposite to that of the misclosure. Table 2.1 shows a misclosure of −0.006 m, hence a total adjustment of +0.006 m must be distributed. Since there are three instrument positions, +0.002 m is added to the reduced levels found from each instrument position. The distribution is shown in the *Adj* (adjustment) column of table 2.1, in which the following cumulative adjustments have been applied. Levels A, B and C, +0.002 m; levels D and E, + (0.002 + 0.002) = +0.004 m; and the TBM, +(0.002 + 0.002 + 0.002) = +0.006 m. No adjustment is applied to the initial TBM since this level cannot be altered.

(7) The adjustments are applied to the *Initial RL* values to give the *Adj*

TABLE 2.2
Height of Collimation Method
(all values in metres)

BS	IS	FS	HPC	Initial RL	Adj	Adj RL	Remarks
2.191			52.064	49.873	—	49.873	TBM 49.873
	2.505			49.559	+0.002	49.561	A
	2.325			49.739	+0.002	49.741	B
3.019		1.496	53.587	50.568	+0.002	50.570	C (CP)
	2.513			51.074	+0.004	51.078	D
1.752		2.811	52.528	50.776	+0.004	50.780	E (CP)
		3.824		48.704	+0.006	48.710	TBM 48.710
6.962	7.343	8.131		300.420			

$$7.343 + 8.131 + 300.420 = 315.894$$
$$52.064 \times 3 + 53.587 \times 2 + 52.528 = 315.894$$

6.962				48.704			
8.131				49.873			
−1.169				− 1.169			

(*adjusted*) *RL* values as shown in table 2.1. These adjusted RL values are used in any subsequent calculations.

The Height of Collimation Method

The level book for the reduction of the levelling of figure 2.23 is shown in the height of collimation form in table 2.2. This method of reducing levels is based on the HPC being calculated for each instrument position and proceeds as follows.

(1) If the BS reading taken to the first TBM is added to the RL of this bench mark, then the HPC for the instrument position I_1 will be obtained. This will be $49.873 + 2.191 = 52.064$ m and is entered in the appropriate column.

(2) To obtain the initial reduced levels of A, B and C the staff readings to those points are now subtracted from the HPC. The relevant calculations are

$$RL \text{ of } A = 52.064 - 2.505 = 49.559 \text{ m}$$
$$RL \text{ of } B = 52.064 - 2.325 = 49.739 \text{ m}$$
$$RL \text{ of } C = 52.064 - 1.496 = 50.568 \text{ m}$$

(3) At point C, a change point, the instrument is moved to position I_2 and a new HPC is established. This collimation level is obtained by adding the BS at C to the RL found for C from I_1. For position I_2, the HPC is $50.568 + 3.019 = 53.587$ m. The staff readings to D and E are now subtracted from this to obtain their reduced levels.

(4) The procedure continues until the initial reduced level of the final TBM is calculated and the misclosure found as before. With the *Initial RL* column in the table completed, the following checks can be applied:

$$\Sigma(BS) - \Sigma(FS) = \text{LAST RL} - \text{FIRST RL} \qquad (2.5b)$$

and

$$\Sigma IS + \Sigma FS + \Sigma RLs \text{ except first} = \Sigma(\text{each HPC} \times \text{number of applications}) \qquad (2.6)$$

Equation (2.5b) only checks reduced levels calculated using BS and FS readings and is shown at the bottom of table 2.2. Equation (2.6) checks reduced levels calculated from IS readings and is added as follows.

Table 2.2 gives

$$\Sigma IS + \Sigma FS + \Sigma RLs \text{ except first} = 7.343 + 8.131 + 300.420$$
$$= 315.894$$

The first HPC, 52.064, was used to calculate the levels of A, B and C and was therefore used three times. The second HPC, 53.587, was used twice to calculate the levels of D and E, and the last HPC of 52.528 was used only once to close the levels onto the final TBM. This gives the second part of the check as

$$52.064 \times 3 + 53.587 \times 2 + 52.528 \times 1 = 315.894$$

After applying the check, any misclosure is distributed as for the rise and fall method.

Summary of the Two Methods

The rise and fall method is quicker to reduce where a lot of back sights and fore sights have been taken and very few intermediate sights taken. For this reason, the rise and fall method tends to be used when establishing control when no intermediate sights would normally be taken.

However, the collimation method is quicker to reduce where a lot of intermediate sights have been taken since fewer calculations are required and it is a good method to use when setting out levels where, usually, many readings are taken from each instrument position. A disadvantage of this method is that the check can be lengthy.

2.16 Precision of Levelling

For normal engineering work and site surveys the *allowable misclosure* for any levelling sequence is given by

$$allowable\ misclosure = \pm5\sqrt{n}\ \text{mm} \qquad (2.7)$$

where n is the number of instrument positions. For example, the allowable misclosure for tables 2.1 and 2.2 is $\pm5\sqrt{3} = \pm9$ mm.

When the actual and allowable misclosures are compared and it is found that the actual value is greater than the allowable value, the levelling should be repeated. If, however, the actual value is less than the allowable value, the misclosure should be distributed equally between the instrument positions as already described. The precision of levelling is also discussed in sections 6.10 and 6.11.

2.17 Errors in Levelling

Many sources of error exist in levelling and those most commonly met in practice are discussed.

Errors in the Equipment

Collimation error

This can be a serious source of error in levelling if sight lengths from one instrument position are not equal, since the collimation error is proportional to the difference in sight lengths. Hence, in all types of levelling, sights should be kept equal, particularly back sights and fore sights. Also, before using any level it is advisable to carry out a two-peg test to ensure that the collimation error is as small as possible (see section 2.11).

Compensator not working

The compensator is discussed in section 2.6 and is checked by moving a footscrew slightly off-level or by tapping the instrument gently to ensure that a reading remains constant as described in section 2.7. If the reading changes to a different position each time the footscrew is moved or the instrument tapped, the compensator is not working properly and the instrument should be returned to the manufacturer for repair.

Parallax

This effect, described in section 2.5, must be eliminated before any readings are taken.

Defects of the staff

It is possible that staff graduations may be incorrect and new or repaired staves should be checked against a steel tape. Particular attention should be paid to the base of the staff to see if it has become badly worn. If this is the case then the staff has a *zero error*. This does not affect height differences if the same staff is used for all the levelling but introduces errors if two staves are being used for the same series of levels. When using a multisection staff, it is important to ensure that it is properly extended by examining the graduations on either side of each joint. If these joints become loose, the staff should be returned for repair.

Tripod defects

The stability of tripods should be checked before any fieldwork commences by testing to see if the tripod head is secure, that the metal shoes at the base of each leg are not loose and that, once extended, the legs can be tightened sufficiently. When fitted, the wing nuts must be tightened before readings are taken.

Field Errors

Staff not vertical

Since the staff is used to measure a vertical difference between the ground and the line of collimation, failure to hold the staff vertical will result in incorrect readings. As stated in section 2.12, the staff is held vertical with the aid of a circular bubble, or it is rocked. At frequent intervals the circular bubble should be checked against a plumb line and adjusted if necessary.

Unstable ground

When the instrument is set up on soft ground and bituminous surfaces on hot days, an effect often overlooked is that the tripod legs may sink into the ground or rise slightly while readings are being taken. This alters the height of collimation and it is therefore advisable to choose firm ground on which to set up the level and tripod, and to ensure that the tripod shoes are pushed well into the ground.

Similar effects can occur with the staff and for this reason it is particularly important that change points should be at stable positions such as manhole covers, kerbstones, concrete surfaces, and so on. This ensures that the base of the staff remains at the same level during all observations to its position. If a stable point cannot be found for a change point, a *change*

Figure 2.24   *Ground plate (courtesy Leica UK Ltd)*

plate or *ground plate* should be used (see figure 2.24) on soft ground. Alternatively, a large stone firmly pushed into the ground can be used.

For both the level and staff, the effect of soft ground is greatly reduced if readings are taken in quick succession.

Handling the instrument and tripod

As well as vertical displacement, the HPC may be altered for any set-up if the tripod is held or leant against. When levelling, avoid contact with the tripod and only use the level by light contact through the fingertips. If at any stage the tripod is disturbed, it will be necessary to relevel the instrument and to repeat all the readings taken from that instrument position.

Instrument not level

For automatic levels this source of error is unusual but, for a tilting level in which the tilting screw has to be adjusted for each reading, this is a common mistake. The best procedure here is to ensure that the main bubble is centralised before and after a reading is taken.

Reading and Booking Errors

Many mistakes are made during the booking of staff readings, and the general rule is that staff sightings must be carefully entered into the levelling table or field book immediately after reading.

Another source of reading error is sighting the staff over too long a distance, when it becomes impossible to take accurate readings. It is, therefore, recommended that sighting distances should be limited to 50 m but, where absolutely unavoidable, this may be increased to a maximum of 100 m.

The Effects of Curvature and Refraction on Levelling

At point A in figure 2.25 where a level has been set up it can be seen that the level and horizontal lines through the instrument diverge (see also figure 2.3). This is caused by level lines following the curvature of the Earth which is defined as mean sea level and is a possible source of error in levelling since all readings taken at A are observed along the horizontal line instead of the level line. The difference between a horizontal and level line is given by

$$c = 0.0785 \, D^2 \tag{2.8}$$

where
c = curvature in metres
D = sighting distance in kilometres.

As shown in figure 2.25, the effect of curvature is to cause staff readings to be too high.

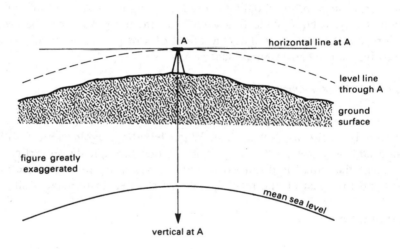

Figure 2.25 *Curvature*

The effect of atmospheric refraction on a line of sight is to bend it towards the Earth's surface causing staff readings to be too low. This is a variable effect depending on atmospheric conditions but for ordinary work refraction is assumed to have a value 1/7 that of curvature but is of opposite sign. The combined curvature and refraction correction is thus

$$c + r = 0.0673 \, D^2 \tag{2.9}$$

The combined correction for a length of sight of 120 m amounts to −0.001 m and the effect of both is thus negligible when undertaking levelling if sightings are less than 120 m, as should always be the case. If longer sight lengths must be used, it is worth remembering that the effects of curvature and refraction will cancel if the sight lengths are equal.

However, curvature and refraction effects cannot always be ignored when calculating heights using theodolite methods and this is discussed in section 3.11.

Weather Conditions

Windy conditions cause the level to vibrate and give rise to difficulties in holding the staff steady. Readings cannot be recorded accurately under these circumstances unless the instrument is sheltered and the minimum number of sections of the staff used.

In hot weather, the effects of refraction are serious and produce a shimmering effect near ground level. This makes it impossible to read accurately the bottom metre of the staff which, consequently, should not be used.

2.18 Summary of the Levelling Fieldwork

When levelling, the following practice should be adhered to if many of the sources of error are to be avoided.

(1) Levelling should always start and finish at points of known reduced level so that misclosures can be detected. When only one bench mark is available, levelling lines must be run in loops starting and finishing at the bench mark.
(2) Where possible, all sights lengths should be below 50 m.
(3) The staff must be held vertically by suitable use of a circular bubble or by rocking the staff and noting the minimum reading.
(4) BS and FS lengths should be kept equal for each instrument position. For engineering applications, many IS readings may be taken from each set-up. Under these circumstances it is important that the level has no more than a small collimation error.
(5) Readings should be booked immediately after they are observed and important readings, particularly at change points, should be checked.
(6) The rise and fall method of reduction should be used when heighting reference or control points and the HPC method should be used for contouring, sectioning and setting out applications.

2.19 Additional Levelling Methods

Inverted Staff

Occasionally, it may be necessáry to determine the reduced levels of points
such as the soffit of a bridge, underpass or canopy. Generally, these points
will be above the line of collimation. To obtain the reduced levels of such
points, the staff is held upside down in an inverted position with its base
on the elevated points. When booking an inverted staff reading it is entered
in the levelling table with a *minus sign*, the calculation proceeding in the
normal way, taking this sign into account.

An example of a levelling line including inverted staff readings is shown
in figure 2.26, table 2.3 showing the reduction of these readings.

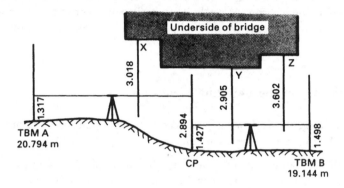

Figure 2.26 *Inverted staff levelling*

TABLE 2.3
Inverted Staff Readings
(all values in metres)

BS	IS	FS	Rise	Fall	Initial RL	Adj	Adj RL	Remarks
1.317					20.794	—	20.794	TBM A 20.794
	−3.018		4.335		25.129	−0.001	25.128	X
1.427		2.894		5.912	19.217	−0.001	19.216	CP
	−2.905		4.332		23.549	−0.002	23.547	Y
	−3.602		0.697		24.246	−0.002	24.244	Z
		1.498		5.100	19.146	−0.002	19.144	TBM B 19.144
2.744		4.392	9.364	11.012	19.146			
4.392			11.012		20.794			
−1.648			−1.648		−1.648			

Each inverted reading is denoted by a minus sign and the rise or fall computed accordingly. For example, the rise from TBM A to point X is $1.317 - (-3.018) = 4.335$ m. Similarly, the fall from point Z to TBM B is $-3.602 - 1.498 = -5.100$ m.

An inverted staff position must *not* be used as a change point since there is often difficulty in keeping the staff vertical and in keeping its base in the same position for more than one reading.

Reciprocal Levelling

True differences in height are obtained by ensuring that BS and FS lengths are equal when levelling. This eliminates the effect of any collimation error that may be present in the level used and also eliminates the effects of curvature and refraction.

There are certain cases, however, when it may not be possible to take readings with equal sight lengths as, for instance, when a line of levels has to be taken over a wide gap such as a river. In these cases, the technique of *reciprocal levelling* can be adopted.

Figure 2.27 shows two points A and B on opposite sides of a wide river. The line of collimation has been assumed to be elevated above the horizontal plane. This may not be the case but does not affect the calculations. To obtain the true difference in level between A and B a level is placed at I_1, about 5 m from A, and a staff is held vertically at A and B. Staff readings are taken at A (a_1) and B (b_1). The level is next taken to position I_2 where readings a_2 and b_2 are recorded.

If AB is a considerable distance, several sets of readings are taken with the instrument being relevelled in a slightly different position for each set. Average values of a_2 and b_1 are then recorded.

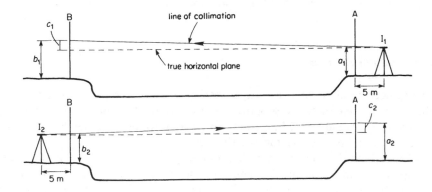

Figure 2.27 *Reciprocal levelling*

Since the observations are taken over the same sighting distances with the same level, the effects of the collimation error will be the same for both cases. Hence the true difference in level ΔH_{AB} is the mean of the two observed differences at I_1 and I_2 and

$$\Delta H_{AB} = \frac{1}{2}\left[(a_1 - b_1) + (a_2 - b_2)\right] \qquad (2.10)$$

When reciprocal levelling with one level, the two sets of observations must follow each other as soon as possible so that refraction effects are the same and are therefore eliminated. Where this is not possible, two levels have to be used simultaneously. It must be realised that the levels should have the same collimation error or the true height difference will not be obtained.

2.20 Applications of Levelling: Sectioning

Levelling has many uses in civil engineering construction. Levels are needed principally in setting out, sectioning and contouring. The applications of levelling in setting out are fully described in chapter 14 and contouring is described in detail in sections 2.22 to 2.25.

Sectioning is usually undertaken for construction work such as roadworks, railways and pipelines. Two types of section are often necessary and these are called *longitudinal* and *cross-sections*.

On many road schemes, longitudinal and cross-sections can be generated by a computer interrogating a Digital Terrain Model (DTM). This is discussed further in sections 9.12 and 13.24 and only the fieldwork required to produce sections by levelling is discussed here.

Longitudinal sections

In engineering surveying, a *longitudinal section* (or *profile*) is taken along the complete length of the proposed centre line of the construction showing the existing ground level. Levelling can be used to measure heights at points on the centre line so that the profile can be plotted.

Generally, this type of section provides data for determining the most economic *formation level*, this being the level to which existing ground is formed by construction methods. The optimum position for the formation level is usually found by using a computer-aided design package but the longitudinal section is sometimes drawn by hand and a mass-haul diagram prepared (see chapter 13).

The fieldwork in longitudinal sectioning normally involves two operations.

Firstly, the centre line of the section must be set out on the ground and marked with pegs. For most works, this is done by theodolite and some form of distance measurement so that pegs are placed at regular intervals (frequently 20 m) along the centre line. Further details of the techniques involved in this stage are given in chapters 10, 11 and 14. Secondly, as soon as the centre line has been established levelling can commence.

The levelling techniques adopted should all conform with the general rules already put forward and these will dictate where the level is to be set up, what bench marks are used and when change points are necessary. For longitudinal sections, it is usually sufficiently accurate to record readings to the nearest 0.01 m. Levels are taken at the following points, the object being to survey the ground profile as accurately as possible.

(1) At the top and ground level of each centre line peg, noting the through chainage of the peg.
(2) At points on the centre line at which the ground slope changes.
(3) Where features cross the centre line, such as fences, hedges, roads, pavements, ditches and so on. At points where, for example, roads or pavements cross the centre line, levels should be taken at the top and bottom of kerbs. At ditches and streams, the levels at the top and bottom of any banks as well as bed levels are required.
(4) Where necessary, inverted staff readings to underpasses and bridge soffits would be taken.

In order to be able to plot levels obtained in addition to those taken at the centre line pegs, the position of each extra point on the centre line must be known. These distances are recorded by measurement with a tape, the tape being positioned horizontally between appropriate centre line pegs.

The method of booking longitudinal sections should always be by the height of collimation method since many intermediate sights will be taken. Distances denoting chainage should be recorded for each level and most commercially available level books have a special column for this purpose. Careful booking is required to ensure that each level is entered in the level book with the correct chainage. Good use should be made of the 'remarks' column in this type of levelling so that each point can be clearly identified when plotting.

When all the fieldwork has been completed and the level book checked, the results can be plotted. The longitudinal section for a small valley is shown in figure 2.28b and its associated level book in figure 2.28a. A longitudinal section is also shown in chapter 13, figure 13.13.

66

BS	IS	FS	HPC	RL	DIST	REMARKS

LEVELS TAKEN FOR PIPELINE CROSSING OF BRAMBER VALLEY
JU/WPP   19 November 1994   Level N° 27 Staff LS14

<table>
<tr><td>BS</td><td>IS</td><td>FS</td><td>HPC</td><td>RL</td><td>DIST</td><td>REMARKS</td></tr>
<tr><td>0.55</td><td></td><td></td><td>35.58</td><td>35.03</td><td></td><td>OBM 35.03 m AOD</td></tr>
<tr><td></td><td>0.48</td><td></td><td></td><td>35.10</td><td>0</td><td>fence</td></tr>
<tr><td></td><td>0.42</td><td></td><td></td><td>35.16</td><td>20</td><td></td></tr>
<tr><td></td><td>0.50</td><td></td><td></td><td>35.08</td><td>24.3</td><td>change of slope (CS)</td></tr>
<tr><td></td><td>1.13</td><td></td><td></td><td>34.45</td><td>40</td><td></td></tr>
<tr><td></td><td>1.57</td><td></td><td></td><td>34.01</td><td>60</td><td></td></tr>
<tr><td></td><td>1.65</td><td></td><td></td><td>33.93</td><td>65.2</td><td>fence</td></tr>
<tr><td></td><td>1.82</td><td></td><td></td><td>33.76</td><td>69.6</td><td>CS</td></tr>
<tr><td></td><td>2.49</td><td></td><td></td><td>33.09</td><td>74.9</td><td>CS</td></tr>
<tr><td></td><td>2.50</td><td></td><td></td><td>33.08</td><td>80</td><td></td></tr>
<tr><td></td><td>2.60</td><td></td><td></td><td>32.98</td><td>100</td><td></td></tr>
<tr><td></td><td>2.61</td><td></td><td></td><td>32.97</td><td>102.1</td><td>bank</td></tr>
<tr><td></td><td>3.47</td><td></td><td></td><td>32.11</td><td>102.2</td><td>bed</td></tr>
<tr><td></td><td>2.78</td><td></td><td></td><td>32.80</td><td>103.0</td><td>water level</td></tr>
<tr><td></td><td>3.40</td><td></td><td></td><td>32.18</td><td>107.6</td><td>bed</td></tr>
<tr><td></td><td>2.58</td><td></td><td></td><td>33.00</td><td>107.8</td><td>bank</td></tr>
<tr><td></td><td>2.56</td><td></td><td></td><td>33.02</td><td>120</td><td></td></tr>
<tr><td>2.20</td><td></td><td>2.40</td><td>35.38</td><td>33.18</td><td>128.0</td><td>edge of footpath</td></tr>
<tr><td></td><td>2.21</td><td></td><td></td><td>33.17</td><td>130.1</td><td>edge of footpath</td></tr>
<tr><td></td><td>2.03</td><td></td><td></td><td>33.35</td><td>140</td><td></td></tr>
<tr><td></td><td>2.00</td><td></td><td></td><td>33.38</td><td>150.4</td><td>edge of road</td></tr>
<tr><td></td><td>1.92</td><td></td><td></td><td>33.46</td><td>152.6</td><td>℄</td></tr>
<tr><td></td><td>1.98</td><td></td><td></td><td>33.40</td><td>155.2</td><td>edge of road</td></tr>
<tr><td></td><td>1.87</td><td></td><td></td><td>33.51</td><td>160</td><td></td></tr>
<tr><td></td><td>1.68</td><td></td><td></td><td>33.70</td><td>173.5</td><td>CS</td></tr>
<tr><td></td><td>1.40</td><td></td><td></td><td>33.98</td><td>180</td><td></td></tr>
<tr><td></td><td>0.81</td><td></td><td></td><td>34.57</td><td>191.8</td><td>CS</td></tr>
<tr><td></td><td>0.66</td><td></td><td></td><td>34.72</td><td>200</td><td></td></tr>
<tr><td></td><td>0.48</td><td></td><td></td><td>34.90</td><td>219.3</td><td>fence</td></tr>
<tr><td></td><td></td><td>0.22</td><td></td><td>35.16</td><td></td><td>OBM 35.16 m AOD</td></tr>
<tr><td>2.75</td><td></td><td>2.62</td><td></td><td>35.16</td><td></td><td></td></tr>
<tr><td></td><td></td><td>2.62</td><td></td><td>35.03</td><td></td><td></td></tr>
<tr><td></td><td></td><td>+0.13</td><td></td><td>+0.13</td><td></td><td></td></tr>
</table>

(a)

N

ploughed — 0m, 24.3
grass — 65.2, 69.6, 74.9
102.1, 102.2 — stream
103.0, 107.6, 107.8
128.0 grass — conc. footpath
130.1
150.4
TM road — 152.6
155.2
grass — 173.5, 191.8, 219.3

levels also taken at 20m intervals

LONGITUDINAL SECTION — BRAMBER VALLEY

(b)

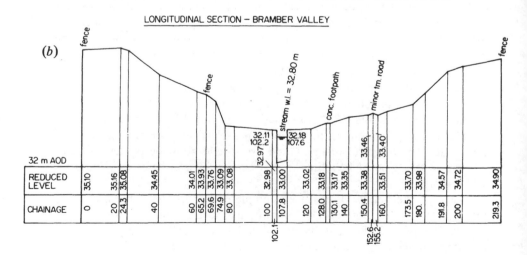

	32 m AOD																								

stream w.l. = 32.80 m
conc. footpath
minor tm. road
fence ... fence

32.11 / 102.2   32.18 / 107.6   32.97   33.46   33.40

REDUCED LEVEL: 35.10 35.16 35.08 34.45 34.01 33.93 33.09 33.08 32.98 33.00 33.02 33.18 33.17 33.35 33.38 33.51 33.70 33.98 34.57 34.72 34.90

CHAINAGE: 0 20 24.3 40 60 65.2 69.6 74.9 80 100 107.8 120 128.0 130.1 140 150.4 160. 173.5 180 191.8 200 219.3   (102.1) (152.6 / 155.2)

Figure 2.28 *Longitudinal section: (a) level book; (b) format for drawing section*

Cross-sections

A longitudinal section provides information only along the centre line of a proposed project. For works such as sewers or pipelines, which usually are only of a narrow extent in the form of a trench cut along the surveyed centre line, a longitudinal section provides sufficient data for the construction to be planned and carried out. However, in the construction of other projects such as roads and railways, existing ground level information at right angles to the centre line is required. This is provided by taking *cross-sections*. These are sections taken at right angles to the centre line such that information is obtained over the full width of the proposed construction.

For the best possible accuracy in sectioning a cross-section should be taken at every point levelled on the longitudinal section. Since this would involve a considerable amount of fieldwork, this rule is generally not observed and cross-sections are, instead, taken at regular intervals along the centre line usually where pegs have been established. A right angle is set out at each cross-section either by eye for short lengths or by theodolite for long distances or where greater accuracy is needed. A ranging rod is placed on either side of the centre line to mark each cross-section.

The longitudinal section and the cross-sections are usually levelled in the same operation. Starting at a TBM or OBM, levels are taken at each centre line peg and at intervals along each cross-section. These intervals may be regular, for example, 10 m, 20 m, 30 m on either side of the centre line peg or, where the ground is undulating, levels should be taken at all changes of slope such that a good representation of existing ground level is obtained over the full width of the construction. The process is continued taking both longitudinal and cross-section levels in the one operation and the levelling is finally closed on another known point.

Such a line of levels can be very long and can involve many staff readings and it is possible for errors to occur at stages in the procedure. The result is that if a large misclosure is found, all the levelling will have to be repeated, often a soul destroying task. Therefore, to provide regular checks on the levelling it is good practice to include points of known height such as traverse stations or TBMs at regular intervals in the line of levels and then, if a large discrepancy is found, it can be isolated into a short stretch of the work.

Examples of plotted cross-sections are shown in figure 2.29 and figure 13.14, and the applications of the results in earthwork calculations are considered in chapter 13.

2.21 Use of the Digital Level in Sectioning

Although the Wild digital level (see section 2.10) has a recording module which makes fieldwork and calculations simpler, electronic levelling is even

BS	IS	FS	HPC	RL	DIST	REMARKS

BADGER LANE – EXISTING LEVELS   CHAINAGE 420m
JU/WFP   26 November 1994   Level N°16 StaffLS B

1·32			150·47	149·15		TBM 149·15m ROADNAIL on ₵ AT
						CH 420
						LEFT
	1·43			149·04	2·3	channel
	1·26			149·21	2·3	kerb
	1·29			149·18	3·4	footpath edge
	1·13			149·34	4	⎫
	1·20			149·27	5	⎬ ground levels
	1·42			149·05	6	⎪
	1·68			148·79	7	⎪
	2·11			148·36	8	⎭
						RIGHT
	1·45			149·02	2·1	channel
	1·28			149·19	2·1	kerb
	1·32			149·15	3·2	footpath edge
	1·15			149·32	4	⎫
	1·10			149·37	5	⎬ ground levels
	1·01			149·46	6	⎪
	0·63			149·84	7	⎪
	0·46			150·01	8	⎭
		1·32		149·15		TBM 149·15m on ₵ at CH420

Note: LEFT and RIGHT viewed when
facing direction of increasing
chainage

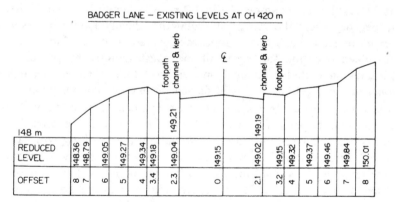

BADGER LANE – EXISTING LEVELS AT CH 420 m

Figure 2.29 *Cross-section drawing and associated field book*

more sophisticated when a digital level is connected to a field computer. Optimal Solutions, who specialise in surveying software, have developed a levelling program for a Husky Hunter field computer (see section 5.15) that can take data directly from a Wild digital level. This set-up is shown in figure 2.30 and with this, readings are recorded automatically in the field computer by the Optimal levelling program but the user can be prompted to

Figure 2.30 *Digital level with field computer (courtesy Leica UK Ltd.)*

input data manually if automatic readings cannot be taken for some reason. The program takes in standard levelling information such as bench mark values and, since it was designed for road construction surveys, allows each point surveyed to be identified by its chainage, offset, point number and remarks. The Husky Hunter is capable of displaying all of these for the current and five previous points together with an instant calculation of the reduced level of each point. Full on-screen editing of all data is possible and a search facility allows the user to locate previously observed data. There is also an option to edit the starting bench mark and recalculate all levels. Checks on fieldwork are automatic and the Husky will highlight any misclosure obtained between bench marks.

The program also has the facility to output data to MOSS, a computer-based highway design package (see section 11.23). MOSS produces such items as longitudinal and cross-sections much more quickly than is possible by hand, and the data for preparing these sections is obtained from the digital level which transfers readings directly to the Optimal levelling program which in turn transfers reduced levels plus other information to MOSS. Although there are no significant gains in time spent on site when using a digital level combined with the levelling program, a considerable amount of time can be saved by being able to transfer reduced levels from the field directly to a computer-aided design package such as MOSS without having to record or calculate anything by hand.

The levelling program also has the facility to download levels from MOSS into the Husky, together with chainage and offset. This enables an instant comparison to be made with observed data on site since the program can display the difference between these.

2.22 Contouring

A *contour* is defined as a line joining points of the same height above or below a datum. These are shown so that the relief or topography of an area can be interpreted, a factor greatly used in civil engineering design and construction.

The difference in height between successive contours is known as the *contour* or *vertical interval* and this interval dictates the accuracy to which the ground is represented. The value chosen for any application depends on

(1) the intended use of the plan
(2) the scale of the plan
(3) the costs involved
(4) the nature of the terrain.

Generally, a small vertical interval of up to 1 m is required for engineering projects, for large-scale plans and for surveys on fairly even sites. In hilly or broken terrain and at small scales, a wider vertical interval is used. Very often, a compromise has to be reached on the value chosen since a smaller interval requires more fieldwork time, thus increasing the cost of the survey.

Electronic instruments such as total stations are normally used to collect data for contouring and contours are generated and plotted making use of computer software and hardware. Such a process, from field to finished plan, can be fully automated if electronic data capture and transfer is used. The instruments and methods used in these surveys are described in chapters 5 and 9. The remainder of this chapter is concerned with how levelling can be used to obtain contours and how they can be plotted by hand.

If drawn manually, contours can be obtained either directly or indirectly using mathematical or graphical interpolation techniques. Once plotted, in addition to indicating the relief of an area, contours can be used to provide sectional information.

2.23 Direct Contouring

In this method the positions of contours are located on the ground by levelling.

A level is set up in the area so that as much ground as possible can be covered by staff observations from the instrument position. A back sight is taken to a bench mark or other point of known reduced level and the height of collimation calculated. For example

$$
\begin{aligned}
\text{RL of bench mark} &= 51.87 \text{ m AOD} \\
\text{BS} &= 1.78 \text{ m} \\
\text{HPC} &= 53.65 \text{ m}
\end{aligned}
$$

To locate each contour the required staff readings are

$$\text{At } 50 \text{ m contour} = 53.65 - 50.00 = 3.65 \text{ m}$$
$$51 \text{ m contour} = 53.65 - 51.00 = 2.65 \text{ m}$$
$$52 \text{ m contour} = 53.65 - 52.00 = 1.65 \text{ m}$$
$$53 \text{ m contour} = 53.65 - 53.00 = 0.65 \text{ m}$$

Considering the 52 m contour, the surveyor directs the person holding the staff to move until a staff reading of 1.65 m is obtained. At this point a signal is given by the surveyor so that the staff position can be marked with a peg or chain arrow. With the staff in other positions the procedure is repeated until the complete 52 m contour is clearly marked on the ground. Only one contour is set out at any one time.

When other contours are located, care must be taken to ensure that the pegs or chain arrows of different contours are coded so that one set cannot be mistaken for another. As soon as all the contours have been marked on the ground the plan positions of all the pegs or chain arrows have to be established. This can be done by some convenient detail surveying method.

Since two operations are involved the method takes longer than others but the advantage of the technique is that it is accurate.

2.24 Indirect Contouring

This involves the heighting of points that do not, in general, coincide with the contour positions. Instead, the points levelled are used as a framework on which contours are later interpolated on a drawing.

Two of the more common methods of indirect contouring involve taking levels either on a regular grid pattern or at carefully selected points.

Grid Levelling

The area to be contoured is divided into a series of lines forming squares and ground levels are taken at the intersection of the grid lines.

The sides of the squares can vary from 5 to 30 m, the actual figure depending on the accuracy required and on the nature of the ground surface (see table 9.1). The more irregular the ground surface the greater the concentration of grid points. Methods of setting out the grid are numerous and one such method is considered here.

Four lines of ranging rods are set out by taping, as shown in figure 2.31, such that each ranging rod marks a grid point. Stepping of the tape will be necessary to establish a horizontal grid. To obtain the ground level at each grid point the person holding the staff lines the staff in with the two ranging rods in each direction that intersect at the point being levelled, and

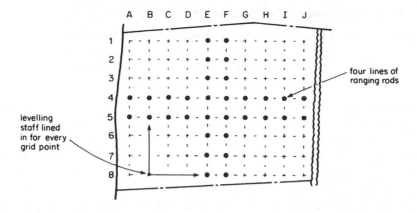

Figure 2.31 *Grid levelling*

a reading is taken. The procedure is repeated at all grid points. Where a ranging rod marks a grid point the staff is placed against the rod and the reading taken.

When taking each reading, a suitable reference system should be adopted, for example, B8 as shown in figure 2.31, and rigorously maintained during the location of each point and the booking of each reading.

Following the fieldwork, the levels are reduced, the grid is plotted and the contours interpolated either graphically or mathematically, taking into account the general shape of the land as observed during the fieldwork.

This method of contouring is ideally suited to gently sloping areas but the setting out of the grid on a large area can take a considerable time. Furthermore, if visibility is restricted across the site, difficulties can occur when locating grid points.

Contours from Selected Points

For large areas or areas containing a lot of detail, contours can be drawn from levels taken at points of detail or at prominent points on open ground such as obvious changes of slope. These points will have been plotted on the plan by one of the methods discussed in chapter 9 and hence the position of each level, or *spot height* as it is called, is known. These spot heights will form a random pattern but the contours are drawn by interpolation as in grid levelling.

This technique is obviously well suited to detail surveying and is the usual method of contouring such surveys. As in all methods, a sufficient number of levels must be recorded so that the ground surface can be accurately represented on the site plan.

2.25 Interpolating Contours

In the direct method of contouring, spot heights are located at exact contour values, plotted on a plan and individual contours are drawn by joining spot heights of equal value with a smooth curve.

In the indirect methods, the plotted spot heights will not be at exact contour values and it is necessary to locate points between them on the plan which do have exact contour values. This is known as *interpolation* and it can be carried out either mathematically or graphically.

The assumption is made when undertaking interpolation that the surface of the ground slopes uniformly between the spot heights. Hence, careful positioning of spot heights in the field is essential if accurate contours are to be produced.

Mathematical Interpolation

This can be a laborious process when there are a large number of spot heights.

The height difference between each spot height is calculated and used with the horizontal distance between them to calculate the position on the line joining the spot heights at which the required contour is located.

With reference to figure 2.32, in which the positions of the 36 m and 37 m contours are to be located between two spot heights A and B of reduced level 37.2 m and 35.8 m respectively. By simple proportion

$$\frac{0.2}{x} = \frac{1.2}{y} = \frac{1.4}{28.7}$$

from which

$$x = 4.1 \text{ m}$$
$$y = 24.6 \text{ m}$$

Horizontal distances x and y are scaled along line BA on the plan to fix the positions of the 36 m and 37 m contours respectively.

When all the exact contour positions have been plotted, they are joined by smooth curves as in the direct method.

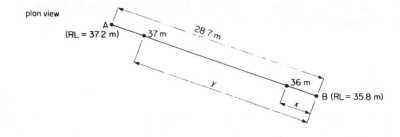

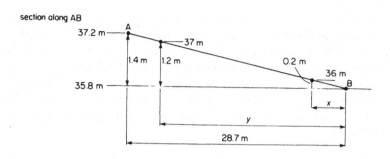

Figure 2.32 *Mathematical interpolation of contours*

Graphical Interpolation

This is a much quicker method where there are large numbers of spot heights. The procedure is as follows.

(1) A piece of tracing paper is prepared with a series of equally spaced horizontal lines as shown in figure 2.33a. Every tenth line is drawn heavier than the others.
(2) The tracing paper is then laid between pairs of spot heights and rotated until the horizontal lines corresponding to the known spot height values pass through the points as shown in figure 2.33b.
(3) The heavy lines indicate the positions of the contour lines where they pass over the line joining the spot heights and these positions are pricked through on to the drawing paper using a sharp point.
(4) The reduced level of each contour is written lightly next to its position. When all the exact contour positions have been located they are joined by smooth curves.

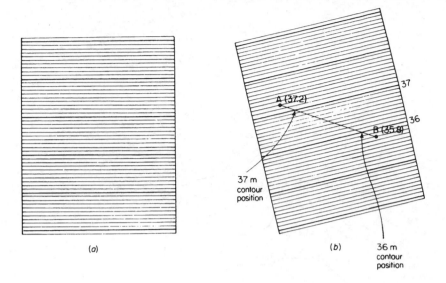

Figure 2.33 *Graphical interpolation of contours*

2.26 Obtaining Sections from Contours

It is possible to use contours to obtain sectional information for use in the initial planning of such projects as roads, pipelines, earthworks and reservoirs.

Figure 2.34 shows part of a contoured plan of an area. The line XX is the proposed route for a straight section of a road centre line and relevant cross-sections are shown at chainages 525 m to 625 m. Using the contours, the approximate shape of the longitudinal and cross-sections can be obtained by scaling height and distance information from the plan at points where the section lines cut contours as shown in figure 2.35.

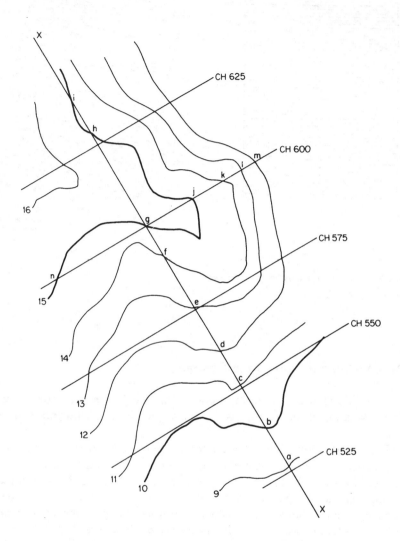

Figure 2.34 *Contoured plan with sections*

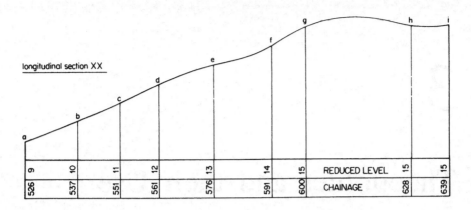

longitudinal section XX

	9	10	11	12	13	14	15	REDUCED LEVEL	15	15
	526	537	551	561	576	591	600	CHAINAGE	628	639

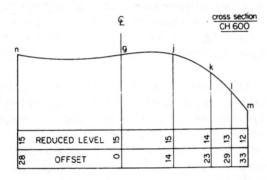

cross section
CH 600

15	REDUCED LEVEL	15	15	14	13	12
28	OFFSET	0	14	23	29	33

Figure 2.35 *Longitudinal and cross-sections from contours*

3

Theodolites and their Use

Theodolites are precision instruments used extensively in construction work for measuring angles in the horizontal and vertical planes. Two types of theodolite are used and these are either optical theodolites which need to be read manually or electronic theodolites which are capable of displaying readings automatically.

Many different theodolites are available for measuring angles and they are often classified according to the smallest reading that can be taken with the instrument. This can vary from 1' to 0.1" and, for example, a 1" theodolite is one which can be read to 1" directly without any estimation.

At this point, it is worth noting that a full circle is 360° or 400^g and a reading system capable of resolving to 1" directly shows the degree of precision in the manufacture of theodolites. The gons angular unit is widely used in Europe instead of degrees, minutes and seconds for the measurement and setting out of angles. In order to relate the type of theodolite to its intended application, the size an angle subtends over a distance is given in table 3.1 for some of the reading precisions of theodolites. Using table 3.1 it is evident that, if a 5 mm tolerance was specified on site for work over distances of up to 100 m, a 6" reading theodolite would be needed to meet this requirement. At distances of 50 m for the same tolerance, a 20" theodolite would be adequate. On the majority of civil engineering projects, it

TABLE 3.1
Angular Precisions

1'	subtends	10 mm	at	34 m
20"	subtends	10 mm	at	103 m
6"	subtends	10 mm	at	344 m
1"	subtends	10 mm	at	2063 m

is not necessary to use 1″ theodolites for most setting out, although they are used when establishing control.

3.1 Principles of Angle Measurement

Figure 3.1*a* shows two points S and T and a theodolite set up on a tripod over a ground point R. The RL of S is greater than that of R which, in turn, is greater than that of T.

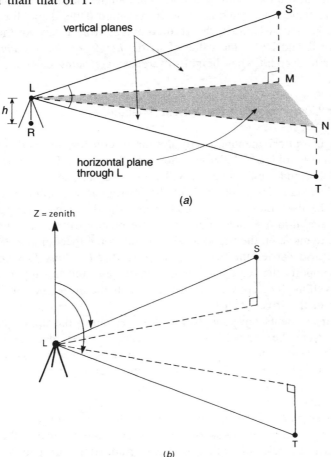

(a)

(b)

Figure 3.1   *(a) Horizontal and vertical angles; (b) zenith angles*

The theodolite is mounted at point L, a vertical distance h above R for ease of observation.

The *horizontal angle* at L between S and T is angle MLN, where M and N are the vertical projections of S and T on to the horizontal plane through L.

The *vertical angles* to S and T from L are angle SLM (an *angle of elevation*) and angle TLN (an *angle of depression*).

Another angle often referred to is the *zenith angle*. This is defined as the angle between the direction vertically above the theodolite and the line of sight, for example angles ZLS and ZLT in figure 3.1*b*.

In order to measure horizontal and vertical angles, the theodolite must be *centred* over point R using a plumbing device and must be *levelled* to bring the angle reading systems of the theodolite into the appropriate planes. Although centring and levelling ensure that horizontal angles measured at point L are the same as those that would have been measured if the theodolite had been set on the ground at point R, the vertical angles from L are not the same as those from R and hence the value of *h*, the *height of the instrument*, must be taken into account when height differences are being calculated.

3.2 Basic Components of an Optical Theodolite

All types of optical theodolite are similar in construction and the general features of the Sokkia TM20H are shown in figure 3.2. The various parts of the theodolite and their functions will now be described.

The *trivet stage* forms the base of the instrument and in order to be able to attach the theodolite to the tripod, most tripods have a *clamping screw* which locates into a $\frac{5}{8}$ inch threaded centre on the trivet. This enables the theodolite to move on the tripod head and allows the theodolite to be centred. The trivet also carries the feet of three *threaded levelling footscrews*. The *tribrach* supports the remainder of the instrument and is supported in turn by the levelling footscrews. The tribrach can, therefore, be levelled independently of the trivet stage.

Many instruments have the facility for detaching the upper part of the theodolite from the tribrach. A special target or other piece of equipment can then be centred in exactly the same position occupied by the theodolite, as shown in figures 3.3 and 3.8. This ensures that angular and linear measurements are carried out between the same positions, thereby reducing errors, particularly *centring errors* (see section 3.10).

The *lower plate* of the theodolite carries the *horizontal circle*. The term *glass arc* has been used to describe optical theodolites because the horizontal and vertical circles on which the angle graduations are photographically etched are made of glass. Many types of optical theodolite are available, varying in reading precision from 1′ to 0.1″ although 20″ and 6″ reading theodolites are most commonly used in engineering surveying.

An *upper plate* or *alidade* is recessed into and can be free to rotate within the lower plate. The upper plate carries the horizontal circle reading system. The various circle reading systems are described in section 3.3.

The *plate level* is also fixed to the upper plate and this is identical to the

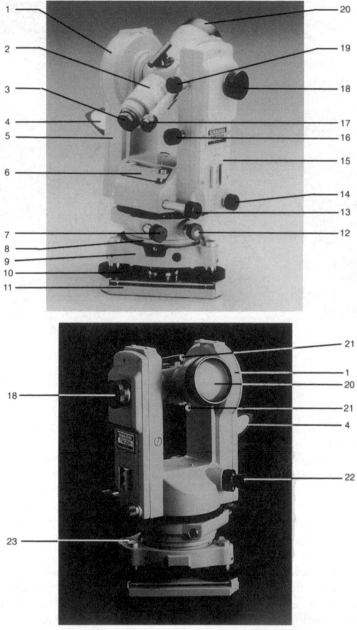

Figure 3.2   *Sokkia TM20H optical theodolite : 1. vertical circle; 2. main telescope focus; 3.
eyepiece; 4. reflecting mirror for reading system; 5. standard; 6. plate level; 7. lower
plate tangent screw; 8. horizontal circle; 9. tribrach; 10. levelling footscrews; 11. trivet;
12. lower plate clamp; 13. and 14. upper plate tangent screw and clamp; 15. standard;
16. telescope clamp; 17. circle reading eyepiece; 18. micrometer screw; 19. telescope
tangent screw; 20. telescope objective; 21. sight; 22. optical plummet; 23. circular
bubble (courtesy Sokkia Ltd)*

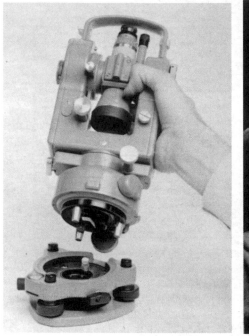

(a)

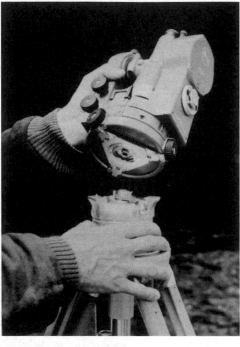

(b)

Figure 3.3   *Forced centring: (a) Wild system; (b) Kern system (courtesy Leica UK Ltd)*

level vial of an optical level as shown in figure 2.13 and is mounted on the upper plate.

On earlier models, the upper and lower plates each have a separate clamp and *slow motion* or *tangent screw* and, to distinguish these, the upper plate screws are milled and the lower plate screws are serrated. For this type of theodolite, if the lower plate is clamped and the upper plate free, rotation in azimuth gives different readings on the horizontal scale. If the lower plate is free and the upper plate clamped, rotation in azimuth retains the horizontal scale reading, that is, the horizontal circle rotates.

Many theodolites do not have a lower plate clamp and tangent screw. There is a facility for altering the position of the horizontal circle within the instrument and this can be achieved using a *horizontal circle setting screw*, as shown in figure 3.33, or by use of a *repetition clamp*, as shown in figure 3.32.

The upper plate also supports two frames called the *standards*. Supported in bearings carried on the standards is the *trunnion* or *transit axis* of the theodolite. Attached to the trunnion axis are the *main telescope*, the *circle reading telescope*, the *micrometer screw* and the *vertical circle*. The mi-

crometer screw is used when horizontal and vertical circle readings are being taken (see section 3.7).

The *focusing screw* of the telescope is fitted concentrically with the bar-
rel of the telescope and the diaphragm (and also the circles) can be illuminated for night or tunnel work. When the main telescope is rotated in altitude about the trunnion axis from one direction to face in the opposite direction it has been *transitted*. The side of the main telescope, viewed from
the eyepiece, containing the vertical circle is called the *face*. The construction of the main telescope is similar to those used in optical levels as described in section 2.4 and it can be clamped in the vertical plane, a *tangent
screw* being provided for fine vertical movement. Fine horizontal movement
is achieved using the upper plate tangent screw (and lower plate tangent
screw, if fitted).

The vertical circles of theodolites are not all graduated in the same way
and it is necessary to reduce the readings to obtain the required vertical
angles (see section 3.8). Some of the graduation systems in use are shown
in figure 3.4.

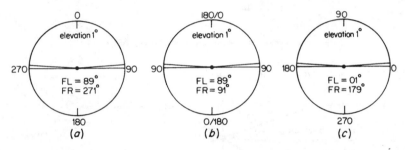

Figure 3.4 *Vertical circle graduations*

Built into the standard containing the vertical circle is a device called an
automatic vertical index (see figure 3.5). This is similar to the compensator
in an automatic level described in section 2.6. Once the theodolite has been
levelled using the plate level and footscrews, the compensator ensures that
the theodolite vertical circle reading system is set properly and the theodolite can be used to read vertical angles.

The arrangement of the axes of the theodolite is shown in figure 3.6.
When the instrument is *levelled*, the vertical axis is made to coincide with
the vertical at the point where the instrument is set up. This is achieved by
using the levelling footscrews and plate level as described in section 3.6.

Centring the theodolite involves setting the vertical axis directly above a
particular point. A hook is provided so that a plumb line can be suspended
underneath the tribrach or centring clamp in order to roughly centre the
instrument within 5 mm. Fine centring is done using the *optical plummet*.

Figure 3.5   *Automatic vertical index (courtesy Sokkia Ltd)*

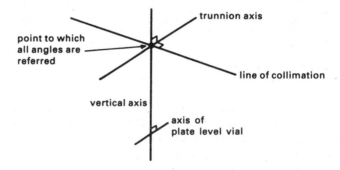

Figure 3.6   *Theodolite axes*

This consists of a small eyepiece, either built into the tribrach or the alidade, the line of sight of which is deviated by 90° so that a point corresponding to the vertical axis can be viewed on the ground. The two types of optical plummet are shown in figure 3.7.

    Some instrument tripods can be fitted with a *centring rod* as a further method of improving centring accuracy. The rod either forms part of the tripod or is detachable. As shown in figure 3.8, the top of the rod is at-

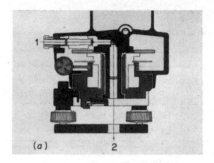
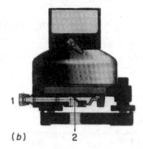

Figure 3.7 *Sections through lower halves of theodolites showing: (a) optical plummet mounted on alidade; (b) optical plummet mounted on tribrach. 1. Eyepiece; 2. line of sight along vertical axis (courtesy Leica UK Ltd)*

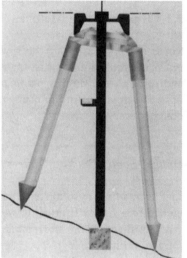

Figure 3.8 *Kern centring system (courtesy Leica Uk Ltd)*

tached to an adaptor plate which, when the rod is moved, slides on the tripod head. A circular level fixed to the rod enables it to be set vertically. When the rod is placed in a vertical position with its base centred over a station mark, the rod and hence the adaptor plate is clamped and the theodolite is centred automatically by fixing it to the adaptor plate as shown in figure 3.3*b*.

Table 3.2 shows a comparison of three centring methods.

TABLE 3.2

Comparison of Centring Methods

Method	Advantages	Disadvantages	Accuracy of centring over point (mm)
Suspended plumb bob	Cheap	Difficult to use in windy conditions	1–2
Optical plummet	Not affected by weather	Must be in good adjustment Takes longer to use	1
Centring rod	Quicker than optical plummet Useful in hilly terrain	Extra piece of equipment to carry	1

3.3 Circle Reading Methods

Since the standards are hollow on optical theodolites and the circles are made of transparent glass, it is possible to direct light into the instrument and through the circles using prisms, and to magnify and read the images using a circle reading telescope. There are three types of reading system in common use.

Optical Scale Reading System

In this reading system, a fixed plate of transparent glass, upon which are etched two scales from 0' to 60', is mounted in the optical path of the light directed through the horizontal and vertical circles, as shown in figure 3.9 for the Wild T16.

When viewed through the circle eyepiece these two scales are seen superimposed on portions of the horizontal and vertical circles and are highly magnified. Readings are obtained directly from the fixed scales, as shown for the Wild T16 instrument in figure 3.10. The length of each scale corresponds exactly to the distance between the images of the circle graduations and there is no possibility of ambiguity.

This system is often referred to as the *direct reading* system since no micrometer adjustment is required (see next sections) to obtain readings.

Only one side of the circle is seen by this method and any circle eccentricity is not eliminated but these errors are likely to be less than the reading accuracy which is direct to 1' with estimation to 0.1'.

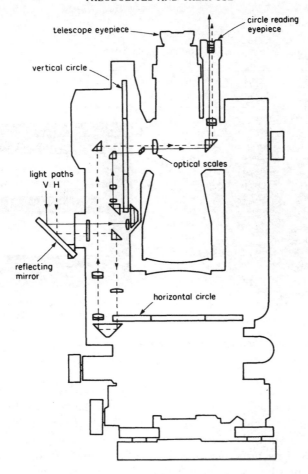

Figure 3.9 *Wild T16 reading system (courtesy Leica UK Ltd)*

Single-reading Optical Micrometer Reading System

This reading system does not have a fixed scale mounted in the optical path. Instead, an optical micrometer is built into the instrument on the standard containing the reading telescope. The micrometer arrangement and optical paths for such a theodolite are shown in figure 3.11 for the Wild T1.

The important part of the optical micrometer is the parallel-sided glass block. This can be rotated by turning the micrometer screw to which the block is geared. If light from the circles enters the block at a right angle it will pass through undeviated, as shown in figure 3.12a. If, however, the

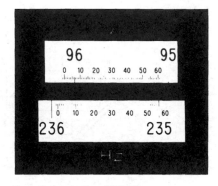

vertical circle reading 96°06.5'
horizontal circle reading 235°56.4'

Figure 3.10 *Wild T16 and reading examples (courtesy Leica UK Ltd)*

block is rotated through an angle θ, the light will be deviated by an amount, d, parallel to the incident ray (see figure 3.12b). It can be shown that the amount of this shift, d, is directly proportional to the angle of rotation of the block, θ. This principle is used to obtain angle readings using index marks built into the optical path which are seen superimposed on the circle images. Suppose the horizontal scale is set as in figure 3.13; the reading will be $62° + x$. By turning the micrometer screw and hence the parallel-sided glass block, the $62°$ graduation can be displaced laterally until it appears to coincide with the index marks as in figure 3.13. The horizontal scale has, therefore, effectively been moved an amount x proportional to the rotation of the glass block. This angular rotation is recorded on a micrometer scale attached to the glass block, the relevant portion of which is seen in the circle eyepiece.

The circle readings are made up of two parts, as shown in figure 3.14. As with the optical scale system, only one side of the circle is read, hence the term *single-reading* optical micrometer. The reading accuracy of the Wild T1 is direct to 6″ with estimation to 3″.

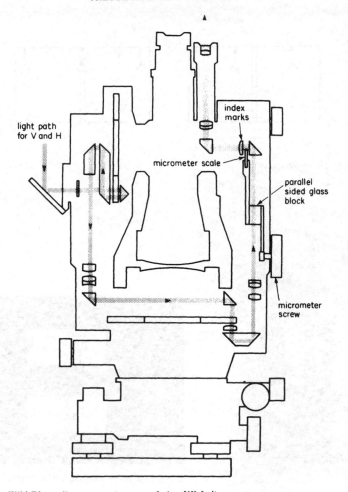

Figure 3.11 *Wild T1 reading system (courtesy Leica UK Ltd)*

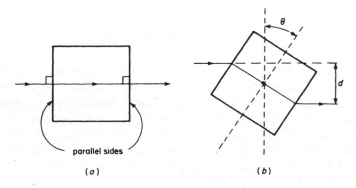

Figure 3.12 *Parallel-sided glass block*

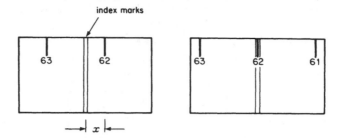

Figure 3.13   *Micrometer setting*

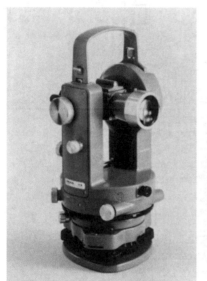

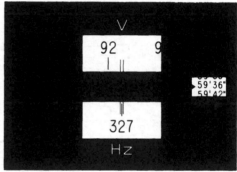

horizontal circle reading 327°59'36"

Figure 3.14   *Wild T1 and reading example (courtesy Leica UK Ltd)*

Double-reading Optical Micrometer Reading System

When reading horizontal (or vertical) angles on a theodolite, if the opposite
sides of the circle are read simultaneously and meaned, the effects of any
circle eccentricity errors are eliminated. This is demonstrated in figure 3.15,
which shows the horizontal circle in plan view (the same theory can be
applied to the vertical circle). If the line joining two diametrically opposed
points A and B on the upper plate corresponds with the centre of the horizontal circle on the lower plate (figure 3.15a) the readings at A (11° 40′
+ x) and B (191° 40′ + y) would be recorded with a mean of 11° 40′ + $\frac{1}{2}$
(x + y). The B degrees are not taken into account. In this case, x = y and,

in theory, only one of the readings, A or B, need be taken. In most cases, however, there is a small displacement between the centre of the horizontal circle and the line joining the two points A and B (see figure 3.15b). The readings here would be A $(11° 40' + x_1)$ and B $(191° 40' + y_1)$, where $x_1 = x + \delta$ and $y_1 = y - \delta$, δ being the eccentricity error. If only one side of the circle is read, the error δ will be included, but if the mean is taken, this gives

$$11° 40' + \tfrac{1}{2}(x_1 + y_1) = 11° 40' + \tfrac{1}{2}((x + \delta) + (y - \delta))$$

$$= 11° 40' + \tfrac{1}{2}(x + y)$$

which is the same value obtained when assuming that $\delta = 0$.

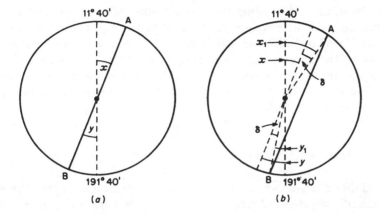

Figure 3.15 *Circle eccentricity errors*

If only one side of the circle is read, circle eccentricity errors will be eliminated by reading on both faces *provided the value of δ remains constant*. Reading only one side of the circle is an acceptable practice when using 20″ or 6″ theodolites but, when using theodolites reading to 1″ or less, the effects of a variable circle eccentricity must be accounted for and this is achieved with the double-reading optical micrometer system. However, instead of noting two separate readings at opposite ends of the circles and calculating the mean, an arrangement is used whereby only one reading is necessary for each setting. With reference to figure 3.15b, imagine that the readings at the opposite ends of the circle were to be deviated through optical paths so that they were viewed simultaneously in the circle eyepiece, as shown in figure 3.16.

If the 11° 40′ and 191° 40′ graduations are made to appear to coincide using parallel-sided glass blocks, they will be optically deviated by amounts

x_1 and y_1 respectively. The mean reading of $11° \ 40' + (x_1 + y_1)$ will be free from circle eccentricity errors provided a suitable optical micrometer system is used to record the mean of the lateral displacement of the two circle images.

The optical arrangement of the double reading Wild T2 is shown in figure 3.17. The optical micrometer basically consists of two parallel-sided glass plates which rotate equally in opposite directions. The images of each side of the circle are brought into coincidence by rotating a single micrometer screw geared to these plates, the amount of rotation being recorded, via a cam, on the moving micrometer scale. The optics are designed so that the micrometer scale reading is the mean of the two circle displacements, free of eccentricity error. The method of reading the Wild T2 is illustrated in figure 3.18.

Double-reading theodolites can use slightly differing optical micrometer systems. In some there is only one parallel plate and this displaces an image of one side of the circle so that it coincides with the other side which is stationary. The parallel plates are replaced by wedges in some designs.

Very often, these instruments do not show the horizontal and vertical scales together in the same field of view. A change-over switch is provided to switch from one scale to the other.

3.4 Electronic Theodolites

A theodolite that produces a digital output of direction or angle is known as an electronic theodolite. These are very similar in appearance to optical theodolites but when using such an instrument the operator does not have to look into a circle reading telescope or set a micrometer screw to obtain a reading. Instead, readings are displayed automatically using a liquid crystal display as shown in figure 3.19 for the Sokkia DT5.

As shown in section 3.3, optical theodolites use glass circles and a series of prisms mounted in the standards to measure angles. Developments in microelectronics have enabled these reading systems to be replaced by electronic components and a microprocessor. All circle measuring systems fitted into electronic theodolites still use a glass circle but this is marked or coded in a special way. Within the theodolite, light is passed through the encoded circle and the light pattern emerging through the circle is detected by photodiodes. Two measurement systems are used to scan this light pattern and these are known as *incremental* and *absolute*. When the horizontal or vertical circle of an electronic theodolite is rotated in an incremental reading system, the amount of incident light passing through to the photodiodes varies in proportion to the angle through which the theodolite has been rotated. This varying light intensity is converted into electrical signals by the photodiodes and these in turn are passed to a microprocessor which con-

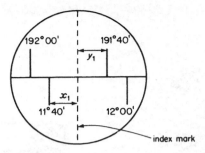

Figure 3.16 *Double reading optical system*

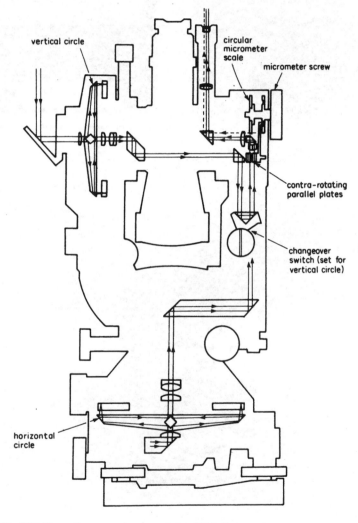

Figure 3.17 *Wild T2 reading system (courtesy Leica UK Ltd)*

horizontal or vertical circle reading 94°12'44"

Figure 3.18 *Wild T2 and reading example (courtesy Leica UK Ltd)*

Figure 3.19 *Sokkia DT5 electronic theodolite (courtesy Sokkia Ltd)*

verts the signals into an angular output. In an absolute reading system, the light pattern emerging through the circle is unique at every point around the circle. This is detected by an array of photodiodes and processed electronically to give the required reading. An example of both reading systems is given in the following sections.

Wild T2002 Electronic Theodolite

The angle measuring system of the Wild T2002 (figure 3.20) consists of a glass circle which is divided into 1024 equally spaced intervals and two pairs of photodiodes mounted diametrically opposite each other. This is shown in figure 3.21 but only for one set of photodiodes. In figure 3.21 one of the photodiodes is fixed and this corresponds to the 'zero mark' on the circle and the movable photodiode corresponds to the direction in which the telescope points. When the theodolite is rotated, the movable photodiode is also rotated but this does not cause the circle itself to rotate.

During the measurement of an angle such as ϕ in figure 3.21, the circle is rotated by a drive motor built into the instrument and a complete rotation is made for each measurement. As light passes through the circle, the circle graduations cause a square wave to be generated by each photodiode, as

Figure 3.20   *Wild T2002 electronic theodolite (courtesy Leica UK Ltd)*

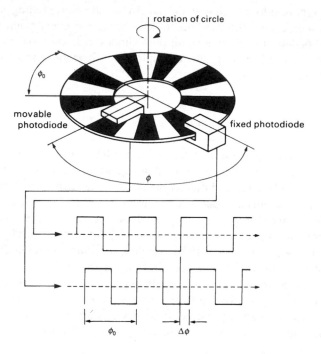

Figure 3.21   *Wild T2002 reading system (courtesy Leica UK Ltd)*

shown in figure 3.21, and these are processed to give a coarse measurement $n\phi_0$ and fine measurement $\Delta\phi$ of the angle where $\phi = n\phi_0 + \Delta\phi$. This method of measurement is directly analogous to that used in electromagnetic distance described in section 5.2.

The coarse measurement is made by an electronic counter which determines the whole number of graduations n between the fixed and movable photodiodes. This is achieved by the photodiodes recognising reference marks on the circle. As soon as one of the photodiodes recognises a reference mark, the counter begins to count graduations until the other photodiode recognises the same reference mark.

The fine measurement is made by determining the phase difference $\Delta\phi$ between the signals generated by a fixed and movable pair of photodiodes. During a single rotation of the circle, this is carried out over 500 times and because the measurements are taken at opposite sides of the circle and are taken at all points on the circle by rotating it, circle eccentricity and circle graduation errors are eliminated. This enables the T2002 to measure angles with a precision of 0.5″ although it has a resolution of 0.1″.

Wild T1010 and T1610 Electronic Theodolites

These instruments (see figure 3.22) have glass circles with 1152 graduation marks encoded into 128 sectors each sector consisting of one sector mark, seven identification marks from 0000000 to 1111111 in binary and a parity check mark. The circle is illuminated by a light emitting diode (LED) and red light passing through the circle from this is directed through optics such that about 1 per cent of the circle is projected onto a photodiode array. The array is made up of 128 separate photodiodes mounted on a small chip some 3.2 mm long.

For both theodolites, the horizontal circle remains in a fixed position and the photodiode array rotates with the alidade whereas the vertical circle is attached to the telescope and the photodiode array is fixed.

As with the T2002, an angle is measured in two parts: the coarse and fine components. However, the T1010 and T1610 use very different methods compared with the T2002: the coarse measurement is a matter of identifying the sector in which the theodolite is pointing and the fine measurement is proportional to that fraction of a sector in which the theodolite is pointing. Both of these are obtained by processing the output from the 128 photodiodes. Since this is different for each position around the circle, the T1010 and T1610 use an absolute system for measurement of angles.

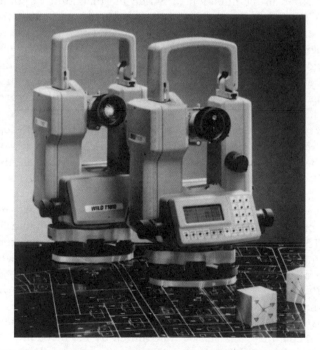

Figure 3.22   *Wild T1010 and T1610 electronic theodolites (courtesy Leica UK Ltd)*

The T1010 has an angular precision of 3″ since it only scans the circle at one point whereas the T1610, which also scans the circle at one point, has a precision of 1.5″. This is made possible by measuring the circle eccentricity errors in the factory and individually programming these into each T1610's software and applying appropriate corrections. Both theodolites display angles to 1″.

Features of Electronic Theodolites

In construction and operation, the electronic theodolite is very similar to the traditional optical theodolite but has a number of additional useful features. Apart from the automatic display of angles from 0.1″ to 20″ many electronic theodolites have some or all of the following features.

Single or dual-axis compensation. This is such an important feature of electronic theodolites (and total stations) that it is described separately in the next section.

Hold/release key. When pressed, this causes the horizontal display to be locked. This is used when setting the horizontal circle to a particular value.

OSET or reset key. This enables a reading of 00° 00′ 00″ to be set on the horizontal circle in the direction in which the theodolite is pointing.

R/L key. Conventionally, the horizontal circle of an optical theodolite is graduated clockwise when viewed from above. This means the readings increase when the telescope is rotated clockwise. By pressing the R/L key, this direction can be reversed to make readings increase when the telescope is rotated anticlockwise. This is a very useful feature for setting out left-handed road curves (see chapters 10 and 11) and in other types of setting out (see chapter 14).

Per cent key. The vertical circle of some electronic theodolites can be made to read percentage of slope instead of a vertical angle. This is accessed by pressing the per cent key. In addition, the vertical circle can also be made to read differently: some of the commonly used options for this are shown in figure 3.23.

Battery power. All electronic theodolites use some form of battery which is usually clipped into one of the standards, as shown in figure 3.24 for the Sokkia DT5. Many different types of battery are used in electronic theodolites, the most popular being a nickel–cadmium rechargeable with an operating time of up to 20 hours in some instruments.

Display illumination. All electronic theodolites have a built-in illumination function for both the display and cross hairs.

Data transfer. Since angle information is generated in a digital format by electronic theodolites, this can be transmitted by the theodolite to a suitable storage device for subsequent processing by a computer. Data is transferred

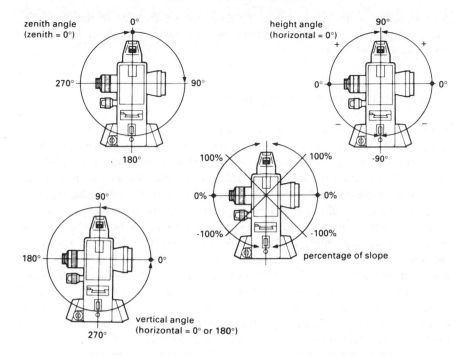

Figure 3.23   *Vertical angle measurement modes (courtesy Sokkia Ltd)*

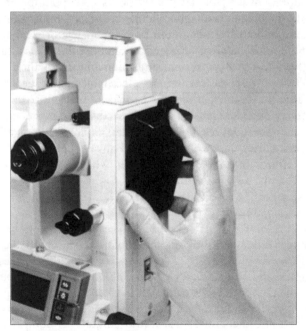

Figure 3.24   *Sokkia DT5 battery (courtesy Sokkia Ltd)*

by connecting a storage device to the data port on the side of the theodolite. Electronic storage devices and methods of data capture are described in more detail in section 5.15.

3.5 Single and Dual-axis Compensators

All theodolites, whether electronic or optical, have to be levelled manually using the footscrews and the plate level. The exact procedure for this is described in section 3.6. However, no matter how carefully a theodolite is levelled, it is unusual for the vertical axis of the instrument to coincide exactly with the vertical through the theodolite and this tilt, even though it is small, can give rise to errors in displayed horizontal and vertical angles.

In the case of optical theodolites, a compensator is built into the light path of the vertical angle reading system to correct for vertical axis tilt in the direction in which the telescope is pointing. This ensures that the vertical circle reading system is set to some multiple of 90° when the telescope is horizontal even if the instrument is not levelled exactly. Consequently, all readings taken using the vertical circle are corrected for vertical axis tilt before reading takes place. It is not possible to correct horizontal circle readings taken with an optical theodolite for the effects of vertical axis tilt until after readings have been taken. This involves a separate calculation and the recording of the plate level position for each pointing of the telescope.

Electronic theodolites and total stations correct for the effects of vertical axis tilt using a liquid or pendulum type compensator. Unlike optical theodolites, the compensator of an electronic theodolite is not mounted in the optical path of the reading system and compensation values are calculated separately from circle readings using electrical signals generated by electronic tilt sensors. In such instruments, the compensator can sometimes be switched off and the amounts of tilt can be displayed: this enables the theodolite to be digitally levelled. However, this is not recommended and under normal circumstances the theodolite should always be levelled using the plate level. This ensures that the compensator is within its working range.

Single-axis Compensation

The effect of vertical axis tilt in the direction in which the telescope is pointing is shown in figure 3.25. As can be seen, this causes an error in vertical angles and this is compensated automatically by some electronic theodolites which apply a correction to the vertical angle. This is known as single-axis compensation.

Some electronic theodolites use liquid single-axis compensators and the type installed by Sokkia in some of their instruments uses a magnetic level

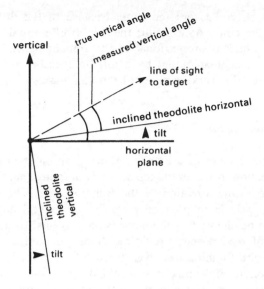

Figure 3.25 *Effect of vertical axis tilt in direction of telescope*

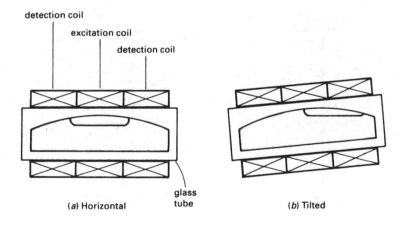

Figure 3.26 *Single-axis compensator (courtesy Sokkia Ltd)*

vial. Shown in figure 3.26, the magnetic tilt sensor consists of a level vial filled with a liquid in which magnetic particles are dispersed. Surrounding the vial are three coils, two detection coils and an excitation coil. When an alternating current is passed through the excitation coil, this causes currents to flow in the detection coils. If the sensor is level as in figure 3.26a, each detection coil generates an equal voltage and the differential output from both coils is zero. When the sensor is tilted as in figure 3.26b, the voltage generated by each coil is not the same because of the different magnetic paths between the coils. As a result, there is a differential output from the detection coils which is proportional to the amount of tilt. This voltage is converted into an angular output by a microprocessor which also corrects the vertical angle. The working range of this single-axis compensator is ±10′.

Dual-axis Compensation

This type of compensator measures the effect of an inclined vertical axis not only in the direction in which the telescope is pointing (single-axis compensation) but also in the direction of the trunnion axis. The effect of an inclined trunnion axis is to produce errors in horizontal angles (see figure 3.27) and it can be shown that the error is proportional to the tangent of the vertical angle of the telescope pointing. If the theodolite was poorly levelled such that the trunnion axis was tilted 60″, for vertical angles of 10° and 50°, horizontal circle readings would be in error by 11″ and 72″ respectively. Clearly, for precise work and for steep sightings, this error could be significant. A dual-axis compensator will measure trunnion axis dislevelment and correct horizontal circle readings automatically for this error.

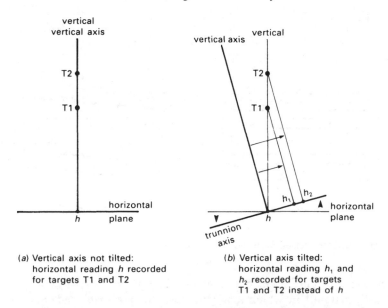

(a) Vertical axis not tilted:
horizontal reading h recorded
for targets T1 and T2

(b) Vertical axis tilted:
horizontal reading h_1 and
h_2 recorded for targets
T1 and T2 instead of h

Figure 3.27 *Effect of vertical axis tilt in direction of trunnion axis*

Most dual-axis compensators are of the liquid type and the Sokkia dual-axis tilt sensor is based, like their single-axis tilt sensor, on a level vial. However, in this case, a circular level vial is used. As shown in figure 3.28,

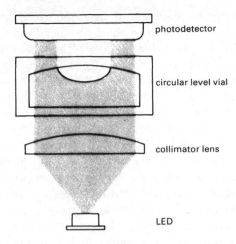

photodetector

circular level vial

collimator lens

LED

Figure 3.28   *Dual-axis compensator (courtesy Sokkia Ltd)*

light from an LED is collimated and passed through the vial which is highly sensitive to movement and mounted in clear glass. This causes a shadow to be projected onto a photodiode. The photodiode is divided into four sections and when the theodolite is levelled properly, the bubble shadow is projected evenly across all four sections. When the instrument is tilted, the position of the shadow changes and alters the amount of incident light falling on each section of the photodiode. This changes the electrical output of the photodiode which is passed to a microprocessor and converted into tilt angles. The working range of this dual-axis compensator is ±3′.

3.6 Setting Up a Theodolite

The process of setting up a theodolite is carried out in three stages: centring the theodolite, levelling the theodolite and elimination of parallax.

Centring the Theodolite

It is possible to centre a theodolite using a number of different methods and each engineer has his or her own preferred method. Any method used to centre a theodolite is perfectly acceptable as long as it is quick, accurate and is not likely to damage the theodolite. A commonly used method is described below where it is assumed that the theodolite is to be centred on its tripod over a ground mark which is a peg driven into the ground. A nail driven into the top of the peg defines the exact position for centring. The mark is referred to as station W.

(1) Leaving the instrument in its case, the tripod is first set up over station W. The legs are placed an equal distance from the peg and are extended to suit the height of the observer. The tripod head should be made as level as possible by eye. The tripod legs are not, at this stage, pushed into the ground and because of further adjustments to be made to the tripod legs, they are not fully extended when setting up initially and about 100 mm of the leg is left above the clamp.

(2) Standing back a few paces from the tripod, the centre of the tripod head is checked to see if it is vertically above the peg at W. This should be done by eye from two directions at right angles. If the tripod is not centred, it is moved in the appropriate direction keeping the head level. This process is repeated until the tripod is centred and levelled. At this point, the tripod legs are pushed firmly into the ground.

(3) The theodolite is carefully taken out of its case, its exact position being noted to assist in replacement, and is securely attached to the tripod head. Whenever carrying a theodolite, always hold it by the standards and *not* the telescope. Never let go of the theodolite until it is firmly screwed onto the tripod.

(4) By looking through the optical plummet, it is focused onto the peg at station W by moving the plummet eyepiece in and out or by rotating it. An image of the peg and a reference mark should be seen in the plummet after focusing.

(5) By adjusting the footscrews on the theodolite, the image of the reference mark seen in the plummet is moved until it coincides with the nail head in the peg at W.

(6) The upper plate is undone and the theodolite is rotated until the plate level lies on a line parallel to the imaginary line joining any two of the tripod legs. By undoing the clamp on one of these legs and by moving the tripod leg up or down, the plate level bubble is brought as near to the centre of its run as possible. No attempt should be made to centre the bubble exactly and care must be taken not to lift the tripod foot out of the ground if the tripod leg is moved upwards.

(7) The theodolite is turned through approximately 90° and the bubble is centred by raising or lowering the third leg. Both (6) and (7) are made much easier if the theodolite has a circular bubble fitted to the tribrach where this is used instead of the plate level. Steps (6) and (7) are repeated until the bubble is more or less centred in both positions.

This completes what is known as the rough centring of the theodolite. When using an optical plummet for fine centring prior to measuring angles, it is essential that the theodolite is properly levelled before centring takes place. If the theodolite is not levelled, the axis of the plummet will not be vertical and even though it may appear to be centered in the plummet eyepiece, the theodolite will be miscentred. Therefore, the fine centring is carried out

after the theodolite has been levelled using the footscrews (see next section). To complete the centring after levelling

(8) The position of the optical plummet on the station mark is checked. It should
   be close enough to the nail head in the peg to enable step (9) to be carried
   out. If it is not, the rough centring should be repeated from step (5).

(9) If the centring is close, the clamping screw on the tripod is undone and
   the theodolite moved by sliding it on the tripod head until it is centred.

(10) After step (9), the theodolite should be checked to ensure that it is
    still level as the act of fine centring can slightly upset the fine level-
    ling. If it is not level, it should be re-levelled and step (9) repeated.

Levelling the Theodolite

The *fine levelling* procedure for a theodolite is as follows.

(a) The alidade is rotated until the plate level is parallel to two footscrews
   as in figure 3.29*a*. These footscrews are turned until the plate level
   bubble is brought to the centre of its run. The levelling footscrews should
   be turned in opposite directions simultaneously, remembering that the
   bubble will move in a direction corresponding to the movement of the
   left thumb.

(b) The alidade is turned through 90° clockwise (see figure 3.29*b*) and the
   bubble centred again using the third footscrew only.

(c) The above operations are repeated until the bubble is central in positions (a) and (b).

(d) The alidade is now turned until it is in a position 180° clockwise from
   (a) as in figure 3.29*c*. The position of the bubble is noted.

(e) The alidade is turned through a further 90° clockwise as in figure 3.29*d*
   and the position of the bubble again noted.

(f) If the bubble is still in the centre of its run for both conditions (d) and
   (e) the theodolite is level and no further adjustment is needed. If the
   bubble is not central it should be off centre by the same amount in both
   conditions (d) and (e). This may be, for example, two divisions to the
   left.

(g) To remove the error, the alidade is returned to its initial position (figure
   3.29*a*) and, using the two footscrews parallel to the plate level, the bubble
   is placed in such a position that *half* the error is taken out; for example,
   in the case quoted, so that it is one division to the left.

(h) The alidade is then turned through 90° clockwise as in figure 3.29*b* and
   half the error again taken out such that, for the example quoted, it is
   again one division to the left.

(i) Conditions (g) and (h) are repeated until half the error is taken out for
   both positions.

(j) The alidade is now slowly rotated through 360° and the plate level bubble
   should remain in the same position.

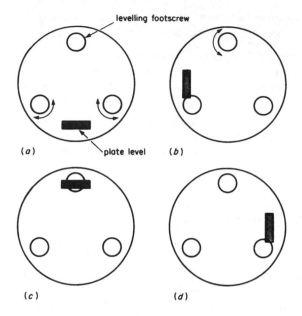

Figure 3.29   *Fine levelling*

The theodolite has now been finely levelled and the vertical axis of the instrument is truly vertical.

From this procedure, it can be seen that the plate bubble is not necessarily in the centre of its run when the theodolite is level. However, as long as the bubble is always set up at the position found by this procedure the theodolite can be used perfectly satisfactorily until a *permanent adjustment* of the plate level can be carried out (see section 3.12).

Once fine levelling has been completed, steps (8), (9) and (10) of the centring procedure should be carried out as described in the previous pages.

Elimination of Parallax

When the theodolite has been levelled and centred, parallax is eliminated by accurately focusing the cross hairs against a light background and focusing the instrument on a distant target (see section 2.4).

At this stage the theodolite is ready for reading angles and this procedure is described in section 3.7.

3.7 Measuring Angles

This section assumes that the theodolite has been set up over a point W as described in section 3.6 and that the horizontal and vertical angles to three distant points X, Y and Z are to be measured (see diagram, table 3.3). In order to be able to measure the directions to X, Y and Z, targets have to be set up at these points.

Targets

All survey targets, whatever type of survey they are being used for, should be set up vertically and should be centred exactly over ground or station marks.

If a target is not vertical then a *centring error* will be introduced into angular observations, even though the base of the target may be centred accurately over, for example, a peg (see figure 3.30). This applies to a target such as a pole-mounted reflector (see section 5.6) which may be held on a control station or point of detail during a survey. From figure 3.30, it can be seen that the lower the point of observation on the target, the smaller will be the centring error. For this reason, the lowest visible point on long targets should always be observed when measuring angles. The effect of miscentring a target or theodolite is discussed further in section 3.10.

When considering the width or diameter of a target, it is a waste of time trying to observe a direction to, say, a ranging rod when the line of sight is short, since accurate bisection is difficult. The width of a target should be proportional to the length of sight and, ideally, should be about the same size as the theodolite cross hairs.

Simple targets often used in control surveys and setting out include the following.

(1) The station mark should be observed directly if possible. This can often be the case over short lines if the mark is a nail in the top of a wooden peg.
(2) If a station cannot be seen directly, a pencil held on it can be used for convenience.
(3) A tripod can be set up such that a plumb bob can be suspended from it directly over the station. The plumb line can then be observed. Care must be exercised to ensure that the plumb bob does not rest on a nail in a peg as the string will then no longer be vertical, as shown in figure 3.31.
(4) For longer lines, ranging rods can be used. These must be carefully centred over stations and must be held vertically by hand or in a ranging rod stand. The lowest part of the rod must be observed.

The targets used for the majority of observations are specially manufac-

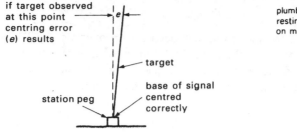

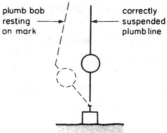

Figure 3.30 *Signal not vertical*   Figure 3.31 *Plumb line not vertical*

tured and are often combined with an EDM prism to enable distances to be measured at the same time as angles. This type of target is described in section 5.6.

Horizontal Angles

Assuming suitable targets are used at X, Y and Z, the observation procedure starts with the selection of one station as the *reference object* (RO). This point may be the most reliable and preferably the most distant of all the stations to be sighted. All the horizontal angles are referred to this point as shown in table 3.3, in which the horizontal angles XWY and XWZ are required.

Following this, a reading is set on the theodolite in the FL position along the direction to the RO. This can be done by one of four methods depending on whether the theodolite is fitted with a lower plate clamp and slow motion screw, a repetition clamp, a circle setting screw or whether the theodolite is electronic.

As an example, the methods for setting a reading of 00°05′00″ are described but these can be adapted to set any reading.

For instruments such as the Sokkia TM20H that has a lower plate clamp and tangent screw (see figure 3.2), the procedure is as follows.

(1) The micrometer is set to read 5′00″.
(2) The upper plate is unclamped and rotated until 0° appears in the theodolite's horizontal angle display. The upper plate is clamped and the upper plate tangent screw used to index 0° exactly. This gives a reading of 00°05′00″.
(3) The lower plate is unclamped and the theodolite turned until it points towards the RO. The reading of 00°05′00″ remains fixed.
(4) The lower plate is clamped and the lower plate tangent screw used to bisect the target at the RO with the vertical hair.
(5) The reading of 00°05′00″ should now be set on the RO and the horizontal circle is read to confirm this. All subsequent pointings should be done using the upper plate clamp and tangent screw.

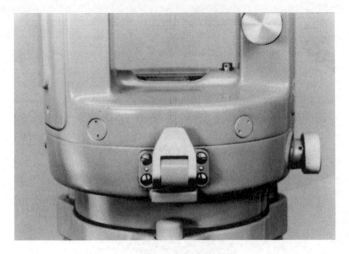

Figure 3.32 *Wild T16 repetition clamp (courtesy Leica UK Ltd)*

For instruments with a repetition clamp such as the Wild T16 (see figures 3.10 and 3.32), the procedure for setting a reading is as follows.

(1) The horizontal clamp is undone and the theodolite rotated until 0° appears in the horizontal angle display.
(2) The clamp is tightened and 00°05'00" is indexed exactly using the horizontal slow motion screw.
(3) The repetition clamp is pulled out and this locks the horizontal circle of this reading. The horizontal clamp is released and the theodolite pointed at the RO. The horizontal clamp is tightened and the RO is bisected using the horizontal slow motion screw.
(4) The reading of 00°05'00" should now be set to the RO. To enable readings to other points to be taken, the repetition clamp should be pushed up (down on some instruments) to release the horizontal circle.

For instruments with a circle setting screw such as the Wild T2 (see figures 3.18 and 3.33) and Sokkia TM1A an exact reading can never be set easily since there is no fine adjustment to the circle setting screw. This is only a feature of 1" theodolites because their main function is to read angles rather than set them out exactly. If an angle is to be set, the following procedure can be used.

(1) The vertical hair of the theodolite is made to bisect the target at the RO using the horizontal clamp and slow motion screw.
(2) The micrometer is set to read 05'00" and the circle setting screw is used to set coincidence at 00°00' on the circle reading system. As soon as this has been done, it is important that the circle setting screw is covered.

Figure 3.33 *Wild T2 horizontal circle setting screw (courtesy Leica UK Ltd)*

(3) The theodolite is moved off target using the horizontal slow motion
screw and the target bisected once again. The horizontal reading should
now be very close to 00°05′00″. If there is a small difference, it might
be difficult to correct this and it is often better to simply read the
theodolite and continue with this as the first reading.

When using an electronic theodolite such as the Sokkia DT5 (see figure
3.19), a reading of exactly 00°00′00″ can be set by pressing the OSET key
when the theodolite is pointing in the required direction. If some other
reading is required to an RO (such as 00°05′00″), the following procedure
is used.

(1) The theodolite is rotated until a horizontal reading close to 0° is displayed. The horizontal slow motion screw is used to obtain an exact
reading of 00°05′00″.
(2) The hold key is pressed and the RO sighted in the usual manner. The
reading of 00°05′00″ should now be set along the direction to the RO.
(3) The hold key is pressed again. Readings will now change as the theodolite is rotated.

Having set the required direction to the RO on face left, the procedure
for measuring angles continues as follows.

(1) Swinging the telescope to the *right*, Y and Z are sighted in turn and the horizontal circle readings are recorded at both sightings.

(2) The telescope is transitted so that the theodolite is now on *face right* (FR), Z is sighted and the horizontal circle reading recorded.

(3) Swinging *left* to Y and X, the horizontal circle readings are recorded.

(4) At this stage, *one round* of angles has been completed. The theodolite is changed to face left and *the zero* changed by setting the horizontal circle to read something different from the reading set for the first round when sighting X, the RO. It is not necessary to set an exact reading on the second round but it is important to realise that as well as changing the degrees setting to the RO, the setting of the minutes and seconds should also be different from that of the first round.

(5) Repeat steps (1) to (3) inclusive to complete a second round of angles.

At least two rounds of angles should be taken at each station in order to detect errors when the angles are computed since each round is independently observed. Both rounds must be computed and compared *before* the instrument and tripod are moved. When sighting targets at other stations, it is better to use approximately the same point on the vertical hair rather than the intersection of the cross hairs since setting coincidence here is time consuming and unnecessary.

Vertical Angles

These should be read after the horizontal angles to avoid confusion when booking. Vertical angles can be observed in any order of the stations. General points in the procedure are given below for the booking shown in table 3.3.

(1) It is usual to take all face left readings first. The horizontal hair is used for sighting targets in this case and it is again not necessary to use the intersection of the cross hairs but it is important that approximately the same point on the horizontal hair is used on both faces.

(2) Readings should again be taken on both faces but in this case only *one* round of angles need be taken.

(3) When reading the vertical circle it is necessary for the recorded angles to be reduced. This is shown in table 3.3.

Having completed all the angular observations, the theodolite is carefully removed from the tripod head and returned to its case. Before removing the theodolite from the tripod head, the three footscrews should be set central in their runs.

If other stations are to be occupied, the theodolite must *never* be left on the tripod when moving between stations since this can distort the axes and, if the operator trips and falls, the instrument may be severely damaged.

TABLE 3.3
Angle Booking

STATION W

HORIZONTAL CIRCLE

POINT	FACE LEFT			FACE RIGHT			MEAN			REDUCED DIRECTION			FINAL ANGLES
X (RO)	00	03	50	180	04	30	00	04	10	00	00	00	
Y	17	22	10	197	23	10	17	22	40	17	18	30	XŴY =
Z	83	58	50	264	00	00	83	59	25	83	55	15	17° 18′ 35″
X (RO)	45	12	30	225	13	30	45	13	00	00	00	00	XŴZ =
Y	62	31	10	242	32	10	62	31	40	17	18	40	83° 55′ 10″
Z	129	07	30	309	08	40	129	08	05	83	55	05	

VERTICAL CIRCLE

POINT	TARGET HEIGHT	FACE LEFT			FACE RIGHT			REDUCED FACE LEFT			REDUCED FACE RIGHT			FINAL ANGLE		
X	1.16 m	88	10	30	271	51	20	01	49	30	01	51	20	+01	50	25
Y	1.52 m	89	34	50	270	27	30	00	25	10	00	27	30	+00	26	20
Z	1.47 m	92	48	20	267	13	40	02	48	20	02	46	20	−02	47	20

DIAGRAM

REMARKS

SURVEY TITLE	A3 Road Improvement
DATE	11 July 1993
THEODOLITE HEIGHT	1.48 m
THEODOLITE NUMBER	816421
OBSERVER	J.U.
BOOKER	W.F.P.

3.8 Booking and Calculating Angles

Table 3.3 shows the horizontal and vertical angle booking and calculation for points X, Y and Z observed from station W. Many different formats exist for recording and calculating angles and only one method is shown in table 3.3.

The mean horizontal circle readings are obtained by averaging the FL and FR readings. To simplify these calculations, the degrees of the FL readings are carried through and only the minutes and seconds values are meaned. These mean horizontal circle readings are then reduced to the RO in the Reduced Direction column to give the horizontal angles. The final horizontal angles are obtained by meaning the values obtained from each round.

From the readings obtained, it can be seen that the vertical circle is graduated as shown in figure 3.4a and therefore it is necessary to reduce the FL and FR readings to ascertain whether the angles are either elevation (+) or depression (−). The final vertical angles are obtained by meaning these reduced FL and FR readings.

In addition, the following procedures should be adopted.

(1) For both types of angle, the stations are booked in clockwise order. This should be the order of observation.
(2) If a single figure occurs in any reading, for example, a 2 or a 4, this should be recorded as 02 or 04. If a mistake is made the number should always be rewritten, for example, if a 4 is written and should be 5, this should be recorded as ̶4̶5, not ̶4̶.
(3) Never copy out observations from one field sheet or field book to another.
(4) The booker, as readings are entered, should be checking for consistency in *horizontal collimation* on horizontal angles and *vertical collimation* on vertical angles. These effects are described in section 3.12 and, referring to table 3.3, the checks are as follows.

For horizontal angles, the difference (FL − FR) is computed for each sighting considering minutes and seconds only. This gives the following results for the first round:

Station	(FL − FR)
X	−00′ 40″
Y	−01′ 00″
Z	−01′ 10″

Assuming a 20″ theodolite was used to record the two rounds shown in table 3.3, this shows the readings to be satisfactory since (FL − FR), for a 20″ theodolite, should agree to within 1′ for each point observed considering the magnitude and the sign of the difference. If, for example, the difference for station Z was −11′ 10″ then an *operator error* of 10′ is immediately apparent. In such a case, the readings for station Z would be checked. For a 1″ theodolite, (FL − FR) should agree within a few seconds, depending on the length of sight and the type of target used. A similar process is applied to the vertical circle readings to check for consistency in vertical collimation. In this case FL + FR *should* = 360° and, for station X, FL + FR = 88°10′30″ + 271°51′20″ = 360°01′50″. For stations Y and Z, 360°02′20″ and 360°02′00″ are obtained. All three values agree very closely which shows the readings to be consistent and therefore acceptable.

3.9 Importance of Observing Procedure

The method of reading angles may be thought to be somewhat lengthy and repetitious but it is necessary to use this so that certain instrumental errors are eliminated (see section 3.12).

(1) By taking the mean of FL and FR readings for horizontal and vertical angles, the effects of *horizontal collimation, vertical collimation* and *trunnion axis dislevelment* are all eliminated.
(2) Observing on both faces also removes any errors associated with an *inclined diaphragm* provided the same positions are used on each cross hair for observing.
(3) In addition to circle eccentricity (see section 3.3), the horizontal circle axis may not coincide with the vertical axis. Furthermore, the graduations may be irregular. These effects are very small and are reduced by changing the zero between rounds. However, two rounds of angles would not be sufficient to reduce these errors significantly: the reason for observing two rounds is to provide a check on observations.
(4) The effect of an *inclined vertical axis* (plate level not set correctly) is not eliminated by observing on both faces but any error arising from this is negligible if the theodolite is carefully levelled. Since this error is proportional to the tangent of the vertical angle of the sighting, care should be taken when recording angles to points at significantly different elevations as is often the case on construction sites. However, when using an electronic theodolite with a dual-axis compensator, the effect of improper levelling is corrected provided the theodolite is levelled such that the tilt sensor is within its working range.

3.10 Effect of Miscentring a Theodolite

Suppose a horizontal angle ABC (θ) is to be measured but, owing to miscentring, the theodolite is set up over B$'$ instead of B as in figure 3.34. As a result, horizontal angle AB$'$C is measured.

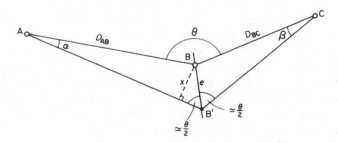

Figure 3.34 *Miscentring*

The miscentring distance, e, is equal to distance BB′ and the maximum error in θ will occur when distance e bisects the observed angle AB′C as shown in figure 3.34.

The total error in angle ABC will be $(\alpha + \beta)$.

With reference to figure 3.34 since α is very small it can be assumed that

$$x = D_{AB}\, \alpha \; (\alpha \text{ in radians})$$

But

$$\sin (\theta/2) = (x/e)$$

Hence

$$x = e \sin (\theta/2)$$

Therefore

$$\alpha = (e/D_{AB}) \sin (\theta/2) \; (\alpha \text{ in radians})$$

since α (in radians) $= \alpha'' \sin 1''$ for small angles.
Then

$$\alpha'' = \frac{e}{D_{AB}} \sin (\theta/2) \operatorname{cosec} 1'' \; (\operatorname{cosec} 1'' = 206\,265) \quad (3.1a)$$

Similarly

$$\beta'' = \frac{e}{D_{BC}} \sin (\theta/2) \operatorname{cosec} 1'' \qquad\qquad (3.1b)$$

The significance of this is that for relatively small values of D, α and β will be large. Therefore, care must be taken in centring when sighting over short distances. The worked example in section 3.13 illustrates this point.

3.11 Height Measurement by Theodolite (Trigonometrical Heighting)

If the reduced levels of several points some distance apart in hilly terrain are required then levelling can be a very tedious task. However, if an accuracy in the order of ± 100 mm is acceptable and the points are visible from other points of known elevation, an alternative and much quicker method is to use a theodolite.

The technique of using a theodolite to obtain heights is known as *trigonometrical heighting* and involves the measurement of the vertical angle between a known point and the point of unknown height. Since the slope distance between the points is required in the calculation, trigonometrical heighting is best undertaken with the aid of an EDM system fitted to the theodolite.

The EDM reflector is set up over the point being sighted and the vertical angle to it is measured with a theodolite reading to 1″ or better. Several measurements of the vertical angle are taken and the mean value used.

Because the slope length of the line of sight between the points may be in the order of kilometres, it is necessary to take into consideration the effects of the curvature of the Earth and the refraction of light by the atmosphere.

In *single ended trigonometrical heighting*, the observations are taken from one end of the line only and curvature and refraction must be allowed for in the calculations.

In *reciprocal trigonometrical heighting*, observations are taken from each end of the line but not at the same time. Curvature and refraction must again be taken into account.

In *simultaneous reciprocal trigonometrical heighting*, the observations are taken from each end of the line at exactly the same time in order that the curvature and refraction effects will cancel each other out in the calculations. The simultaneous method also provides a means of measuring the coefficient of atmospheric refraction.

If care is taken, the accuracy of heights obtained by each method over distances of several kilometres can be as follows

single ended method	±200 mm
reciprocal method	±100 mm
simultaneous method	± 50 mm

The accuracy of heights obtained by single ended and reciprocal methods depends to a great extent on the value of the coefficient of atmospheric refraction used in the calculations.

Single Ended Observations

Figure 3.35 shows a single ended observation carried out between points A and B. In this case, the theodolite is reading an angle of elevation, θ, between the horizontal at T and the direction of the telescope pointing along TS.

If the height difference between A and B was calculated using θ, an incorrect result would be obtained because of curvature and refraction effects. Between A and B the curvature of the Earth is represented by vertical distance FG which is the difference between the level and horizontal lines through T over distance AB. Refraction causes the theodolite line of sight to be deviated along TP although vertical angles are measured along TS.

From figure 3.35 the height of B (H_B) relative to the height of A(H_A) is given by

$$H_B = H_A + i + L \sin [(+\theta) + (\gamma - \alpha)] - b$$

where

i = height of theodolite trunnion axis above point A
b = height of EDM reflector above point B
L = slope distance between A and B obtained from the EDM readout
θ = vertical angle obtained from the theodolite T
γ = curvature angle between A and B
α = refraction angle between A and B.

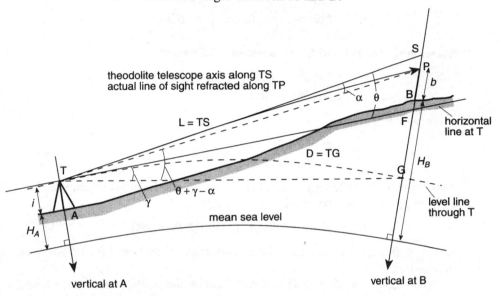

Figure 3.35 *Single ended observation*

The angle $\delta\theta = (\gamma - \alpha)$ can be considered as a correction to the observed vertical angle to account for curvature and refraction. The correction $\delta\theta$ is obtained as follows with reference to figure 3.35

$$\gamma \simeq \frac{FG}{D} \text{ rad and } \alpha \simeq \frac{SP}{D} \text{ rad}$$

where $D = L \cos\theta$ = horizontal distance between A and B. Therefore

$$\delta\theta = \frac{1}{D} (FG - SP) \text{ radians} \qquad (3.2)$$

It can be shown that

$$FG = D^2/2R$$

where R = average radius of the Earth between A and B, and from figure 3.35

$$SP = \alpha D$$

The coefficient of atmospheric refraction, k, is given by

$$k = (\alpha/\beta)$$

Where β = the angle subtended by AB at the centre of the Earth. Therefore

$$SP = k\beta D = \frac{k D^2}{R} \quad (\text{since } \beta \simeq D/R)$$

Substituting for FG and SP in equation (3.2) gives

$$\delta\theta = \frac{1}{D} \left[\frac{D^2}{2R} - \frac{kD^2}{R} \right] \text{ radians}$$

From which

$$\delta\theta = \frac{D (1 - 2k)}{2R (\sin 1'')} \text{ seconds} \tag{3.3}$$

This leads to the following general equation for single ended trigonometrical heighting, which can be applied to all cases:

$$H_B = H_A + i - b + L \sin [(\pm \theta) + \delta\theta] \tag{3.4}$$

where $+ \theta$ is used for an angle of elevation and $- \theta$ is used for an angle of depression.

When using equation (3.3) in Great Britain, the value of R is often taken as 6375 km and the value of k is usually assumed as 0.07. However, the value of k is open to some doubt because of atmospheric uncertainties and, as a result, it is recommended that for any particular survey the simultaneous method should be used wherever possible.

If any single ended observations are necessary, the value of k which should be applied can be calculated from simultaneous readings taken at approximately the same time.

The worked example in section 3.13 shows how single ended observations are used to calculate the heights of points.

Reciprocal Observations

Although each direction is not necessarily observed on the same day, the accuracy obtained from this method will be improved if the same time of day is used for each observation since the k values should be comparable.

The two observations are each computed as for the single ended method, the final height difference being obtained by meaning the two individual height differences.

Simultaneous Reciprocal Observations

In this method two theodolites are required in order that observations can be taken from each end at exactly the same time to eliminate the effect of refraction. Since the sighting distances in each direction are also exactly the same, the effect of curvature is also eliminated.

The worked example in section 3.13 shows how the simultaneous reciprocal method can be used both to calculate heights of points and to calculate a value of k for use in single ended observations.

3.12 Adjustments of a Theodolite

There are two types of adjustment necessary, these being the *station* or *temporary adjustments* and the *permanent adjustments*.

The station adjustments are carried out each time the theodolite is set up and have been described in section 3.6. These adjustments are centring, levelling and removing parallax.

Figure 3.6 shows the arrangement of the axes of the theodolite when it is in *perfect adjustment*. This configuration is rarely achieved in practice and the purpose of the permanent adjustments is to set the instrument so that the axes take up positions as close as possible to those shown in figure 3.6. The permanent adjustments should be carried out when first using an unknown instrument and periodically thereafter since the setting of the axes tends to alter with continual use of the theodolite.

Plate Level Adjustment

The aim of this test is to check and, if necessary, set the vertical axis truly vertical when the plate level bubble is central. In other words, the plate level axis is to be set perpendicular to the vertical axis. In order to check and adjust the plate level, the following procedure is used.

(1) Level the theodolite as described in section 3.6 until the plate level bubble is in the same position for a complete 360° rotation of the alidade. The bubble is not necessarily in the middle of its run.
(2) In this position the vertical axis is truly vertical and only the bubble is out relative to its main divisions.
(3) Bring the bubble back to the centre of its run using the adjustment provided.

Horizontal Collimation Adjustment

The aim of this test is to set the line of sight (line of collimation) perpendicular to the trunnion axis.

The error is caused by a displaced vertical hair and it can be detected as soon as face left and face right readings have been taken as, for example, in table 3.3 where the theodolite used has a horizontal (FL − FR) difference of about 1′ 00″. To check the horizontal collimation of any theodolite the following procedure is used.

(1) Set the instrument in the face left position.
(2) Turn the theodolite and sight, using the vertical hair, a well defined target preferably about 100 m distant from and at about the same height as the theodolite.
(3) Read the horizontal circle.
(4) Transit the telescope to face right and sight the target again and read the horizontal circle.
(5) Subtract the face right reading from the face left reading. This difference, less 180°, equals the horizontal collimation error in the theodolite. As stated in section 3.9, this error is cancelled by taking the mean of face left and face right readings.

If adjustment is necessary it is always better to return the theodolite to the manufacturer or supplier for adjustment in the laboratory but if an onsite adjustment is absolutely necessary, the horizontal collimation error can be removed using the following procedure.

(1) Set up and level the theodolite on reasonably flat ground such that there is a clear view of approximately 100 m on either side. A marking arrow is placed at point A, approximately 100 m from the instrument, and the vertical hair aligned to it on *face left* (see figure 3.36a).
(2) The telescope is transitted and a second arrow placed at point B, again approximately 100 m from the instrument. The theodolite is now on *face right* (see figure 3.36b).
(3) Keeping *face right*, the telescope is rotated in azimuth and exact coincidence obtained at A (figure 3.36c).
(4) The telescope is again transitted so that it is now *face left*. If there is no collimation error, B will be intersected. Usually, however, B is not intersected and a third arrow is placed on the line of sight next to B, at C (figure 3.36d). The distance BC represents four times the collimation error and, if it is small, it is usually ignored.
(5) If the error is to be removed, a fourth arrow is placed at D such that CD = DF.
(6) The vertical hair is moved using the diaphragm adjusting screws until point D is intersected.
(7) To check the adjustment, transit the telescope, reintersect A and retransit. The vertical hair should exactly intersect F.

Diaphragm Orientation

In carrying out the horizontal collimation adjustment, the diaphragm is moved. This may alter the setting of the vertical hair in a plane perpendicular to the trunnion axis so that it no longer sweeps out a vertical plane when the trunnion axis is horizontal.

Assuming that a horizontal collimation adjustment has just been completed, the following procedure should be adopted to check the orientation of the diaphragm.

(1) Relevel the instrument carefully and sight A (as in figure 3.36) on either face.
(2) Move the telescope up and down while observing A. If the vertical hair stays on point A then it is set correctly.
(3) If adjustment is necessary, the diaphragm is moved until the vertical hair remains on point A while moving the telescope in altitude.

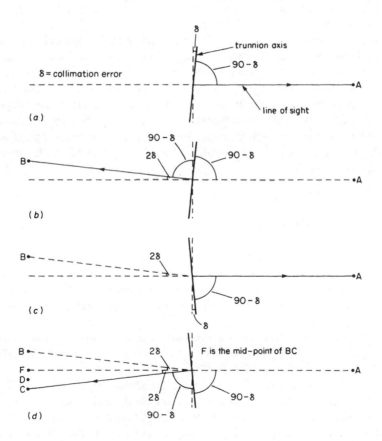

Figure 3.36 *Plan view of horizontal collimation test*

The horizontal collimation and diaphragm orientation are interdependent and both are undertaken consecutively until a satisfactory result is obtained for each.

The diaphragm is constructed by the instrument manufacturer so that the horizontal and vertical hairs are perpendicular. Setting the vertical hair vertical therefore sets the horizontal hair in a horizontal plane.

Adjustment of the Vertical Indices (Index Error or Vertical Collimation)

The aim of this test is to check if the vertical circle is set to some multiple of 90° when the line of sight is horizontal and the theodolite has been levelled.

This error is shown in table 3.3 where the theodolite used has a (FL + FR) consistency of about 2'00". To check the vertical collimation of any theodolite the following procedure is used.

(1) Set the instrument in the face left position.
(2) Using the horizontal hair, sight a well defined target about 100 m distant and read the vertical circle.
(3) Transit to face right, resight the target and read the vertical circle again.
(4) Add the two vertical circle readings. The difference between this sum and 360° (or 180° in some instruments) is caused by the vertical collimation error of the theodolite. As with horizontal angles, this error is cancelled by taking the mean of face left and face right readings after reducing them. Vertical collimation is caused by the compensator built into the automatic vertical index being out of adjustment and if adjustment is necessary, the theodolite should be returned to the manufacturer or supplier.

Trunnion Axis Dislevelment

The purpose of this test is to check if the trunnion axis is perpendicular to the vertical axis. The trunnion axis will then be horizontal when the instrument is levelled. If the trunnion axis is not horizontal the telescope will not define a vertical plane and this will give rise to incorrect vertical and horizontal angles.

In all theodolites, it is rare for this not to be the case and, consequently, none provide for this adjustment. However, satisfactory results will be obtained by meaning FL and FR readings.

To check the trunnion axis, set up the theodolite adjacent to a tall structure and sight the top (A). Depress the telescope and read a scale rule or mark a point (B) at the base of the structure. Change face and sight A again and depress the telescope. If the reading on the scale is the same or point B is intersected, the theodolite is in adjustment. If it is out of adjustment, it should be returned to the manufacturer or supplier.

Adjustment of the Optical Plummet

The line of collimation of an optical plummet must coincide with the vertical axis of the theodolite when it is levelled. Two tests are possible, depending on the type of instrument used.

If the optical plummet is on the alidade and can be rotated about the vertical axis (figure 3.7a).

Secure a piece of paper on the ground below the instrument and make a mark where the optical plummet intersects it. Rotate the alidade through 180° in azimuth and make a second mark. If the marks coincide, the plummet is in adjustment. If not, the correct position of the plummet axis is given by a point midway between the two marks.

   Consult the instrument handbook and adjust either the diaphragm (cross hairs) or objective lens on the optical plummet.

If the optical plummet is on the tribrach and cannot be rotated without disturbing the levelling (figure 3.7b).

Set the theodolite on its side on a bench with its base facing a wall and mark the point on the wall intersected by the optical plummet. Rotate the tribrach through 180° and again mark the wall. If both marks coincide, the plummet is in adjustment. If not, the plummet diaphragm should be adjusted to intersect a point midway between the two marks.

In both cases, it is difficult to adjust the plummet precisely under site conditions. If the plummet needs adjusting, it is best to return the theodolite to the manufacturer or supplier.

3.13 Worked Examples

(1) Miscentring a Theodolite

Question
From a traverse station Y, the horizontal angle between two stations X and Z was measured with a 1″ theodolite as 123°18′42″.

   The theodolite at Y was miscentred by 9 mm and the horizontal distances YX and YZ were measured as 69.41 m and 47.32 m respectively.

Calculate the maximum angular error in angle XYZ owing to the theodolite being miscentred.

Solution
For the maximum angular error, equation (3.1) gives

$$\alpha = \frac{0.009}{69.41} \sin\left[\frac{123°}{2}\right] 206\,265 = 23.5''$$

$$\beta = \frac{0.009}{47.32} \sin\left[\frac{123°}{2}\right] 206\,265 = 34.5''$$

Therefore

maximum angular error = ($\alpha + \beta$) = 58''

(2) Simultaneous Reciprocal Trigonometrical Heighting

Question

Simultaneous reciprocal trigonometrical heighting observations were taken from station A to station B and from station B to station A as follows

At station A
Instrument height   = 1.49 m
Target height   = 1.50 m
Vertical angle to B = −01°17′26″

At station B
Instrument height   = 1.53 m
Target height   = 1.75 m
Vertical angle to A = +01°17′03″

Immediately after these simultaneous observations, the following single ended observation was taken from station A to a station C:

Target height at C = 1.96 m
Vertical angle to C = −02°24′53″

The height of station A was 117.43 m AOD and the slope distances AB and AC were measured using EDM equipment as 1863.12 m and 1543.28 m, respectively. The radius of the Earth is 6375 km. Calculate

(1) the height of station B
(2) the value of the coefficient of atmospheric refraction which prevailed during the observations
(3) the height of station C.

Solution
(1) The height of station B
From the observation at station A, equation (3.4) gives

$$H_B = H_A + i - b + L \sin\left[(-\theta) + \delta\theta\right] \qquad (3.5)$$

But $\delta\theta$ can be ignored when simultaneous observations are taken, therefore

$$H_B = 117.43 + 1.49 - 1.75 + 1863.12 \sin(-01°17'26'')$$

From which

$$H_B = 75.208 = 75.21 \text{ m}$$

From the observation at station B, equation (3.4) gives

$$H_A = H_B + i - b + L \sin[(+\theta) + \delta\theta] \qquad (3.6)$$

Which, again ignoring $\delta\theta$, gives

$$117.43 = H_B + 1.53 - 1.50 + 1863.12 \sin(+01°17'03'')$$

From which

$$H_B = 75.646 = 75.65 \text{ m}$$

Therefore

$$\textbf{Height of B} = \frac{75.21 + 75.65}{2} = \textbf{75.43 m AOD}$$

(2) The value of k

From the two height differences calculated above, the true height difference is obtained from

$$H_A - H_B \text{ from the observation at A} = 42.222 \text{ m}$$
$$H_A - H_B \text{ from the observation at B} = 41.784 \text{ m}$$

Hence

$$\text{True height difference} = \frac{42.222 + 41.784}{2} = 42.003 \text{ m}$$

This is substituted into equation (3.5) and (3.6) in turn to calculate first $\delta\theta$ and then k.

Substitution into equation (3.5) gives

$$H_B - H_A = -42.003 = 1.49 - 1.75 + 1863.12 \sin[(-01°17'26'') + \delta\theta]$$

From which

$$\delta\theta = 24.27''$$

From equation (3.3)

$$\delta\theta = \frac{D(1 - 2k)}{2R(\sin 1'')} \text{ seconds}$$

where

$$D = L \cos\theta = 1863.12 \cos(01°17'26'') = 1862.65 \text{ m}$$

Hence

$$24.27 = \frac{1.86265}{2(6375)} (1 - 2k) \, 206 \, 265$$

From which

$$k = 0.0973$$

Substituting into equation (3.6) and solving for $\delta\theta$ and then for k gives

$$\delta\theta = 24.20'' \text{ and } k = 0.0984$$

Hence

mean value of k $= \dfrac{0.0973 + 0.0984}{2} = \mathbf{0.098}$

(3) The height of station C
Since the observation to C was taken immediately after the simultaneous reciprocal observations, $k = 0.098$ can be used, therefore

$$D_{AC} = 1543.28 \cos(02°24'53'') = 1541.91 \text{ m}$$

And

$$\delta\theta = \frac{1.541 \, 91}{2(6375)} (1 - 2(0.098)) \, 206 \, 265 = 20.06''$$

Substituting into equation (3.4) gives

$$H_C = 117.43 + 1.49 - 1.96 + 1543.28 \sin \left[(-02°24'53'') + 20.06'' \right]$$

From which

$$H_C = 52.09 = \mathbf{52.1 \text{ m AOD}}$$

4

Distance Measurement: Taping and Stadia Tacheometry

In engineering surveying, three types of distance are used: slope distance, horizontal distance and vertical distance (or height difference).

With reference to figure 4.1

$$\text{slope distance} = AB = L$$
$$\text{horizontal distance} = AB' = A'B = D$$
$$\text{vertical distance} = AA' = BB' = V = \Delta H$$

Horizontal and vertical distances are used in survey drawings, setting out plans and engineering design work. Slope distances and vertical distances are used when setting out design points on construction sites.

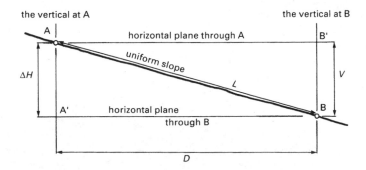

Figure 4.1 *Types of distance*

Distances can be measured and set out either directly using tapes or indirectly using optical equipment or electronic equipment. This chapter deals

127

with taping and stadia tacheometry (an optical method of distance measurement). Electromagnetic distance measurement is covered in chapter 5.

4.1 Direct Distance Measurement

This is carried out with tapes and usually involves laying the tape on the surface of the ground as shown in figure 4.2*a*. However, when measuring over very steep or undulating ground, the tape may be held horizontally as shown in figure 4.2*b*, this technique being known as *stepping*. Occasionally, it may be necessary to suspend a tape between two supports in *catenary* as shown in figure 4.2*c*.

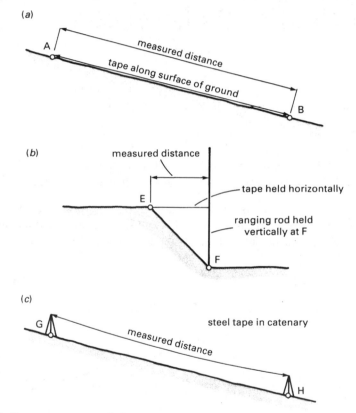

Figure 4.2   *Tape measurement methods*

Height difference is obtained by allowing a tape to hang freely with a weight attached to its free end. A common application of this is in the transference of height from floor to floor in a multi-storey building by measuring up or down vertical columns.

4.2 Steel Tapes

Steel tapes are available in various lengths up to 100 m, 20 m and 30 m being the most common. They are encased in steel or plastic boxes with a recessed winding lever (closed case steel tapes) or are mounted on open frames with a folding winding lever (open frame steel tapes). Most tapes incorporate a small loop or grip at the end of the tape, this marking the zero point, although it is possible to have a different zero marked on the steel band itself. It is essential to find where the zero point is marked before using any tape. Examples of steel tapes are shown in figure 4.3.

Various methods are used for graduating tapes and the UK metric and EC styles are shown in figure 4.4. Since these are different, it is advisable to inspect tape graduations before fieldwork commences.

Figure 4.3   *Steel tapes (courtesy Fisco Products Ltd)*

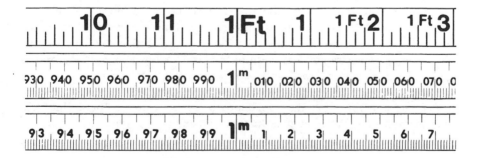

Figure 4.4   *Steel tape graduations (courtesy Fisco Products Ltd)*

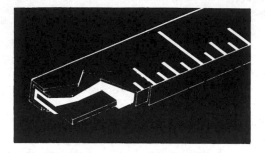

Figure 4.5 *Cross-section through steel tape (courtesy Fisco Products Ltd)*

A section through a steel tape is shown in figure 4.5 in which it can be seen that the steel band is protected by covering it with coats of polyester or nylon. The second coat gives a tape its characteristic colour (usually yellow or white) and since the graduations are printed on this another transparent coat covers these and gives added protection to the tape. All steel tapes are manufactured so that they measure their nominal length at a specific temperature and under a certain pull. These standard conditions, 20°C and 50 N, are printed somewhere on the first metre of the tape. The effects of variations from the standard conditions are discussed in section 4.4.

4.3 Steel Taping: Fieldwork

Distance measurement using steel tapes involves determining the straight-line distance between two points.

When the length to be measured is less than that of the steel tape, measurements are carried out by unwinding and laying the tape along the straight line between the points. The zero of the tape (or some convenient graduation) is held against one point, the tape is straightened, pulled taut and the distance read directly on the tape at the other point.

When the length of the line between two points exceeds that of the tape, some form of alignment is necessary to ensure that the tape is positioned along the straight line required. This is known as *ranging* and is achieved using *ranging rods* and *marking arrows* (see figure 4.6).

For measuring long lines two people are required, identified as the *leader* and the *follower*, the procedure being as follows for a line AB. This method of measurement is known as *ranging by eye*.

(1) Ranging rods are erected as vertical as possible at the points A and B and, for a measure in the direction of A to B, the zero point or some convenient graduation of the tape is set against A by the follower.

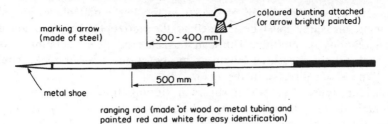

Figure 4.6 *Ranging rod and marking arrow*

(2) The leader, carrying a third ranging rod, unwinds the tape and walks towards point B, stopping just short of a tape length, at which point the ranging rod is held vertically.

(3) The follower removes the ranging rod at A and, stepping a few paces behind point A, lines up the ranging rod held by the leader with point A and with the rod at B. This lining-in should be done by the follower sighting as low as possible on the poles.

(4) The tape is now straightened and laid against the rod at B by the leader, pulled taut and the tape length marked by placing an arrow on line.

(5) For the next tape length the leader and the follower move ahead simultaneously with the tape unwound, the procedure being repeated but with the follower now at the first marking arrow. Before leaving point A, the follower replaces the ranging rod at A as this will be sighted on the return measurement from B to A, which should always be taken as a check for gross errors.

(6) As measurement proceeds the follower picks up each arrow and, on completion, the number of arrows held by the follower indicates the number of whole tape lengths measured. This number of tape lengths plus the section at the end less than a tape length gives the total length of the line.

4.4 Steel Taping: Corrections

The following corrections are applied to taped distances in order to improve their precision: slope, standardisation, tension, temperature and sag.

Slope Measurements and Slope Corrections

The method of ranging described in section 4.3 can be carried out for any line, either sloping or level. Since all surveying calculations, plans and

setting-out designs are based or drawn in the horizontal plane, any sloping length measured must be reduced to the horizontal before being used
for calculations or plotting. This can be achieved by calculating a *slope
correction* for the measured length or by measuring the horizontal equivalent of the slope directly in the field.

Consider figure 4.7a which shows a sloping line AB. To record the
horizontal distance D between A and B, the *method of stepping* may be
employed in which a series of horizontal measurements is taken. To measure
D_1 the tape zero or a whole metre graduation is held at A and the tape
then held horizontally and on line towards B against a previously lined-in
ranging rod. The horizontality of the tape should, if possible, be checked
by a third person viewing it from one side some distance away.

At another convenient tape graduation (preferably a whole metre mark
again) the horizontal distance is transferred to ground level using a *plumb
line* (a string line with a weight attached) or a *drop arrow* (a marking
arrow to which a weight is attached).

The tape is now moved forward to measure D_2 in a similar manner. It
is recommended that the maximum length of an unsupported tape should
be 10 m and that this should be considerably shorter on steep slopes since
the maximum height through which a distance is transferred should be 1.5 m.

As an alternative to stepping, the slope angle, θ, can be determined
and the horizontal distance D calculated from the measured slope distance
L as shown in figure 4.7b. Alternatively, a correction can be computed
from

$$\text{slope correction} = -L \ (1 - \cos\theta) \qquad (4.1)$$

This correction is always negative and is applied to the measured length L.

The slope angle can be measured using a theodolite which is set up
(say) at A and the slope angle measured along A'B' (see figure 4.7b). In
this case, h will be the height of the theodolite trunnion axis above ground
level.

Comparing these methods of obtaining a horizontal distance, stepping
is more useful when the ground between points is very irregular, whereas
the theodolite is suitable for measurements taken on uniform slopes.

A third method is available if the height difference between the two
points is known and the slope between them is again uniform. In figure
4.7c, if Δh is the height difference between A and B, then

$$\text{slope correction} = -\frac{\Delta h^2}{2L} - \frac{\Delta h^4}{8L^3}$$

For slopes less than 10 per cent the last term in this expression can be
ignored and

$$\text{slope correction} = -\frac{\Delta h^2}{2L} \qquad (4.2)$$

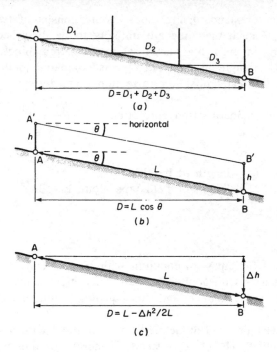

Figure 4.7 *Slope measurements*

Standardisation

Under given conditions a tape has a certain nominal length. However, with a lot of use, a tape tends to stretch and this effect can produce serious errors in length measurement. Therefore, standardisation of steel tapes should be carried out frequently against a reference tape or baseline. If using a reference tape, standardisation should be done on a smooth, flat surface such as a surfaced road or footpath. The reference tape should not be used for any fieldwork and should be checked by the manufacturer as often as possible. From standardisation measurements a correction is computed as follows

$$\text{standardisation correction} = \frac{L\,(l' - l)}{l} \tag{4.3}$$

where

L = recorded length of a line
l = nominal length of field tape (say 30 m)
l' = standardised length of field tape (say 30.011 m).

The sign of the correction depends on the values of l and l'.

A baseline for standardising tapes should consist of two fixed points, located on site such that they are unlikely to be disturbed. These points could be nails in pegs, but marks set into concrete blocks or pillars are preferable. The length of the field tape is compared to the length of the baseline and the standardisation correction given by

$$\text{standardisation correction} = \frac{L\ (l_B - l_F)}{l_B} \qquad (4.4)$$

where

l_B = length of baseline

l_F = length of field tape along baseline.

Tension

The steel used for tapes, in common with many metals, is elastic and the tape length varies with applied tension. This effect tends to be overlooked by an inexperienced engineer and, consequently, errors can arise in measured lines.

Every steel tape is manufactured and calibrated with a standard tension of 50 N applied. Therefore, instead of merely pulling the tape taut, an improvement in precision is obtained if the tape is pulled at its standard tension. This can be achieved using *spring balances* specially made for use in ground taping together with a device called a *roller grip*. When measuring, one end of the tape is held firm near the zero mark, the spring balance and roller grip are hooked to the other end of the tape and the spring balance handle is pulled until its sliding index indicates that the correct tension is applied, as shown in figure 4.8. This tension is then maintained while measurements are taken.

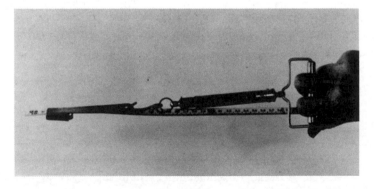

Figure 4.8 *Tensioning equipment (Building Research Establishment: Crown copyright)*

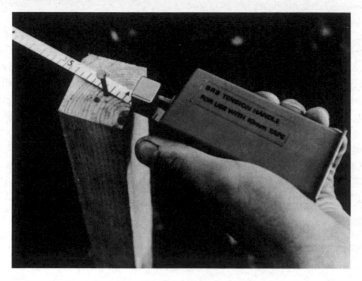

Figure 4.9   *Constant tension handle (Building Research Establishment: Crown copyright)*

When setting out, this method of tensioning can be difficult and a *constant tension handle* can be used to minimise errors. The use of the constant tension handle is shown in figure 4.9 and since the correct tension is always applied to the tape it is particularly suitable for use by unskilled operatives.

Should a tape be subjected to a pull other than the standardising value, it can be shown that a correction to an observed length is given by

$$\text{tension correction} = \frac{L\,(T_{\text{F}} - T_{\text{S}})}{AE} \qquad (4.5)$$

where

$T_{\text{F}} =$ tension applied to the tape (N)
$T_{\text{S}} =$ standard tension (N)
$A =$ cross-sectional area of the tape (mm^2)
$E =$ modulus of elasticity for the tape material (N mm^{-2})
(for steel tapes, typically 200 000 N mm^{-2}).

The sign of the correction depends on the magnitudes of T_{F} and T_{S}.

Temperature Variations

In addition to the effects of standardisation and tension, steel tapes contract and expand with temperature variations and are, therefore, calibrated at a standard temperature of 20°C.

In order to improve precision, the temperature of the tape has to be recorded since it will seldom be used at 20°C, and special surveying thermometers are used for this purpose. When using the tape along the ground, measurement of the air temperature can give a different reading from that obtained close to the ground, so it is normal to place the thermometer alongside the tape at ground level. For this reason, the thermometers are usually metal-cased for protection. When in use they should be left in position until a steady reading is obtained since the metal casing can take some time to reach a constant temperature. It is also necessary to have the tape in position for some time before readings are taken to allow it also to reach the ambient temperature. It is bad practice to measure a distance in the field in winter with a tape that has just been removed from a heated office.

The temperature correction is given by

$$\text{temperature correction} = \alpha L \, (t_F - t_S) \tag{4.6}$$

where

α = the coefficient of expansion of the tape material (for example 0.000 011 2 per °C for steel)

t_F = mean field temperature (°C)

t_S = temperature of standardisation (20°C).

The sign of this correction is given by the magnitudes of t_F and t_S.

Sag (Catenary)

When the ground between two points is very irregular, surface taping can prove to be a difficult process and it may be necessary to suspend the tape above the ground between the points in order to measure the distance between them. This can be done by holding the tape in tension between tripods or wooden stakes, the stakes being driven in approximately 1 m above ground level. For long lines, these tripods or stakes must be aligned by theodolite before taping commences. When measuring distances less than a tape length on site between elevated points on structures, the tape may be suspended for ease of measurement.

Whatever the case, the tape will sag under its own weight in the shape of a catenary curve as shown in figure 4.10.

Since the distance required is the chord AB, a sag correction must be applied to the catenary length measured. This correction is given by

$$\text{sag correction} = -\frac{w^2 L^3 \cos^2\theta}{24 T_F^2} \tag{4.7}$$

where

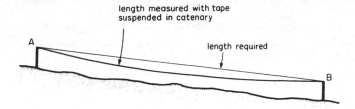

Figure 4.10   *Measurement in catenary*

θ = the angle of slope between tape supports
w = the weight of the tape per metre length ($N\ m^{-1}$)
T_F = the tension applied to the tape (N).

This correction is always negative.

Combined Formula

The corrections discussed in the preceding sections are usually calculated separately and then used in the following equations:

For horizontal measurements

$$D = L - \text{slope} \pm \text{standardisation} \pm \text{tension}$$

$$\pm \text{ temperature } - \text{ sag} \qquad (4.8)$$

For vertical measurements

$$V = L \pm \text{standardisation} \pm \text{tension} \pm \text{temperature} \qquad (4.9)$$

Both of these can be used when measuring and setting out distances, as shown in the worked examples in section 4.6.

4.5 Steel Taping: Precision and Applications

The general rules for the precision of steel taping can be summarised as follows.

For a maximum precision of 1 in 5000 (that is ±6 mm per 30 m), measurements can be taken over most ground surfaces if only standardisation and slope corrections are applied.

If the tape is tensioned correctly and temperature variations are taken into account, the precision of taping can be increased to the order of 1 in

TABLE 4.1

Applications of Steel Taping with Precisions

Type of work	Precision required
Location of spoil heaps and soft detail	1 in 500 to 1 in 5000
Setting out sewer pipelines. Location of hard detail	1 in 5000 to 1 in 10 000
Measuring traverse legs. Setting out road centre lines, grids, baselines, offset pegs General site setting out, setting out buildings, establishing secondary control	1 in 10 000 to 1 in 20 000
Setting out primary control	1 in 20 000 upwards

10 000 (that is ±3 mm per 30 m). On specially prepared surfaces or over spans less than a tape length, the precision may be improved further.

To obtain the best precision of 1 in 20 000 (that is ±1.5 mm per 30 m), sag corrections should be applied in addition to all the other corrections.

Some of the common applications of steel taping in engineering surveying are shown in table 4.1 together with an indication of the precisions normally required for site work. Table 4.2 summarises taping corrections and the effect of precision on these.

4.6 Steel Taping: Worked Examples

(1) Measuring a Horizontal Distance with a Steel Tape

Question
A steel tape of nominal length 30 m was used to measure a line AB by suspending it between supports. The following measurements were recorded.

Line	Length measured	Slope angle	Mean temperature	Tension applied
AB	29.872 m	3°40'	5°C	120 N

The standardised length of the tape against a reference tape was known to be 30.014 m at 20°C and 50 N tension.

TABLE 4.2
Taping Corrections

Correction	Formula	Procedure required to achieve stated precision	
		1:5000	1:10 000
Slope	$-\dfrac{\Delta h^2}{2L}$ or $-L(1 - \cos\theta)$	Slope correction always applied. Usually largest correction	
Standardisation	$\dfrac{L(l' - l)}{l}$ or $\dfrac{L(l_B - l_F)}{l_B}$	Standardise tape and apply correction	
Tension	$\dfrac{L(T_F - T_s)}{AE}$	Negligible effect if tape pulled 'sensibly'	Standard tension must be applied or apply correction
Temperature	$\alpha L(t_F - t_s)$	Only important in hot or cold weather	Measure temperature and apply correction
Sag	$-\dfrac{w^2L^3\cos^2\theta}{24T_F^2}$	Use tape fully supported	Apply correction for suspended measurements

If the tape weighs 0.17 N m^{-1} and has a cross-sectional area of 2 mm^2, calculate the horizontal length of AB.

Young's modulus (E) for the tape material is 200 kN mm^{-2} and the coefficient of thermal expansion (α) is 0.000 011 2 per °C.

Solution
A series of corrections is computed as follows

$$\text{slope correction} = -L(1 - \cos\theta)$$
$$= -29.872(1 - \cos 3°40')$$
$$= \mathbf{-0.0611 \ m}$$

$$\text{standardisation correction} = \frac{L\ (l' - l)}{l}$$

$$= \frac{29.872\ (30.014 - 30.000)}{30.000}$$

$$= +0.0139\ \text{m}$$

$$\text{tension correction} = \frac{L\ (T_F - T_s)}{AE} = \frac{29.872\ (120 - 50)}{2(200\ 000)}$$

$$= +0.0052\ \text{m}$$

$$\text{temperature correction} = \alpha L\ (t_F - t_s)$$

$$= 0.000\ 011\ 2\ (29.872)(5 - 20)$$

$$= -0.0050\ \text{m}$$

$$\text{sag (catenary) correction} = -\frac{w^2 L^3 \cos^2 \theta}{24\ T_F^2}$$

$$= -\frac{(0.17)^2\ (29.872)^3\ \cos^2\ 3°40'}{24(120)^2}$$

$$= -0.0022\ \text{m}$$

The horizontal length of AB is given by substituting the corrections into equation (4.8) as follows

horizontal length AB

$$= 29.872 - 0.0611 + 0.0139 + 0.0052 - 0.0050 - 0.0022$$

$$= 29.8228 = \textbf{29.823 m} \text{ (rounded to the nearest mm)}$$

(2) Setting Out a Slope Distance with a Steel Tape

Question

On a construction site, a point R is to be set out from a point S using a 50 m steel tape. The horizontal length of SR is designed as 35.000 m and it lies on a constant slope of 03°27′.

During the setting out the steel tape is laid on the ground and pulled at a tension of 70 N, the mean temperature being 12°C.

The tape was standardised as 40.983 m at 50 N tension and 20°C on a baseline of length 41.005 m. The coefficient of thermal expansion of the tape material is 0.000 011 2 per °C, Young's modulus is 200 kN mm^{-2} and the cross-sectional area of the tape is 2.4 mm^2.

Calculate the length that should be set out on the tape along the direction SR to establish the exact position of point R.

Solution

Equation (4.8) is again used but in this case D is known and L must be calculated.

The slope, standardisation, tension and temperature corrections must all be calculated. The sag correction does not apply since the tape is laid along the ground.

Although L is not known, for the purposes of calculating the corrections it is sufficiently accurate to use D instead of L in the individual formulae. Therefore

$$\text{slope correction} = -D(1 - \cos\theta)$$

$$= -35.000(1 - \cos 03°27') = -\mathbf{0.0634 \ m}$$

$$\text{standardisation correction} = \frac{D \ (l_B - l_F)}{l_B}$$

$$= \frac{35.000 \ (41.005 - 40.983)}{41.005}$$

$$= +\mathbf{0.0188 \ m}$$

$$\text{tension correction} = \frac{D \ (T_F - T_S)}{AE}$$

$$= \frac{35.000 \ (70 - 50)}{2.4 \ (200 \ 000)}$$

$$= +\mathbf{0.0015 \ m}$$

$$\text{temperature correction} = \alpha D \ (t_F - t_S)$$

$$= 0.000 \ 011 \ 2 \ (35.000)(12 - 20)$$

$$= -\mathbf{0.0031 \ m}$$

The slope length SR is obtained from equation (4.8) as follows

$$D_{SR} = L_{SR} - \text{slope} \pm \text{standardisation} \pm \text{tension}$$

$$\pm \text{temperature}$$

From which

$$35.000 = L_{SR} - 0.0634 + 0.0188 + 0.0015 - 0.0031$$

Therefore

$$L_{SR} = 35.0462 = \mathbf{35.046 \ m} \text{ (rounded to the nearest mm)}$$

(3) Measuring a Vertical Distance with a Steel Tape

Question
A steel tape of nominal length 30 m was used to transfer a level from a reference line near the base of a vertical concrete column to a reference line near its top.

A 100 N weight was attached to its free end and the tape was hung freely down the side of the column such that its 100 mm mark was against the bottom reference line. A reading of 14.762 m was obtained at the top reference line.

The tape used was standardised on the flat as 30.007 m at a tension of 50 N and a temperature of 20°C. It had a cross-sectional area of 1.9 mm^2, the coefficient of thermal expansion of the tape material was 0.000 011 2 per °C and Young's modulus was 210 kN mm^{-2}. During the measurement the mean temperature of the tape was 3°C.

Calculate the vertical distance between the two reference lines.

Solution
For this measurement, only the standardisation, tension and temperature corrections apply.

$$\text{measured length} = L = 14.762 - 0.100 = 14.662 \ m$$

$$\text{standardisation correction} = \frac{14.662 \ (30.007 - 30.000)}{30.000}$$

$$= +0.0034 \ m$$

$$\text{tension correction} = \frac{14.662 \ (100 - 50)}{1.9 \ (210 \ 000)}$$

$$= +0.0018 \ m$$

$$\text{temperature correction} = 0.000 \ 011 \ 2(14.662)(3 - 20)$$

$$= -0.0028 \ m$$

The vertical distance is given by equation (4.9) as

$$V = L \pm \text{standardisation} \pm \text{tension} \pm \text{temperature}$$

$$= 14.662 + 0.0034 + 0.0018 - 0.0028$$

$$= 14.6644$$

$$= \mathbf{14.664 \ m} \text{ (to the nearest mm)}$$

4.7 Other Types of Tape

In addition to steel tapes, the following tapes are sometimes used in engineering surveys.

Fibreglass tapes are available in a variety of lengths and their construction is shown in figure 4.11. Typical graduations are shown in figure 4.12. When compared with steel tapes, fibreglass tapes are lighter, more flexible and less likely to break but they tend to stretch much more when pulled. As a result, this type of tape is used mainly in detail surveying, sectioning and earthworks where precisions in the order of 1 in 1000 are acceptable for linear measurements.

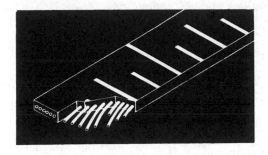

Figure 4.11   *Cross-section through fibreglass tape (courtesy Fisco Products Ltd)*

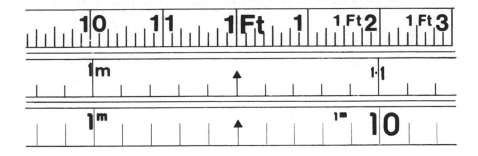

Figure 4.12   *Fibreglass tape graduations (courtesy Fisco Products Ltd)*

Invar tapes are made from an alloy of nickel and steel and have a coefficient of thermal expansion approximately one-tenth or less that of steel. Consequently, these tapes are almost independent of temperature changes and are ideal for use where very precise measurements are required. However, since invar tapes are expensive and must be handled with great care to avoid bends and kinks, they are not used for ordinary work.

4.8 Optical Distance Measurement (ODM)

Optical distance measurement on engineering sites nowadays consists only of the occasional use of stadia tacheometry. The other methods of ODM requiring specialised equipment, such as self-reducing tacheometers and subtense bars, have been superseded almost completely by EDM equipment and are not dealt with here.

Stadia tacheometry can be used to measure horizontal and vertical distances. However, because its relative precision is usually only 1 in 500 the uses of stadia tacheometry in engineering surveying are restricted to the survey of natural features such as trees, hedges, river banks and so on, in the production of site plans and to obtaining spot heights for estimating earthwork quantities and for contouring.

4.9 Stadia Tacheometry

In stadia tacheometry, a levelling staff is held vertically at one end of the line being measured and a level or theodolite is set up above the other.

The staff is read using the stadia lines engraved on the telescope diaphragm as shown in figure 4.13. The vertical angle along the line of sight is also recorded. If a level is used, the line of sight will be horizontal assuming that the level has no collimation error. If a theodolite is used, the line of sight can be either horizontal or inclined as shown in figure 4.13. The vertical compensating system of the theodolite must be in correct adjustment since vertical angles are read on one face only.

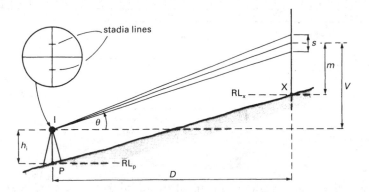

Figure 4.13 *Inclined line of sight in stadia tacheometry*

With reference to figure 4.13

Horizontal distance PX $= D = K s \cos^2\theta + C \cos \theta$ (4.10)

Vertical distance $V = \frac{1}{2}K s \sin 2\theta + C \sin \theta$ (4.11)

Reduced level of X $= RL_x = RL_p + h_i + V - m$ (4.12)

where

K is the multiplying constant of the instrument, usually 100

C is the additive constant of the instrument, usually 0

s is the staff intercept, that is, the difference between the two stadia readings

θ is the vertical angle along the line of sight

h_i is the height of the trunnion axis above point P

m is the middle staff reading at X

$+V$ is used if there is an angle of elevation

$-V$ is used if there is an angle of depression.

4.10 Worked Example: Use of Stadia Tacheometry Formulae

Question
A 1″ reading theodolite having a multiplying constant of 100 and an additive constant of 0 was centred and levelled a height of 1.48 m above a point C of reduced level 46.87 m.

A levelling staff was held vertically at points D and L in turn and the readings shown in table 4.3 were taken.

Calculate (i) the reduced levels of points D and L

(ii) the horizontal length of DL.

TABLE 4.3

Staff position	Staff readings (m)	Vertical circle readings	Horizontal circle readings
D	3.240, 3.047, 2.853	87°38′53″	56°49′31″
L	2.458, 2.230, 2.002	92°21′36″	98°07′18″

Solution
(i) The reduced levels of points D and L

Figure 4.14 shows the data obtained.

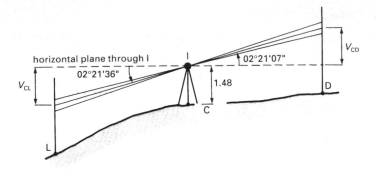

Figure 4.14

From equation (4.11)

$$V_{CD} = \tfrac{1}{2}(100)(3.240 - 2.853)\,\sin(04°42'14'') + 0 = 1.587 \text{ m}$$

$$V_{CL} = \tfrac{1}{2}(100)(2.458 - 2.002)\,\sin(04°43'12'') + 0 = 1.876 \text{ m}$$

Applying equation (4.12) gives

$$\mathbf{RL}_D = RL_C + h_i + V_{CD} - m$$

$$= 46.87 + 1.48 + 1.587 - 3.047 = \mathbf{46.89 \text{ m}}$$

and

$$\mathbf{RL}_L = RL_C + h_i - V_{CL} - m$$

$$= 46.87 + 1.48 - 1.876 - 2.230 = \mathbf{44.24 \text{ m}}$$

(ii) The horizontal length of DL

From equation (4.10)

$$D_{CD} = 100(3.240 - 2.853)\,\cos^2(02°21'07'') + 0 = 38.635 \text{ m}$$

$$D_{CL} = 100(2.458 - 2.002)\,\cos^2(02°21'36'') + 0 = 45.523 \text{ m}$$

Figure 4.15 shows a plan view of points C, D and L. From this

$$\text{Angle DCL} = 98°07'18'' - 56°49'31'' = 41°17'47''$$

In triangle DCL, the cosine formula gives

$$D^2_{DL} = D^2_{CD} + D^2_{CL} - 2(D_{CD})\,(D_{CL})\,\cos 41°17'47''$$

With $D_{CD} = 38.635$ and $D_{CL} = 45.523$, the horizontal length DL is

$$D_{DL} = 30.368 = \mathbf{30.4 \text{ m}} \text{ (rounded to the nearest 0.1 m)}$$

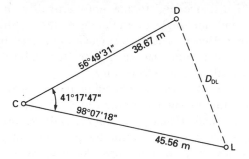

Figure 4.15

4.11 Accuracy and Sources of Error in Stadia Tacheometry

The accuracy of basic stadia tacheometry depends on two categories of error, instrumental errors and field errors.

Instrumental Errors

These include

(1) An incorrectly assumed value for K, the multiplying constant, caused by an error in the construction of the diaphragm of the theodolite or level used.
(2) Errors arising out of the assumption that K and C are fixed when, strictly, both K and C are variable.

The possible errors due to (1) and (2) above limit the overall accuracy of distance measurement by stadia tacheometry to 1 in 1000.

Field Errors

These can occur from the following sources.

(1) When observing the staff, incorrect readings may be recorded which result in an error in the staff intercept, s. Assuming $K = 100$, an error of $\pm$ 1 mm in the value of s results in an error of $\pm$ 100 mm in D. Since the staff reading accuracy decreases as D increases, the maximum length of a tacheometric sight should be 50 m.
(2) Non-verticality of the staff can be a serious source of error. This and poor accuracy of staff readings form the worst two sources of error.

The error in distance due to the non-verticality of the staff is proportional to both the angle of elevation of the sighting and the length of the sighting. Hence, a large error can be caused by steep sightings, long sightings or a combination of both. It is advisable not to exceed $\theta = \pm 10°$ for all stadia tacheometry.

(3) A further source of error is in reading the vertical circle of the theodolite. If the line of sight is limited to $\pm 10°$, errors arising from this source will be small. Usually, it is sufficiently accurate to measure the vertical angle to $\pm 1'$ and, although it is possible to improve this reading accuracy, it is seldom worth doing so owing to the magnitude of all the other errors previously discussed.

Considering all the sources of error, the overall accuracy expected for distance measurement is 1 in 500 and the best possible accuracy is only 1 in 1000.

The vertical component, V, is subject to the same sources of error described above for distances, and the accuracy expected is approximately ± 50 mm.

The precision of stadia tacheometry is also discussed in section 6.10.

4.12 Applications of Stadia Tacheometry

Vertical staff tacheometry is ideally suited for detail surveying by radiation techniques. This is discussed fully in section 9.7.

Since the best possible accuracy obtainable is only 1 in 1000, the method is best restricted to the production of contoured site plans and should not be used to measure distances where precisions better than this are required.

5

Distance Measurement: EDM and Total Stations

In civil engineering and construction the use of electromagnetic distance measurement (EDM) is now so widespread that it would be difficult to imagine contemporary site surveying without it. The rapid development of EDM equipment in recent years has enabled the surveyor and engineer to measure distances much more easily and to a higher precision than is possible using taping or optical methods. As a result of these technical advances, many changes have taken place in surveying techniques. For example, the application of traversing and combined networks in control surveys covering large areas is commonplace; detail surveying using theodolite-mounted EDM and total stations gives rise to more efficient methods of producing maps and plans and many modern setting-out techniques would be impossible without EDM equipment.

To use an EDM system, the instrument is set over one end of the line to be measured and some form of reflector is set over the other end such that the line of sight between the instrument and the reflector is unobstructed. An electromagnetic wave is transmitted from the instrument towards the reflector where part of it is returned to the instrument. By comparing the transmitted and received waves, the instrument is able to compute and display the required distance.

Since there are at present many different EDM systems available, any detailed operating instructions for any particular instrument have been excluded. Such information is available in the handbooks supplied by manufacturers for their respective instruments.

5.1 Electromagnetic Waves

When a length is measured with EDM equipment, no visible linear device is used to determine the length as, for instance, when a tape is aligned in successive lengths along the line being measured. The question often asked is what, then, are electromagnetic waves?

For the simplest treatment they can be considered to be the means by which electrical energy is conveyed through a medium, particularly the atmosphere. If an electric current is fed to an aerial this creates an electrical disturbance in and around the aerial. The disturbance is not confined to the aerial but spreads out into space by varying the electric and magnetic fields in the medium surrounding the aerial. Therefore, energy is propagated outwards and, since the energy is transmitted by varying electric and magnetic fields, the energy is said to be propagated by electromagnetic waves.

The electromagnetic waves so created require no material medium to support them and can be propagated in a vacuum or in the atmosphere. The type of electromagnetic wave generated depends on many factors but, principally, on the nature of the electrical signal used to generate the waves.

Properties of Electromagnetic Waves

Although electromagnetic waves are extremely complex in nature, they can be represented in their simplest form as periodic sinusoidal waves and therefore have predictable properties by which all electromagnetic radiation is defined.

Figure 5.1 shows a sinusoidal waveform which has the following properties. The wave completes a *cycle* when moving between identical points on the wave and the number of times in one second the wave completes a cycle is termed the *frequency* of the wave. The frequency is represented by f hertz, 1 hertz (Hz) being 1 cycle per second. The *wavelength* is the distance which separates two identical points on the wave or is that length traversed in one cycle by the wave and is denoted by λ metres. The *speed of propagation* of the wave is the remaining property. Whereas frequency and wavelength can vary according to the electrical disturbance producing the wave, the speed (v) of an electromagnetic wave depends on the medium through which it is travelling. The speed of an electromagnetic wave in a vacuum is termed the *speed of light* and is given the symbol c. The value of c is known at the present time as 299 792 458 m s^{-1} and an exact knowledge of this constant is essential to EDM.

All of the above properties of electromagnetic waves are related by

$$\lambda = \frac{v}{f} \qquad (5.1)$$

A further term associated with periodic waves is the *phase* of the wave.

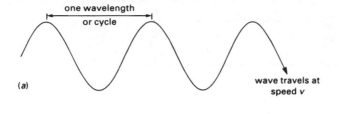

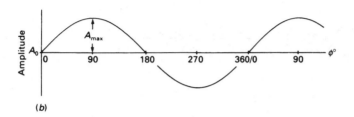

Figure 5.1   *Sinusoidal wave motion: (a) as a function of distance (or time); (b) as a function of phase angle ϕ*

As far as EDM is concerned, this is a convenient method of identifying fractions of a wavelength or cycle. A relationship that expresses the instantaneous amplitude of a sinusoidal wave is

$$A = A_{max} \sin \phi + A_0 \qquad (5.2)$$

where A_{max} is the maximum amplitude developed by the source, A_0 is the reference amplitude and ϕ is the phase angle. Angular degrees or radians are used as units for the phase angle up to a maximum of 360°, or 2π radians, for one complete cycle.

As can be seen, a quoted phase value can apply to the same point on *any* cycle or wavelength. This has importance when measuring lengths using electromagnetic waves.

5.2 Phase Comparison

In an EDM system, distance is determined by measuring the difference in phase angle between transmitted and reflected signals. This phase difference is usually expressed as a fraction of a cycle which can be converted into distance when the frequency and velocity of the wave are known.

The methods involved in measuring by phase comparison are outlined as follows.

In figure 5.2a, an EDM instrument has been set up at A and a reflector at B so that distance AB = D can be measured.

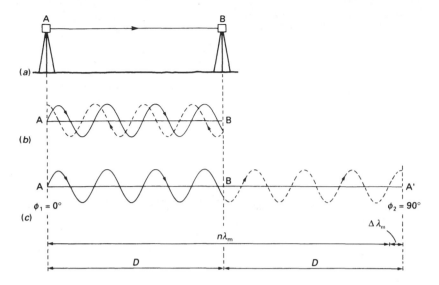

Figure 5.2 *Phase comparison*

Figure 5.2*b* shows the same EDM configuration as in figure 5.2*a*, but only the details of the electromagnetic wave path have been shown. The wave is continuously transmitted from A towards B, is instantly reflected at B and received back at A. For clarity, the same sequence is shown in figure 5.2*c* but the return wave has been opened out. Points A and A' are the same since the transmitter and receiver would be side by side in the same unit at A.

From figure 5.2*c* it is apparent that the distance covered by the wave in travelling from A to A' is given by

$$2D = n\lambda_m + \Delta\lambda_m \tag{5.3}$$

where D is the distance between A and B, λ_m the wavelength of the measuring unit, n the whole number of wavelengths travelled by the wave and $\Delta\lambda_m$ the fraction of a wavelength travelled by the wave. Since the double distance is measured, an EDM instrument has a measuring wavelength of $\lambda_m/2$.

Equation (5.3) shows the distance D to be made up of two separate elements: these are determined by two processes.

(1) The *phase comparison* or $\Delta\lambda_m$ measurement is achieved using electrical phase detectors.

Consider a phase detector, built into the unit at A, which senses or measures the phase of the electromagnetic wave as it is transmitted from A. Let this be ϕ_1 degrees. Assume the same detector also measures the phase of the wave as it returns at A' ($\phi_2°$). These two can be compared to give a measure of $\Delta\lambda_m$ using the relationship

$$\Delta\lambda_m = \frac{\text{phase difference in degrees}}{360} \times \lambda_m$$

$$= \frac{(\phi_2 - \phi_1)°}{360} \times \lambda_m \qquad (5.4)$$

The phase value ϕ_2 can apply to any incoming wavelength at A' and the phase comparison can only provide a means of determining by how much the **wave travels** in excess of a whole number of wavelengths.

(2) Some method of determining $n\lambda_m$, the other element comprising the unknown distance, is required. This is often referred to as *resolving the ambiguity* of the phase comparison and can be carried out by one of three methods.

(a) The measuring wavelength can be increased manually in multiples of 10 so that a coarse measurement of D is made, enabling n to be deduced.

(b) D can be found by measuring the line using three (or more) different, but closely related, wavelengths, to form simultaneous equations of the form $2D = n\lambda_m + \Delta\lambda_m$. These can be solved, making certain assumptions, to give a value for D.

(c) Most modern instruments use electromechanical or electronic devices to solve this problem automatically, the machine displaying the required distance D.

5.3 Analogy with Taping

Referring to the example of figure 5.2, assume the measuring wavelength is 30 m. From the diagram $n = 7$, $\phi_1 = 0°$ and $\phi_2 = 90°$.

The double distance is given by

$$2D = n\lambda_m + \Delta\lambda_m = n\lambda_m + \frac{(\phi_2 - \phi_1)}{360} \times \lambda_m$$

$$= (7 \times 30) + \frac{(90 - 0)}{360} \times 30$$

Hence

$$D = 108.75 \text{ m}$$

Imagine the distance between A and B was to be measured with a tape x metres in length. Following section 4.3, this would involve aligning the tape in successive lengths along the line AB (giving mx where m is the number of whole tape lengths) and noting the fraction of a tape length remaining (Δx) to complete the measurement. Hence $D = mx + \Delta x$.

If $x = 30$ m, measurement of AB would be recorded as

$$D = 3 \times 30 + 18.75 = 108.75 \text{ m}$$

Measurement of a length using electromagnetic waves is, therefore, directly analogous to taping, indeed it can be said that in EDM the electromagnetic wave has replaced the tape as the measuring medium.

5.4 Measurement Requirements

EDM employs the wavelength λ_m of an electromagnetic wave as the basic unit for measuring a distance. The value chosen for λ_m depends to a great extent on the desired accuracy of the EDM instrument and on phase resolution, the smallest fraction of a cycle that the instrument is capable of resolving.

For many EDM instruments an accuracy in measurement of between 1 and 10 mm is specified at short ranges and a phase resolution of 1 in 10 000 is normal. Assuming an accuracy of at least 1 mm, a measuring unit or wavelength of 10 m is required. Using an approximate value for the speed of light of 3×10^8 m s^{-1}, 10 m corresponds to a frequency of 30 MHz. This frequency is in the VHF part of the electromagnetic spectrum and, although it is possible to generate a 30 MHz signal fairly easily, problems exist when such a frequency is to be propagated in the atmosphere. To transmit this order of frequency over distances of many kilometres without significant attenuation of the signal would require either a very large transmitter antenna with dimensions approaching $10\lambda_m$ for effective propagation, or a reasonably sized, very inefficient antenna that requires a considerable power supply to drive it. Both of these alternatives are unacceptable for portable surveying equipment. Another objection to using a 10 m wave is that, although the wavelength of emitted radiation would be constant, it would not be coherent over long ranges. The effect of this is that each cycle slightly overlaps the next over long distances. A solution to these problems might be at first sight to decrease the value of λ_m and therefore increase the frequency of measurement and accuracy of length measurement. This could be done until a suitable compromise is reached for both transmission and measurement. Unfortunately, the phase measurement process tends to become extremely unstable at high frequencies, and use of a very short measuring wavelength would result in difficulties with resolving the ambiguity of measurement.

In order to be able to use a measuring unit of 10 m and combine this with efficient propagation, the process of modulation is used in EDM, in

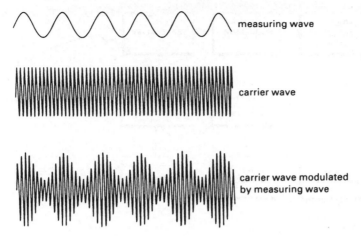

measuring wave

carrier wave

carrier wave modulated
by measuring wave

Figure 5.3   *Amplitude modulation*

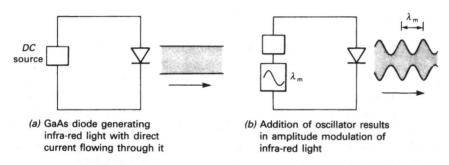

(a) GaAs diode generating
infra-red light with direct
current flowing through it

(b) Addition of oscillator results
in amplitude modulation of
infra-red light

Figure 5.4   *Modulation of GaAs diode*

which the measuring wave is mixed with a carrier wave of much higher
frequency. The type of modulation used in the majority of EDM systems is
amplitude modulation (figure 5.3) in which the measuring wave is used to
vary the amplitude of the carrier wave. The carrier wave used in nearly all
EDM instruments is infra-red and this is due to the carrier source which is
a gallium arsenide (GaAs) infra-red emitting diode. These diodes can be
very easily amplitude modulated at the high frequencies required for EDM
and provide a simple and inexpensive method of producing a modulated
carrier wave, as shown in figure 5.4.

5.5 EDM System

The schematic diagram of figure 5.5 shows the essential parts of an EDM
system. The only carrier source shown is an infra-red beam produced by a
GaAs diode.

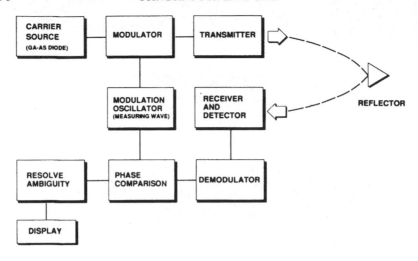

Figure 5.5　*EDM system*

In figure 5.5 the sinusoidal modulation signal or measuring wave is derived from a crystal-controlled frequency oscillator, the frequency value of which is typically 15–40 MHz for reasons already outlined. It is necessary that the frequency of the measuring wave be held at a constant value within a few parts per million (ppm) of the nominal frequency, as this determines the accuracy of the measuring scale of the EDM system. The exact value for the frequency of the measuring wave depends to some extent on the GaAs diode used and also on standard atmospheric conditions.

The amplitude-modulated infra-red beam is transmitted from the EDM unit towards a reflector at the remote end of the line to be measured. The transmitter built into an EDM instrument using continuous-wave infra-red comprises optical components since the GaAs carrier wavelength is in the near visible part of the electromagnetic spectrum. Consequently, mirrors and prisms can be used to direct the infra-red beam through the instrument and, to overcome signal loss, a combination of lenses is used to focus and transmit the infra-red carrier as a highly collimated beam with a divergence of less than 15′ of arc. The latter helps to increase the measuring range of the instrument.

The receiving optics of infra-red distance measurers are usually mounted coaxially or alongside the transmitting optics and they occupy as large an area as possible so as to collect sufficient signal for measurement purposes. Upon re-entering the instrument, the modulated infra-red beam is detected and demodulation takes place (the separation of the measuring and carrier waves).

Modulated signals are often detected by silicon photodiodes since these are small in size and, if used in conjunction with a suitable electronic amplifier, can detect very weak signals especially in the infra-red parts of the spectrum.

From the demodulator, the return signal is fed into the phase comparison stage of the EDM instrument. A second signal, also derived from the original modulation oscillator, is also fed into the phase-comparison stage. These two signals can be processed by several different methods to produce a $\Delta\phi$ or $\Delta\lambda$ value for the relevant line. In addition to this, further measurements may be taken at different frequencies to resolve the ambiguity of measurement.

5.6 EDM Reflectors

Since the infra-red carrier waves used in most EDM systems have wavelengths close to visible light, they can be treated as beams of light and, for simplicity, a plane mirror could be used to reflect them. Unfortunately, this would require very accurate alignment of the mirror because the transmitted beam has a narrow spread. To overcome this problem, a special form of reflector known as a *corner cube prism* (or *retroreflector*) is always used. As shown in figure 5.6, these are constructed from glass cubes or blocks and they will always return a beam along a path exactly parallel to the incident path over a range of angles of incidence of about 20° to the normal of the front face of the prism. As a result, the alignment of the prism is not critical and it is quickly set when on site. A range of *prisms* and *prism* sets (a combination of a prism and an optical target) are shown in figure 5.7 and a *pole-mounted reflector* (or *detail pole*) in figure 5.8.

Associated with all reflecting prisms is a *prism constant* (or *offset*). This is the distance between the effective centre of the prism and the plumbing and pivot point of the prism. Owing to the refractive properties of glass which slows down the carrier wave when it passes through a prism, the

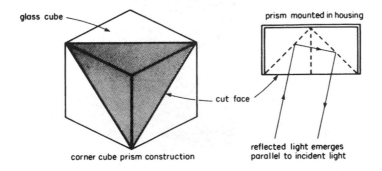

glass cube

cut face

corner cube prism construction

prism mounted in housing

reflected light emerges
parallel to incident light

Figure 5.6   *Corner cube prism (retroreflector)*

(a)   (b)

single and triple prism sets for coaxial EDM

(c)

single prism set (adjustable) for theodolite-mounted EDM

(d)   (e)

single and triple prism sets for yoke-mounted EDM

Figure 5.7   *EDM prism sets (courtesy Sokkia Ltd)*

effective centre of a prism is normally well behind the physical centre, or vertex, as shown in figure 5.9. A prism constant is typically −30 or −40 mm and this value is set into an EDM instrument as a correction that is applied automatically to each distance measured. If ignored, or applied incorrectly, this is systematic error present in all measured distances and is not eliminated by applying any field procedure. It is, therefore, most important that the correct prism constant is identified for the prism in use with an EDM instrument and that this is corrected for.

5.7 EDM Specifications

All EDM systems have an instrumental accuracy quoted in the form

$$\pm (a \text{ mm} + b \text{ ppm})$$

Figure 5.8   *Detail pole*

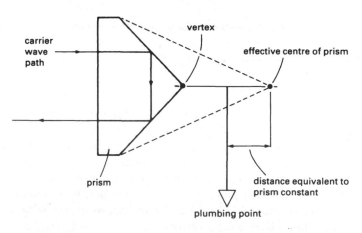

Figure 5.9   *Prism constant*

Constant *a* is made up from internal sources within the EDM instrument, and these are normally beyond the control of the user. This error is an estimate of the individual errors caused by such phenomena as unwanted phase shifts in electronic components, errors in phase measurement and index errors in centring the instrument and reflector.

The systematic error b is proportional to the distance being measured, and depends on the atmospheric conditions at the time of measurement and on the frequency drift in the crystals of the modulation oscillator. Atmospheric conditions are the worst source of error in EDM, and, since these are proportional to distance, extra care should be taken in the recording of meteorological conditions when measuring long lines (see section 5.19).

5.8 Theodolite-mounted EDM Systems

At present, many EDM systems are available, the majority of which use an infra-red carrier source. Without doubt, the most useful facility with infra-red instruments is that they can be combined with a theodolite in some way since the infra-red units are light and compact. This facility enables angles and distances to be measured simultaneously and three types of system can be identified: theodolite-mounted EDM systems, total stations and distancers. This section covers theodolite-mounted EDM: total stations and distancers are covered in the next sections.

In theodolite-mounted EDM, a specially designed lightweight EDM unit is either attached to the telescope of a theodolite or is yoke-mounted. Figure 5.10a shows a Wild DI1600 EDM unit and figure 5.10b shows a Geodimeter 220 EDM unit attached to a theodolite. Figure 5.11 shows a yoke-mounted Sokkia RED2L EDM unit which can also be telescope mounted. A summary of the specifications for these instruments is given in table 5.1.

(b)

(a)

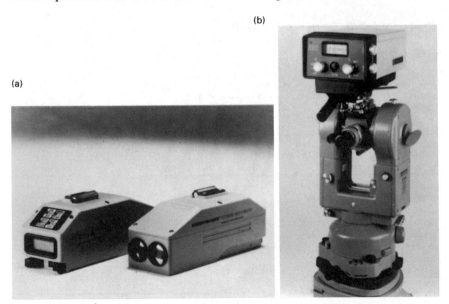

Figure 5.10   (a) Wild DI1600 and (b) Geodimeter 220 (courtesy Leica UK Ltd and Geotronics Ltd)

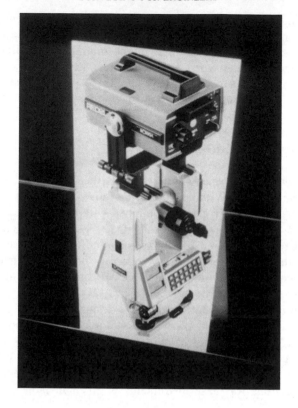

Figure 5.11   *Sokkia RED2L (courtesy Sokkia Ltd)*

TABLE 5.1
Some Examples of Theodolite-mounted EDM Systems

Instrument	Leica DI1600	Geodimeter 220	Sokkia RED2L
Distance measurement			
to one prism	2.5 km	2.3 km	4.0 km
to three prisms	3.5 km	4.0 km	5.2 km
accuracy	$\pm(3 \text{ mm} + 2 \text{ ppm})$	$\pm(3 \text{ mm} + 3 \text{ ppm})$	$\pm(5 \text{ mm} + 3 \text{ ppm})$
measurement time	1.5 seconds	5 seconds	6 seconds
Data displayed	SD	SD,HD,VD	SD,HD,VD,SO
Battery	external	internal	internal
	12 V	12 V	6 V
Weight	1.1 kg	1.8 kg	2.0 kg

Notes: SD = slope distance; HD = horizontal distance; VD = vertical distance; SO = setting out.

On site, the EDM unit and theodolite take measurements independently and, in some cases, horizontal distances are obtained from measured slope distances by observing vertical angles along the direction of the distance measurement (see section 5.22). This angle is then transferred manually into the EDM unit which calculates and displays the horizontal distance.

Many of the features of theodolite-mounted EDM systems are also incorporated into total stations and these are described in the following section.

5.9 Total Stations

The total station, or electronic tacheometer as it is sometimes known, is an instrument that is capable of measuring angles and distances electronically. In common with other electronic surveying equipment, total stations are operated using a multi-function keyboard which is connected to a microprocessor built into the instrument. The microprocessor not only controls both the angle and distance measuring systems but is also used as a small computer that can calculate slope corrections, vertical components, rectangular coordinates and, in some cases, can also store observations directly using an internal memory.

Figure 5.12 shows the Nikon DTM-A5LG, Sokkia SET3C and the Zeiss Elta 5, a sample of total stations from the extensive range now available. Their technical specifications are summarised in Table 5.2.

5.10 Features of Total Stations

Although the total stations currently available have differing technical specifications, they tend to be made to a similar format. Those features common to the majority of total stations are described in this section.

Angle Measurement

This is carried out using an electronic theodolite. All the features associated with these described in section 3.4 are applicable to total stations. Typically, a total station can record angles with a resolution of between 1″ and 20″ and all instruments incorporate some form of compensator, the more expensive using dual-axis and the less sophisticated, single-axis compensation.

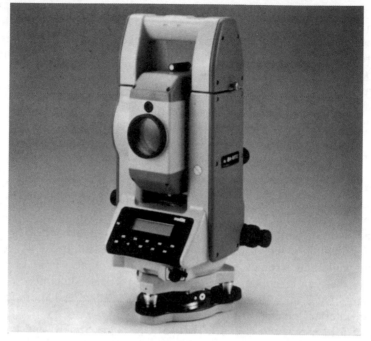

(a)

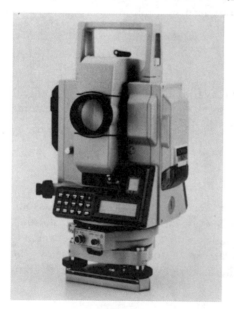

(b)

(c)

Figure 5.12 *(a) Nikon DTM-A5LG; (b) Sokkia SET3C; (c) Zeiss Elta 5 (courtesy Nikon Corporation, Sokkia Ltd and Carl Zeiss Ltd)*

TABLE 5.2
Some Examples of Total Stations

Instrument	Nikon DTM-A5LG	Sokkia SET3C	Zeiss Elta 5
Angle measurement			
H accuracy	±2″	±3″	±5″
V accuracy	±2″	±3″	±5″
Distance measurement			
to one prism	2.3 km	2.2 km	1.0 km
to three prisms	3.1 km	2.9 km	1.5 km
accuracy	±(2 mm + 2 ppm)	±(3 mm + 3 ppm)	±(5 mm + 3 ppm)
measurement time	3.0 seconds	3.2 seconds	3–4 seconds
Data displayed			
H and V angles	yes	yes	yes
SD, HD and VD	yes	yes	yes
X, Y and Z coords	yes	yes	yes
Setting out data	yes	yes	yes
Data recording	data recorder field computer	data recorder field computer memory card	data recorder field computer
Compensator	single-axis	dual-axis	single-axis
Battery	NiCad 7.2 V	NiCad 6 V	NiCad 4.8 V

Distance Measurement

At present, most total stations use a GaAs infra-red carrier source and phase comparison techniques in order to measure distances. However, compared to theodolite-mounted systems, nearly all total stations use coaxial optics in which the EDM transmitter and receiver are combined with the theodolite telescope. This makes the instrument much more compact and easier to use on site. Normally, a total station will measure a slope distance and the microprocessor uses the vertical angle recorded by the theodolite along the line of sight (line of distance measurement) to calculate the horizontal distance. In addition, the height difference between the trunnion axis and prism centre is also calculated and displayed. All instruments use some form of signal attenuation to protect the receiver.

Three modes are usually available for distance measurement.

Standard (or coarse) mode which has a resolution of 1 mm and a measurement time of 1–2 seconds.

Precise (or fine) mode which again has a resolution of 1 mm but a measurement time of 3–4 seconds. This is more accurate than the standard mode since the instrument repeats the measurement and refines the arithmetic mean value.

Tracking (or fast) mode in which the distance measurement is repeated automatically at intervals of less than one second. Normally, this mode has a resolution of 10 mm and is used extensively when setting out since readings are updated very quickly and vary in response to movements of the prism which is usually pole-mounted. Setting out using total stations is discussed in chapter 14.

The range of a total station is typically 1–3 km to a single prism assuming visibility is good and up to a range of 500 m, which covers 90 per cent of the distances measured on site, the precision of a typical total station is about 5 mm. Most instruments allow for the input of temperature and pressure which enables the distance readings to be automatically corrected for atmospheric effects (see section 5.19). In addition, any value of prism constant can be entered into the instrument to suit whatever prism is being used.

Control Panel

A total station is activated through its control panel which consists of a keyboard and multiple line liquid crystal display (often abbreviated to LCD). The LCD of a total station is moisture proof, it can be illuminated and some incorporate contrast controls to accommodate different viewing angles. A number of instruments have two control panels (one on each face of the theodolite) which makes them easier to use. The keyboard enables the user to select and implement different measurement modes, enables instrument parameters to be changed and allows special software functions to be used. Some keyboards incorporate multi-function keys that carry out specific tasks whereas others use keys to activate and display menu systems which enables the total station to be used as a computer might be.

In addition to controlling the total station, the keyboard is often used to code data generated by the instrument. Angles and distances are usually recorded electronically by a total station in digital form as raw data (slope distance, vertical angle and horizontal angle). If a code is entered from the keyboard to define the feature being observed, the data can be processed much more quickly by downloading it into appropriate software. On numeric keyboards, codes are represented by numbers only whereas on alphanumeric keyboards codes can be represented by numbers and/or letters which gives greater versatility and scope. The alphanumeric control panel of the Topcon GTS-6 is shown in figure 5.13. Feature codes and their application to large-scale surveys are discussed in section 9.11.

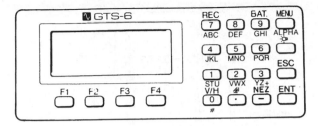

Figure 5.13   *Alphanumeric control panel of Topcon GTS-6 (courtesy Topcon Corporation)*

Power Supply

Rechargeable nickel–cadmium (NiCad) batteries are now standard for surveying instruments and these are connected directly to the total station without using cables. For angle and distance measurements, between two and ten hours' use can be obtained from a battery, depending on the instrument. Most total stations are capable of giving a battery power indication and some have an auto power save feature which switches the instrument off or into some standby mode after it has not been used for a specified time.

It is good practice, no matter what assurances a manufacturer may give about the life of a battery, to have a fully charged spare with the instrument at all times.

Useful Accessories

Geotronics and Nikon manufacture devices known as the *Tracklight* and *Lumi-Guide* respectively. Both of these are similar and the Tracklight is a visible light which enables a pole-mounted prism to be set directly on the line of sight without the need for hand signals from the total station. The device consists of a flashing three colour light: if the prism is to the left of centre of the line of sight, a green light is seen; and if the prism is to the right, a red light is seen – as shown in figure 5.14. When the prism is on line, a flashing white light is seen, the frequency of which doubles when it strikes the prism giving confirmation that the prism is in the correct position.

The *Geotronics Unicom* is a communication system which allows speech to be transmitted from a Geodimeter instrument to a prism. This consists of a small microphone on the control panel which is activated by pressing a key and a receiver with small loudspeaker mounted on the prism pole (see figure 5.15). The usual method of communicating on site, however, is by use of short range VHF hand-held radios.

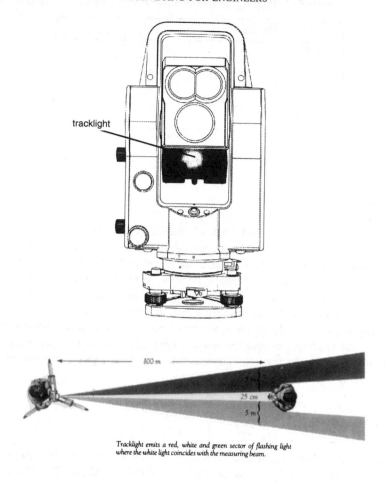

Tracklight emits a red, white and green sector of flashing light
where the white light coincides with the measuring beam.

Figure 5.14   Geodimeter Tracklight (courtesy Geotronics Ltd)

5.11 Onboard Software

As well as controlling the angle and distance functions of a total station,
the microprocessor is also programmed to perform coordinate and other
calculations. Some of these are described below although not all are available on every instrument.

Slope Corrections and Reduced Levels

From raw data (slope distance, vertical angle and horizontal distance), a
total station will calculate and display horizontal distance and vertical distance. If the reduced level of the instrument station, the height of the in-

Figure 5.15   *Geodimeter Unicom (courtesy Geotronics Ltd)*

strument and the height of the prism are entered, the reduced level of the prism station can also be calculated and displayed (see section 5.24). These are the most basic functions of a total station.

Horizontal Circle Orientation

The horizontal circle of a total station can be set to read a known bearing by entering the easting (E) and northing (N) coordinates of the station occupied followed by the E and N of a reference station. The reference station is then sighted and the orientation program is activated to calculate the bearing from the station occupied to the reference station and to set the horizontal circle to display this bearing (figure 5.16). The instrument would now be ready for further coordinate measurements or for setting out.

Coordinate Measurement

Having orientated the horizontal circle of a total station, the coordinates of other points can be determined fairly easily. A new point is sighted, and

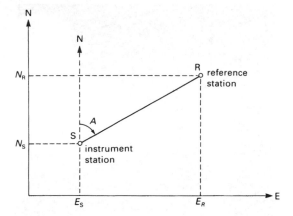

Figure 5.16   *Horizontal circle orientation: E_S, N_S and E_R, N_R are entered into total station which then calculates A and orientates to this*

distance and circle readings taken: when using the coordinate measurement program the instrument will now display the coordinates of the new point (figure 5.17). This can be extended to three dimensions if the reduced level of the instrument station and appropriate instrument and prism heights are also entered into the total station which will then display the reduced level of the new point.

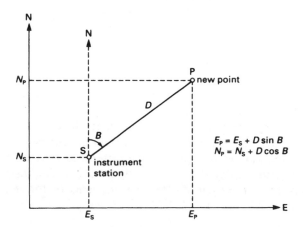

$$E_P = E_S + D \sin B$$
$$N_P = N_S + D \cos B$$

Figure 5.17   *Coordinate measurement: total station determines D and B and calculates E_P, N_P from instrument station*

Traverse Measurements

Traversing consists of a series of distance and angle measurements taken between successive points that enable the coordinates of those points to be calculated. Shown in figure 5.18, this type of coordinate determination is performed by a total station as a series of horizontal circle orientations and coordinate measurements taken at each traverse station. Traversing is discussed in much more detail in chapter 7 and the reader is strongly advised to consult this before attempting to use a total station in control surveys.

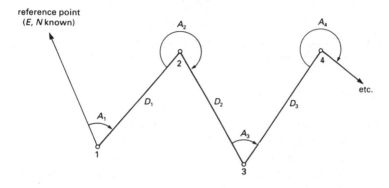

Figure 5.18   *Traverse measurements with a total station: set up at station 1 and orientate onto reference point. Measure A_1 and D_1 and obtain coordinates of station 2. Move to station 2, orientate onto station 1 and measure A_2 and D_2 to obtain coordinates of station 3. Move to other stations and repeat procedure*

Resection (or Free Stationing)

At level two control points (1 and 2 in figure 5.19) are required for free stationing, the coordinates of which have to be entered into the total station. These points are sighted in turn and distances and circle readings measured to each. Using this data, the instrument calculates the coordinates of the instrument station. This can be a useful facility when setting out as a temporary control point can be established in any desired location to suit site needs. As with other coordinate functions, if reduced levels and prism heights at the control points are also fed into the total station as well as the instrument height, the reduced level of the instrument station can also be computed and displayed.

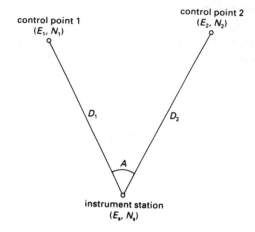

Figure 5.19   *Free stationing (resection): E_1, N_1 and E_2, N_2 entered into total station; D_1, D_2 and A*
            are measured and total station then calculates and displays E_s and N_s

Missing Line Measurement (MLM)

This software option allows a total station, from a single instrument position, to determine the horizontal distance and height difference between a
start point and a series of subsequently selected points. In figure 5.20, points
1 and 2 are sighted and the distances and circle readings to them recorded
from the instrument station. The MLM program then computes the horizontal
distance D_{12} and height difference Δh_{12} between these points. If the distance

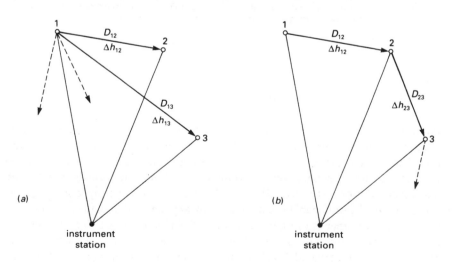

Figure 5.20   *Missing line measurement (MLM): (a) radial; (b) continuous*

and circle reading to a third point are included in the sequence, the total station can display D_{13} and Δh_{13} (radial MLM) or it can display D_{23} and Δh_{23} (continuous MLM). Any number of points can be added to the sequence.

Remote Elevation Measurement (REM)

This function is used to determine heights at inaccessible points where it is not possible to locate a prism. Since measurements are taken along an extended plumb line through the prism, the prism must be positioned vertically above or below the point(s) to be surveyed. The prism height p is entered into the instrument and the horizontal distance D to the prism determined (figure 5.21). In REM mode, the total station will now display the height from the ground at the prism to any point along the vertical through the prism. In figure 5.21, the top of a structure S could be sighted directly above the prism and its height h recorded. REM can also be used to set profile boards at their correct heights.

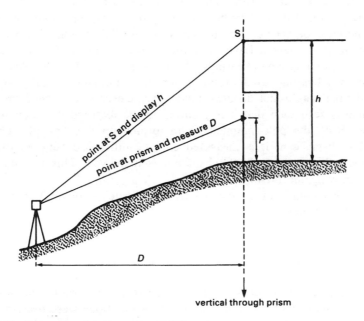

Figure 5.21   *Remote elevation measurement (REM)*

Setting Out Functions

Total stations can be used for setting out with given horizontal angle and distance values or with given coordinates.

When the horizontal angle and distance to be set out are known, these are entered into the instrument which has already had its horizontal circle orientated to a reference station. As soon as the appropriate key(s) are pressed to activate the setting out mode on the total station, it displays the difference (dHA) between the entered and measured horizontal angle values. In order to set the required direction for setting out, the telescope is rotated until a difference of zero is displayed (that is, dHA = 0). Following this, a pole-mounted prism is located on the line of sight as near to the required distance as possible: devices such as the Tracklight or Unicom (see section 5.10) are useful for lining in the prism but VHF radios with hand signals between instrument and prism are often used. Once aligned, the prism is sighted and the distance to it measured by the total station. The difference between the measured and entered distances is displayed and by moving the prism this difference is reduced to zero to locate the point.

When the coordinates of the point to be set out are known, those of the instrument and reference stations should also be known. Prior to measurement, the station coordinates are entered into the instrument as well as those of a reference station and the horizontal circle is orientated to the coordinate grid such that it displays bearings directly. Next, the coordinate values of the point to be set out are entered into the total station. When the setting out mode is selected the instrument displays the difference between calculated and measured bearings. There is no need for the observer to calculate any bearings as the total station does this automatically. As with the horizontal angle and distance mode, the telescope is rotated until this difference is zero such that it is pointing in the required direction for setting out. With the pole-mounted prism located on this line of sight, the horizontal distance to it is measured and the difference between this and the value calculated by the total station is displayed. This is reduced to zero by moving the prism.

For further details of setting out by coordinates with total stations see chapters 10, 11 and 14.

5.12 Total Stations: What to do and what not to do

The previous section discussed the onboard software installed in total stations and a number of different measurement functions were described. As far as site surveying is concerned, some of these have created a situation where the emphasis on surveying is apparently turning away from hand calculations and associated checking procedures to good site practice and field checking procedures.

A note of caution is expressed here. Even though a total station can perform many of the calculations often done manually on site, this does not mean the surveyor or engineer should lose this ability. For this reason,

Surveying for Engineers deals with and strongly emphasises coordinate calculations throughout. A knowledge of these may avoid situations where too great a reliance is placed on the digital readout obtainable from a total station and may avoid obvious mistakes when generating data using setting-out functions. In other words, the ability to process and manipulate coordinates by hand gives a surveyor or engineer the 'feel' for a correct orientation and distance, especially when setting out.

It is accepted that the total station is a very sophisticated instrument but this in itself can create problems. As an example, a traverse (see chapter 7) is a recognised surveying procedure for obtaining the coordinates of control points. Using onboard software, a total station can produce a set of coordinates much more quickly than by traversing since observations can be taken from a single instrument position. Although the methods used by total stations for obtaining coordinates have their applications, they are very dangerous if used in the wrong circumstances, especially when fixing the positions of control points. There are many cases where the time taken to arrive at the end result when using a total station is reduced by such a large factor compared with an established field procedure that the temptations are often too great and mistakes occur. While every opportunity should be made to make full use of a total station, any field procedure involving a total station that does not include an independent check on fieldwork must be treated as incomplete. Throughout *Surveying for Engineers*, recommendations for good practice are given for all procedures covered: there are still good reasons why these must be applied even when the most up-to-date instruments are being used on site or elsewhere when surveying.

5.13 Specialised Total Stations

Geotronics AB of Sweden currently manufacture a number of total stations with unique features.

The *Geodimeter 464* (figure 5.22) has all of the features described in sections 5.10 and 5.11 but is, in addition, a servo-driven total station. In use, all the operator needs to do to align onto a prism is to point the instrument roughly at the prism and then press a key to initiate a measurement. The servo motors then take over pointing the instrument at the prism and readings are recorded automatically. This has some advantages over a manually pointed system since a servo-driven instrument can aim and point much more quickly and with a better precision. For setting out with the Geodimeter 464, a point number is entered and the instrument instantly computes setting-out data and then automatically positions itself on the calculated bearing. If an elevation is stored, the instrument will also position itself in the third dimension.

The *Geodimeter System 500* is a range of total stations which allows users

Figure 5.22   *Geodimeter 464 (courtesy Geotronics Ltd)*

to purchase an instrument to suit their own requirements. Similar in appearance to the Geodimeter 464, instruments in this system can have three different angle and distance specifications: they can be servo-driven or manual, a numeric or alphanumeric keyboard can be selected, and different software options are available. An innovative product made by Geotronics for use with their System 500 total stations is the *RPU 500* (*RPU = Remote Positioning Unit*). This consists of a prism and target for angle and distance measurement, a receiver and a control panel all pole-mounted, as shown in figure 5.23. Using a telemetry link, the RPU controls all of a Geodimeter's functions other than aiming and this enables the surveyor or engineer to act as the prism (or 'staff') holder and at the same time take readings and feature-code a survey at points of detail. This is a much better method of surveying compared with the normal procedure in which the surveyor works at the instrument. Geotronics also manufacture the *RPU 502* which is shown in figure 5.24. Compared with the RPU 500, this is lighter and easier to handle since its components are broken down into three modules: target and prism, control panel, and telemetry link.

   The *Geodimeter System 4000* takes automation a stage further and consists of the Geodimeter 4400, a servo-driven total station known as the station unit and the RPU 4000 and RPU 4002 remote positioning units. All surveying carried out with this system is controlled from its RPU but the

Figure 5.23   *Geodimeter RPU 500 (courtesy Geotronics Ltd)*

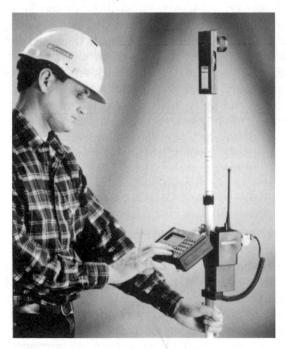

Figure 5.24   *Geodimeter RPU 502 (courtesy Geotronics Ltd)*

Geodimeter 4400, even though it is left unattended, is used as if the operator were standing behind it. As soon as the RPU and station unit are switched on, a signal is transmitted from the RPU to instruct the Geodimeter 4400 to search for and lock onto the RPU. As measurements are taken, the Geodimeter 4400 automatically follows the movements of the RPU and if contact is lost, a search routine is used to restore the link. When collecting data with the System 4000, the operator places the RPU at a point of interest and, by pressing keys on the control panel, the angle and distance are measured to it from the Geodimeter 4400. A feature code is allocated and all of this data is stored in the control panel. When setting out, coordinates are downloaded from a computer into the control panel in the office and when on site, keying in a relevant point number produces setting-out data immediately. Assuming the Geodimeter 4400 has been correctly orientated and is locked onto the RPU, the RPU will display the amounts by which it has to be moved so that the point is set out. The Geodimeter 4000 is, then, a surveying system that is operated by one person since the station unit can be left unattended. It is prudent, however, not to do this simply because the station unit could be damaged or stolen.

The *Leica VIP Survey System* is based on the Wild TC1610 total station. The operating system installed in this instrument allows the standard code function to be replaced by users who can define their own point coding system. In addition, the input and output displays can be changed as desired. All of this is carried out by programming an IBM-compatible microcomputer with the new codes and other instructions and then transferring the program into the total station. As well as the freely definable coding, the large memory of the TC1610 also enables programs to be loaded from the VIP program library. These include free stationing, height transfer, setting out, and so on and are similar to those described in section 5.11 for onboard software. The difference between this and other total stations is, however, their flexibility and only those programs needed on a particular job need to be installed in the TC1610.

5.14 Distancers

These are simply EDM devices that measure distances only and are not combined with a theodolite. Most instruments in this category are specialised in some way. For example, the Leica ME5000 (figure 5.25) is a very high precision EDM instrument and has a standard error of $\pm(0.2$ mm $+ 0.2$ ppm) in distance up to a range of 8 km. This instrument uses a HeNe laser as a carrier wave in order to predict more closely the effects of the atmosphere on a distance measurement and to increase the range.

Also in this category of EDM are microwave instruments such as the Tellumat CMW20 (figure 5.26). Unlike electro-optical instruments, an EDM

Figure 5.25   *Kern ME5000 (courtesy Leica UK Ltd)*

instrument such as the CMW20 uses a microwave carrier and active rather than passive signal reflection. Electronic reflection is achieved by placing at the remote terminal of the line another instrument which is identical to the measuring instrument. This remote instrument receives the transmitted signal, amplifies it and retransmits it back to the measuring (or master) instrument in exactly the same phase as it was received. Phase comparison is thus possible and, since the signal is amplified as well as reflected, a greater working range is obtained. As can be seen, microwave EDM requires two instruments and two operators to measure distances. The CMW20 also has a speech facility between master and remote to help the operators proceed through the measuring sequence. Signals are radiated from the CMW20 using small aerials built into the instrument case and this is shown in the upright (measuring) position in figure 5.26. These aerials produce a directional signal with a beam width of about 3° so alignment of the master and remote units is not critical. The maximum range of the CMW20 is 25 km, the accuracy being ±(5 mm + 3 ppm). In engineering surveying, an instrument such as the CMW20 would be used mainly in the establishment of control for very large projects.

Figure 5.26   *Tellumat CMW20 (courtesy Tellumat Ltd)*

5.15 Electronic Data Recording

One of the earliest and most successful applications of computers in surveying was the creation of software and hardware for plotting detail surveys. These improvements, however, caused problems with data transfer
because at the time computers were first used in surveying, it was only
possible to record observations, by hand, in field books. This meant that
all data collected on site had to be taken to an office and entered manually,
via a keyboard, into a computer if one was to be used for plotting. This is
a relatively slow process prone to error. The need for a better method of
getting information from field to computer was also accentuated with the
introduction of total stations which are capable of generating computer-
compatible angle and distance readings. As a result, the conventional method
of recording surveys was overtaken by developments in computer mapping
and survey instrumentation which made electronic data recording and transfer essential.

Initially, the devices produced to do this were fairly simple data loggers
but major advances were made when it became possible to connect small
portable computers to total stations. These intelligent data loggers could be
programmed to ask the surveyor for information, to record data from an

instrument in a suitable format and, if necessary to perform calculations using data transmitted to them. A number of different methods of recording data electronically have been developed and these include *data recorders*, which are dedicated to a particular instrument and can only store and process surveying observations, *field computers* which are general purpose hand-held computers adapted to survey data collection, *recording modules* which take the form of plug-in cards onto which data is magnetically encoded by a total station and *internal memories*.

Data Recorders

One of the most popular range of data recorders in use at present is the SDR Series manufactured by Sokkia of Japan for use with their SET total stations. The term SDR is an abbreviation for Sokkia Data Recorder but they are often referred to as electronic field books. In common with many other data recorders, they use solid state technology which enables them to store large amounts of data in a device not much bigger or heavier than a pocket calculator. Built with field survey in mind, the SDRs can withstand exposure to some rain, they can operate over a wide range of working temperatures and they are shockproof.

For detail surveys, angle and distance readings are transmitted from the total station directly to the SDR and these are stored together with point numbers generated by the recorder and feature codes which are entered manually on site. The type of feature code entered for each point depends on the software to be used to edit and plot a survey. Observations are normally stored as angles and distances (called raw data) but a data recorder can convert these to three-dimensional coordinates prior to transfer to a microcomputer. The Sokkia SDR33 (figure 5.27) can store 2400 or 7900 surveyed points.

All data recorders have some resident programs installed and these make it possible for them to collect and process data in a variety of ways. For instance, the readings for a three-dimensional traverse network can be entered into the SDR33 and it will compute and check the network, on site, as soon as observations are complete.

After completion of a survey, or at intermediate stages, data collected needs to be transferred from a data recorder to a computer. In the case of the SDR33, this is controlled by a program which allows the SDR33 to transmit the data held to a computer or backup storage device, as shown in figure 5.28. The format of data transmission can be varied but a compressed binary format is normally used by companies such as Sokkia as it allows faster data transmission. This, however, requires a special program to be installed in the receiving computer in order to decode the information. To prevent accidental data loss, the readings stored in an SDR cannot be altered

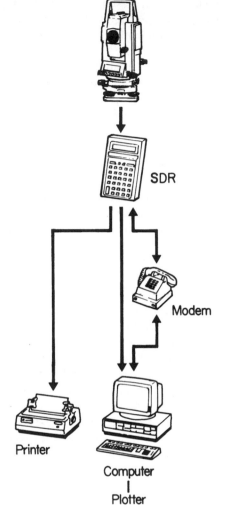

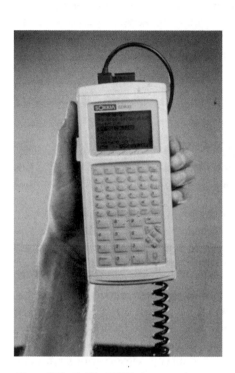

Figure 5.27   *Sokkia SDR33 data recorder*
            (courtesy   Sokkia LTD)

Figure 5.28   *Data transfer with SDRs*

or erased until they have been transmitted to a computer or printer and
protection from transmission errors is provided through checksum values
and parity.

Another well known survey company, Leica, markets two hand-held data
terminals called the GRE4n and GRE4a (figure 5.29). Both models are similar
but the GRE4a is alphanumeric and the GRE4n only accepts numeric data
input. These recorders are similar to and have many of the features described for the SDRs, the most significant difference being their memory
capacity (2000 data blocks). However, both versions of the GRE4 are freely

Figure 5.29 *Wild GRE4 data terminals (courtesy Leica UK Ltd)*

programmable and the 32 kbyte memory set aside for this can be loaded with any BASIC programs. As might be expected, Leica have developed software for this purpose and a suite of programs is available from them.

Field Computers

These are fully functional portable computers made suitable for outdoor use. Compared with a data recorder, they offer a more flexible approach to data collection since they can be programmed for many forms of data entry from any instrument to suit the individual requirements of any user. At present, the surveying field computer market is dominated by Husky Computers although the Psion Organiser is gaining in popularity.

The Husky Hunter 2 is a fully waterproof battery operated field computer with a standard QWERTY keypad and large display area (figure 5.30). The storage capacity of this computer is similar to a data recorder, it has a CPM compatible operating system and for programming Hunter BASIC is installed, although PASCAL, C, FORTRAN and COBOL compilers are available.

The Hunter 16 and 16/80 are improved versions of the Hunter 2 and they have larger memory capacities (up to 4 Mbytes of RAM) as well as MS-DOS 3.3 operating systems with GWBASIC. The FS/2 (figure 5.31) is the latest hand-held computer designed by Husky and this also has a PC-compatible MS-DOS operating system as well as 4 Mbytes of RAM.

The Psion Organiser (figure 5.32), although not made exclusively for surveying, is used as a data recorder by a number of survey manufacturers. These hand-held computers offer a much cheaper alternative to the systems described so far and they can be linked to any electronic instrument and

Figure 5.30   *Husky Hunter 2 field computer (courtesy Husky Computers Ltd)*

Figure 5.31   *Husky FS2 (courtesy Husky Computers Ltd)*

Figure 5.32   *Psion Organiser (courtesy Psion UK PLC)*

computer for data collection and transfer. Of course, they are not as sophisticated as Husky products or as specialised as a data recorder; data cannot be edited easily once entered into them and they are not ruggedised by Psion. However, they can be programmed, a special protective waterproof case can be purchased for field use and the latest versions are capable of storing over 2000 points of detail with feature coding.

Each Psion Organiser consists of a keyboard, a two or four line LCD and all models can accommodate special Psion storage devices, known as datapaks, that fit into two slots in the back of the housing (figure 5.32). These are the equivalent of floppy discs in PCs and provide the user with substantial extra memory capacity. A number of survey companies have developed software packages for the Psion Organiser and these are stored on a datapak which, when inserted into the Organiser, makes it suitable for survey data collection. The other slot is occupied by a memory pack which collects data and these are available in different formats and sizes.

Memory Cards (Recording Modules)

The method of storing and processing information using data recorders and field computers is thought of by some surveyors as inconvenient since it involves the use of an extra piece of equipment that could fail in some way. While the likelihood of anything going wrong with a data recorder or field computer is negligible, alternatives to these have been developed.

Data is collected on a memory card (or recording module) using a total station fitted with a microprocessor that has been programmed to perform the functions normally carried out by an external data recorder or field computer. A typical instrument in this category, the Topcon GTS-6B shown in figure 5.33, has a card holder (see in the open position in this case) and an alphanumeric keyboard to enable feature codes and other information to be entered. Data is transmitted to the memory card using a non-contact magnetic coupling system which eliminates the need to attach sockets or pins to the card, both of which risk being damaged under field conditions.

About the size of a credit card, data can be stored on memory cards quite safely for long periods and as well as offering the advantage of internal data storage, any number of cards could be used on a survey so there is no

Figure 5.33 *Topcon GTS-6B (courtesy Topcon Corporation)*

Figure 5.34 *Geodat 500 (courtesy Geotronics Ltd)*

limit to the number of points that can be measured. At present, card memories vary from storage of about 500 suitably coded points, to 4000 points. For automatic data transfer, the contents of a card are read into a desktop computer via a card reader (see figure 2.17) and information is processed by the computer in much the same way as if it were received from a data recorder or field computer.

Internal Memories

The distinguishing feature of the Geodimeter System 500 (see section 5.13) is that any total station in this series can be fitted with an internal memory of various sizes capable of storing from 900 to 10 000 points. This enables data to be collected without the need for a memory card or data recorder and files can be retrieved, checked and edited in the field using the instrument's display. Each Geodimeter has a serial two-way communication port which allows it to transfer data directly to most computers.

Although their instruments are fitted with internal memories, Geotronics also market a data recording unit known as the Geodat 500 (figure 5.34) and this acts as a portable 'hard-disc' for any total station in the Geodimeter System 500. About the same size as a data recorder, the Geodat 500 does not have a keypad or display and all recording takes place through whatever instrument it is connected to. Capable of storing 3000 points, no special reader is required for transferring data from this to a computer. In use, the

Geodat 500 tends to be used as an additional storage area once the internal memory of the instrument is full or it is used to store data simultaneously with the internal memory if a backup is required directly in the field.

5.16 Timed-pulse Distance Measurement

In this technique, distances are obtained by measuring the time taken for a pulse of laser radiation to travel from an instrument to a reflector (or target) and back. The distance D between instrument and reflector (or target) is given by

$$D = \frac{vt}{2} \tag{5.5}$$

where t is the measured transit time and v is the speed of propagation of electromagnetic radiation in the atmosphere.

For surveying applications, the transit time can be extremely small. Using an approximate value for v of 3×10^8 m s^{-1}, a transit time of 1 μs (10^{-6} s) corresponds to an object distance of 150 m. Consequently, the time duration of the transmitted pulse must be short compared with transit times.

Figure 5.35 shows a simplified block diagram of a timed-pulse distance measurer in which a GaAs laser diode is shown, although other types of

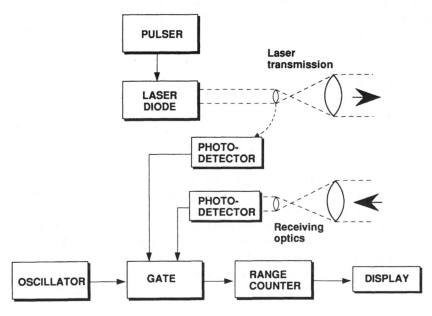

Figure 5.35 *Timed-pulse EDM system*

laser can be used. Just before a laser pulse is transmitted from the instrument, it is detected and this small sample of the energy transmitted is used as a start signal to open an electronic gate which connects an oscillator to a high-speed counter. This gate and therefore the connection between oscillator and counter is closed by the return signal pulse and the number of oscillator cycles fed into the counter during the time taken by the pulse to travel to the reflector (or target) and back is proportional to the required distance.

Although only a single pulse is necessary to measure a distance, the accuracy obtained would be poor and to improve this, a large number of pulses are analysed during each measurement. However, a limitation is imposed on the rate at which pulses can be transmitted owing to the requirement of the system to produce unambiguous measurements over its working range. For example, if an instrument has a maximum range of 15 km, a pulse will travel 30 km to the reflector and back in 0.1 ms. If pulses are transmitted and received by the instrument at a faster rate than one every 0.1 ms then ambiguity of distance up to 15 km will not be resolved. A pulse repetition rate of 0.5 ms is used by some manufacturers which is equivalent to 2000 pulses being transmitted every second.

To improve accuracy further, methods for refining the measurement of the transit time are used. As already stated, the transit time is counted by a reference oscillator and a typical frequency for this is 15 MHz giving a resolution of 10 m in distance. By obtaining a fine measurement within a reference oscillator period, a single pulse resolution approaching 10 mm can be achieved. The reliability of this depends primarily on the stability of the oscillator but repeating the measurement with a large number of pulses produces a result with millimetre accuracy.

A benefit of using pulses to measure distances is that the method relies on pulse detection and is therefore not dependent on signal amplitude: this enables a more efficient optical receiver to be built which in turn increases the measuring range compared with instruments of a similar size using phase comparison. Use of laser radiation instead of ordinary radiation to produce pulses also improves the measuring range since lasers are highly collimated. In addition, the use of lasers with their relatively closely defined wavelengths enables atmospheric effects on measured ranges to be predicted much more closely, thereby reducing the systematic error in the specification of such instruments.

5.17 Timed-pulse Instrumentation

The Wild DI3000S (figure 5.36) is a timed-pulse distancer with an accuracy of ±(3 mm + 1 ppm) in its standard measuring mode. Weighing only 1.7 kg it fits onto any Wild theodolite and, in average atmospheric conditions, can

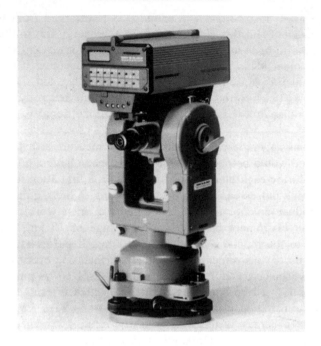

Figure 5.36 *Wild DI3000S (courtesy Leica UK Ltd)*

measure distances up to 9 km with a single prism. This instrument is well suited to control surveys where long distances are to be measured.

The Wild DIOR3002S (figure 5.37) is a special version of the DI3000S and has been designed specifically for distance measurement without reflectors. A maximum range of 350 m can be obtained to uncooperative targets depending on the reflecting surface with an accuracy of 5 mm. The DIOR3002S emits a visible laser beam which is used to mark points along the axis of the instrument in which the distance is measured using an invisible GaAs laser diode. If used with a single prism, the DI3002S has a measuring range of 6 km.

Many other timed-pulse distancers are currently available, the most widely used of which are IBEO products. These have various measuring ranges, the longest being the Pulsar 100 (figure 5.38) which can measure up to 1 km without reflectors with an accuracy of about 20 mm.

When taking reflectorless distance measurements with timed-pulse EDM instruments, many measuring and monitoring tasks in building and construction can be accomplished more easily. The following examples are common applications of reflectorless EDM.

(1) Monitoring of deformation in bridges, cooling towers and large struc-

Figure 5.37 *Wild DIOR3002S (courtesy Leica UK Ltd)*

Figure 5.38 *IBEO Pulsar 100 (courtesy Pulsar Measuring Systems Ltd)*

tures where access is not possible. This is discussed further in chapter 15.

(2) Monitoring waste disposal sites, slurry pits and other sites where access is difficult or even dangerous.

(3) Measuring volumes in quarries and open cast mines.

(4) Controlling and positioning construction equipment.

(5) Mapping the facades of buildings and as-built surveys of large areas of wall without the need to erect large amounts of scaffolding for access.

(6) Dimensional control of steelwork prior to lifting in position. In this example, a large number of readings to different surfaces are required in a short period.

(7) Road profiling from moving vehicles.

5.18 EDM Corrections

When measurements are taken using an EDM instrument, atmospheric and instrumental effects may give rise to errors in the distances displayed and corrections are required to account for these. In addition, it is sometimes necessary to apply a series of geometric corrections to slope distances measured in order that horizontal distances may be obtained.

5.19 Atmospheric Effects

All electromagnetic waves, when travelling in a vacuum, travel at the speed of light, a universal constant, but when travelling in the atmosphere the speed v of an electromagnetic wave is reduced from the free-space value c owing to the retarding action of the atmosphere. Consequently, v will be a variable depending on atmospheric conditions and the modulation wavelength will vary for all EDM measurements since $\lambda_m = v/f$. The significance of this is that the measuring unit λ_m is not constant and the distance recorded by an instrument will include a systematic error. The same effect occurs in timed-pulse EDM where the speed of the pulses is affected. This is analogous to steel taping where variations in temperature cause the tape to contract and expand from some reference value. Atmospheric effects are normally defined in terms of the *refractive index ratio* of the atmosphere n, where $v = c/n$ $(n > 1)$.

To correct for atmospheric effects, some EDM instruments are fitted with an *atmospheric correction switch* which is set according to the atmospheric pressure and temperature prevailing at the time of measurement, these being measured on site. Another method of removing atmospheric effects in EDM measurements is to enter corrections directly into the EDM unit using a dial mounted on the control panel for this purpose. As with the atmospheric correction switch, the atmospheric conditions must be measured and the correction, usually in ppm, is deduced from charts supplied with the instrument. A further alternative to both of these methods is to calculate n and correct measurements directly.

Whatever method is used to correct for atmospheric effects, it is evident that this requires meteorological conditions to be determined at some stage in the measurement of an EDM line. Great care should be taken when recording this data as the main factor that limits the accuracy of any EDM measurements is the uncertainty in the meteorological conditions. In order to maintain a precision of 1 ppm in measured distances, temperature and pressure need to be measured with precisions of 1°C and 3 mm Hg respectively. Also, since atmospheric conditions are proportional to the distance being measured, extra care should be taken in the recording of meteorological conditions when measuring long lines.

5.20 Instrumental Errors

All EDM measurements are subject to the following errors.

Scale Error (or Frequency Drift)

This is caused by variations in the modulation frequency f of the EDM instrument and the error is proportional to the distance measured since $\lambda_m = v/f$. Consequently, the effect is more noticeable on long lines and can sometimes be as high as 20–30 ppm for short-range instruments.

Zero Error (or Index Error)

This occurs if there are differences in the mechanical, electrical and optical centres of the EDM instrument and reflectors, and includes the prism constant discussed in section 5.6. This error is of constant magnitude and is not dependent on range, and care must be taken to eliminate it. The value of a zero error obtained from a calibration procedure usually applies to an instrument and reflector and if the reflector is changed, the zero constant changes.

Cyclic Error (or Instrument Non-linearity)

This error is caused by unwanted interference between electrical signals generated in the EDM unit and can be investigated by measuring a series of known distances spread over the measuring wavelength of the instrument. If a calibration curve of (observed - measured) distances is plotted against distance and a periodic wave is obtained, the EDM instrument has a cyclic error.

5.21 EDM Calibration

The determination of scale, zero and cyclic errors of EDM instruments is known as *calibration* and can be carried out by a number of different methods which have a varying degree of sophistication. For most site work, calibration is carried out using baselines and techniques involving both unknown and known baseline lengths are used.

The advantage of methods based on *unknown baseline lengths* is that the long-term stability of the points marking the baseline need not be known. However, it is not possible to determine scale error using such a baseline. The simplest method for determining the zero error of an EDM instrument involves the use of an unknown three-point baseline as shown in figure 5.39. All three distances are measured with the instrument being calibrated and l_{12}, l_{23} and l_{13} are obtained. If each measured distance has the same zero error z

$$l_{12} = d_1 + z$$
$$l_{23} = d_2 + z$$
$$l_{13} = d_1 + d_2 + z$$

where d_1 and d_2 are the correct distances from 1 to 2 and from 2 to 3. This gives

$$z = l_{13} - (l_{12} + l_{23})$$

This is sometimes called the *three-peg test*.

Figure 5.39   *Three-peg test*

If the baseline consists of multiple sections, z is given by

$$z = \frac{L - \Sigma l_i}{n - 1} \tag{5.6}$$

where

L = total length of baseline
l_i = length of each baseline section
n = number of baseline sections.

An improvement in accuracy can be obtained by measuring all possible baseline lengths and by using a least squares adjustment procedure (see section 6.2) to determine z. The results from such a calibration would also yield a value for any cyclic errors provided these are greater than any random errors present in observations.

The advantage of calibration methods based on *known baseline lengths* is that all three error components can be determined. However, the long-term stability of points along the baseline is important and all distances along the baseline should be measured with a high degree of precision at frequent intervals. As with an unknown baseline, multiple measurements are taken to calibrate an instrument and least squares techniques used to calculate the instrumental errors.

5.22 Geometric Corrections

Slope Correction

When the slope distance L has been obtained from an EDM measurement, a slope correction must be applied to it in order to obtain the equivalent horizontal distance.

If the EDM and theodolite are coaxial as in most total stations and the telescope is tilted and pointed at the centre of the prism (figure 5.40a), the correct slope distance is obtained no matter how the prism is tilted and the horizontal distance D is given by (see section 4.4)

$$D = L \cos\theta = L \sin z \qquad (5.7)$$

where

θ is the vertical angle
z is the zenith angle.

For an integrated total station, this calculation is carried out by the instrument's microprocessor and the result displayed automatically.

For theodolite-mounted EDM systems of the 'add-on' type in which the EDM unit is attached directly to the theodolite telescope and where the prism–target distance is compatible with the EDM–theodolite distance (see figure 5.40b), the correct slope distance is obtained provided the prism is tilted normal to the line of sight of the theodolite telescope. For a yoke-mounted system (figure 5.40c), a correct slope distance is obtained since the prism is centred and tilted along the vertical through the prism set. However, for both telescope and yoke-mounted EDMs, the horizontal distance is only given by $D = L \cos \theta = L \sin z$, as before, provided the telescope is pointed at the correct target on the prism set and the vertical angle measured parallel to the slope distance (see figures 5.40b and 5.40c).

When using an incompatible target and prism with theodolite-mounted EDM, an error could arise in the horizontal distance if the prism–target and

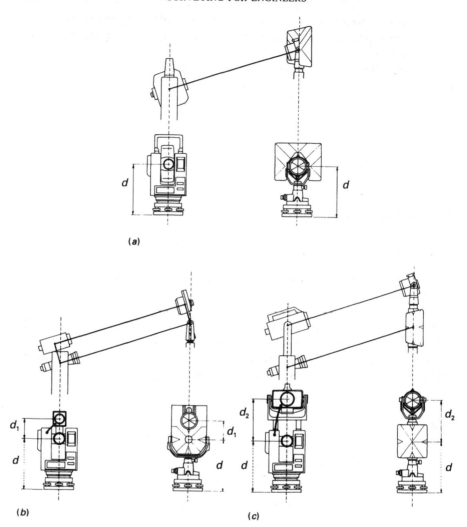

(a)

(b)   (c)

Figure 5.40   *EDM slope correction: (a) coaxial optics; (b) telescope mounted; (c) yoke mounted*
(courtesy Sokkia Ltd)

EDM–theodolite distances are not the same. However, some prism sets are adjustable (see figure 5.7*b*) and the distance from the target to a mounted prism can be altered to suit a range of different EDMs and theodolites.

In the case where the instrument and prism heights are known, the horizontal distance can be obtained as follows. In figure 5.41, the EDM unit is set h_e above station A of reduced level RL_A. The measurement of h_e refers to the height of the EDM unit and not the theodolite, except for a total station where the two are coaxial. At station B (reduced level RL_B), the prism height is h_p above station B. The horizontal distance is given by

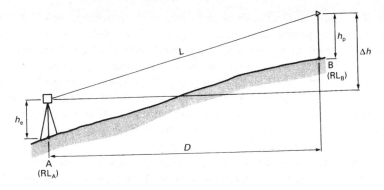

Figure 5.41 *EDM slope correction–heights method*

$$D^2 = L^2 - \Delta h^2 \tag{5.8}$$

where $\quad \Delta h = (RL_B - RL_A) + (h_p - h_e)$

Alternatively, the slope correction is

$$\text{slope correction} = \frac{\Delta h^2}{2L} \tag{5.9}$$

where

$$D = L - \text{slope correction}$$

Height Correction

When a survey is to be based on the National Grid coordinate system (see sections 1.5 and 1.9), the line measured must be reduced to its equivalent length at mean sea level (MSL). The height or MSL correction is given by

$$\text{height correction} = -\frac{Dh_m}{R} \tag{5.10}$$

where

h_m is the mean height of the instrument and reflector above MSL

R is the radius of the Earth.

The correction is negative unless a line below MSL is measured.

5.23 Scale Factor

As outlined in section 1.5, all Ordnance Survey maps and plans in Great Britain are based on a rectangular coordinate system known as the *National*

Grid. The National Grid is derived from a *map projection* which is a Transverse Mercator projection with an origin at 2°W, 49°N. A map projection provides a means of representing the curved surface of the Earth on a plane surface so that coordinate grids can be defined and maps drawn. In forming the National Grid, the relative positions of points on the grid are altered slightly from their ground positions as a result of using the Transverse Mercator projection to account for the curvature of the Earth. Therefore, distances calculated from National Grid coordinates will not, in some cases, agree with their equivalent measured on site.

To convert measured distances to projection (or grid) distances the measured distance is converted to its equivalent at MSL and the scale factor (F) used as follows

$$\text{grid distance} = \text{measured distance (at MSL)} \times F \qquad (5.11)$$

The value of the scale factor varies across the country and for a point P, of National Grid easting E_p, the scale factor is given by

$$F_p = F_0 \left[1 + \frac{(E_p - 400\,000)^2}{2R^2} \right] \qquad (5.12)$$

Using the Transverse Mercator projection and figure of the Earth adopted for the National Grid, $F_0 = 0.999\,601\,3$ and an average value for R can be taken as 6381 km.

The scale factor for any point can be calculated from equation (5.12) remembering that the units for E_p and R must be compatible. For example, given $E_p = 495\,676.241$ m E

$$F_p = 0.999\,601\,3 \left[1 + \frac{(495\,676 - 400\,000)^2}{2(6\,381\,000)^2} \right]$$

$$= 0.999\,713\,7$$

For a point Q with $E_Q = 182\,073.450$

$$F_Q = 0.999\,601\,3 \left[1 + \frac{(182\,073 - 400\,000)^2}{2(6\,381\,000)^2} \right]$$

$$= 1.000\,184\,3$$

Use of Scale Factor in Distance Measurement

Suppose a distance of 122.619 m was recorded by a total station for a line AB. If the scale factor for AB is 0.999 631 2 and the average height of AB is 31.72 m above Ordnance Datum (sea level), the grid distance AB is obtained as follows. Equation (5.10) gives

$$\text{MSL correction} = -\frac{122.619 \times 31.72}{6\,381\,000}$$
$$= -\,0.0006 \text{ m}$$

and

$$\text{MSL distance} = 122.619 - 0.0006 = 122.6184 \text{ m}$$

Applying the scale factor gives (using equation 5.11)

$$\text{grid distance} = \text{measured distance (at MSL)} \times F$$
$$= 122.6184 \times 0.999\,631\,2$$
$$= \textbf{122.573 m}$$

The grid distance would be used in place of the measured distance in any National Grid calculations.

Use of Scale Factor in Setting Out

In a road scheme, let the National Grid coordinates of a point on a road centre line be 612 910.741 m E, 157 062.283 m N. This is to be set out by polar coordinates from a nearby control station with National Grid coordinates 612 963.524 m E, 157 104.290 m N. The setting-out distance is calculated as follows

$$\Delta E = 612\,963.524 - 612\,910.741 = 52.783 \text{ m}$$
$$\Delta N = 157\,104.290 - 157\,062.283 = 42.007 \text{ m}$$
$$\text{grid distance} = (\Delta E^2 + \Delta N^2)^{1/2} = 67.4584 \text{ m}$$

For equation (5.12)

$$F = 1.000\,158$$

Therefore

$$\text{horizontal setting out distance} = \frac{67.4584}{1.000\,158}$$
$$= \textbf{67.448 m}$$

5.24 Measuring Reduced Levels Using EDM

Figure 5.42 shows two points X and Y. An EDM and theodolite unit is set up at point X and a reflector at Y. The slope distance between X and Y is measured by the EDM as L_{xy} and the vertical angle as θ_{xy}. For coaxial instruments, the height of the EDM (and theodolite) above X is h_i and the height of the prism above Y is h_p. For compatible theodolite-mounted EDM

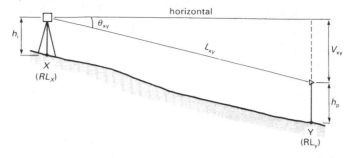

Figure 5.42 *Measurement of reduced levels by EDM*

systems, h_i is taken to be the height of the theodolite above X and h_p the height of the target sighted above Y.

If the reduced level of X (RL_x) is known, the reduced level of Y (RL_y) is given by

$$RL_y = RL_x + h_i - V_{xy} - h_p \qquad (5.13)$$

In figure 5.42, the vertical component V_{xy} is negative since the line of sight is below the horizontal. In general

$$RL_y = RL_x + h_i \pm V_{xy} - h_p \qquad (5.14)$$

The vertical component V_{xy} is obtained from

$$V_{xy} = L_{xy} \sin \theta_{xy} \qquad (5.15)$$

As with horizontal distances, many EDM instruments will calculate and display V_{xy} automatically. When measuring reduced levels with a prism fitted to a detail pole, it is good practice to set the height of the prism to the same height as the EDM system. This simplifies the calculations since h_i and h_p will then cancel each other out and $RL_y = RL_x \pm V_{xy}$.

In this section, no allowance has been made for the effects of the curvature of the Earth and for atmospheric refraction. Over short distances they can be ignored, for example, at 120 m the correction required is only −1 mm in height and at 400 m it is only slightly greater than −10 mm. At longer distances, however, the precision of heights obtained by EDM deteriorates. This is discussed in section 6.10.

5.25 Applications of EDM

The introduction of EDM into surveying has had such an enormous impact that it has replaced or changed many traditional methods of surveying. Generally, the use of EDM in engineering surveying results in a saving in time and, in most cases, an improvement in the accuracy of distance measurement when compared with taping over distances greater than a tape length and when compared with optical methods.

The use of EDM for large-scale route and site surveys (scales of 1:1000 and larger) is now standard practice among all but the smallest survey organisations. Such surveys are normally carried out using total stations with some form of data recorder which can store and then transmit field data directly to a computer for processing and plotting (see sections 5.15 and 9.11). Compared with conventional tape and tacheometric methods, electronic detail surveying is quicker, has a better precision and is more reliable, especially when using electronic data transfer. In addition, EDM is extremely well adapted to forming digital terrain models (see section 9.12) when the data storage unit is directly interfaced with the computer forming the model.

On construction sites, EDM is now used so extensively that many of the methods used for setting out have changed. The most significant of these changes is the emphasis now placed on polar methods of setting out based on coordinates in preference to other methods based on site grids and offsets (this is discussed in chapters 10, 11 and 14). For example, most road centre lines are now set out using distances and angles measured from control stations positioned away from the centre line rather than using the tangential angles method along the centre line itself (see section 11.16). Buildings, traditionally set out by occupying proposed corner positions and establishing right angles, now tend to be set out from control stations again by use of distances and angles. Many of the features associated with total stations, especially the tracking facility and the software functions dealing with coordinates, make it possible for setting out to be completed much more efficiently and with a better reliability than before.

An area where the use of EDM in civil engineering is increasing is deformation monitoring, in which structural movement over a period of time is measured. This is discussed in chapter 15.

6

Measurements and Errors

In the previous chapters the type of measurements fundamental to engineering surveying have been shown to be horizontal distance, vertical distance (or height) and horizontal and vertical angles. As shown throughout this book, many different techniques can be used to measure these quantities and many different instruments and methods have been developed for this purpose. Surveying, then, is a process that involves taking observations and measurements with a wide range of electronic, optical and mechanical equipment some of which is very sophisticated. However, even when using the best equipment and methods, it is still impossible to take observations that are completely free of small variations caused by errors. These effects are very important since they are a property of all measurements and this chapter serves as an introduction to errors and their effects on measurements.

6.1 Types of Error

Gross Errors

These are often called mistakes or blunders, and they are usually much larger than the other categories of error. On construction sites, mistakes are frequently made by inexperienced engineers and surveyors who are unfamiliar with the equipment and methods that they are using. Gross errors are due, then, to carelessness or incompetence and many examples can be given of these. Common mistakes include reading a theodolite micrometer scale or tape graduation incorrectly or writing the wrong value in a field book by transposing numbers (for example 28.342 is written as 28.432). Failure to detect a gross error in a survey or in setting out can lead to serious problems, and for this reason it is vital that all survey work has observational

202

and computational procedures that can be checked so that mistakes can be eliminated. Examples of good practice for the elimination of gross errors are given throughout this book.

Systematic Errors

Systematic errors are those which follow some mathematical law and they will have the same magnitude and sign in a series of measurements that are repeated under the same conditions. If an appropriate mathematical model can be derived for a systematic error, it can be eliminated from a measurement using corrections. For example, in section 4.4 it was shown that the effects of any temperature and tension variations in steel taping can be removed from a measurement by calculation using simple formulae. Another method of removing systematic errors is to calibrate the observing equipment and to quantify the error, allowing corrections to be made to further observations. In section 5.21, it was shown that it is often necessary to calibrate an EDM instrument when measuring distances, where it is expected that a frequency drift will occur giving rise to a systematic error in displayed distances. Observational procedures can also be used to remove the effect of systematic errors and a good example of this is to take the mean direction from face left and face right readings when measuring angles with a theodolite, as shown in sections 3.7, 3.8 and 3.9.

Random Errors

When all gross and systematic errors have been removed, a series of repeated measurements taken of the same quantity under the same conditions would still show some variation beyond the control of the observer. These variations are inherent in all types of measurements and are called random errors, the magnitude and sign of which are not constant. Random errors cannot be removed from observations but methods can be adopted to ensure that they are kept within acceptable limits. In this context, the use of the word error does not always imply that something has gone wrong, it simply tells us that a difference exists between the true value of a quantity and a measured value of that quantity. It is important to realise that, for surveying measurements, the true value of a quantity is usually never known and, therefore, the exact error in a measurement or observation can never be known.

In order to analyse random errors or variables, statistical principles must be used and in surveying it is usual to assume that random variables are *normally distributed* as shown in figure 6.1. This figure demonstrates the general laws of probability that random errors follow, and these are

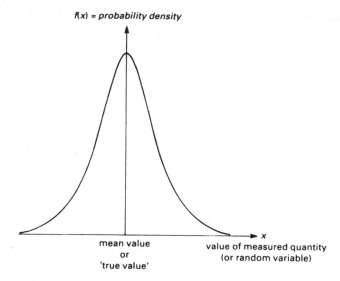

Figure 6.1 *Normal distribution curve*

(1) Small errors occur frequently and are therefore more probable than large ones
(2) Large errors happen infrequently and are therefore less probable; very large errors may be mistakes and not random errors
(3) Positive and negative errors of the same size are equally probable and happen with equal frequency.

Throughout the remainder of this chapter it is assumed that all gross and systematic errors have been removed from observations and that only normally distributed random errors are being dealt with.

6.2 Least Squares Estimation and Most Probable Value

In the absence of gross and systematic errors, a random error is the difference between the true value of a quantity and an observation or measurement of that quantity. Consequently, before any random errors (or simply errors) can be calculated for a set of observations or measurements, the true value of the observed or measured quantity must be known. Since the true value is seldom known in surveying, errors are also unknown but an estimate for a true value can be found using the principle of *Least Squares* which states that

'The best estimate or *most probable value* (MPV) obtainable from a set of measurements of equal precision is that value for which the sum of squares of the residuals is a minimum.'

A *residual* is the difference between any measured value of a quantity and its most probable value and since residuals can be determined, they are used instead of errors to analyse surveying measurements. However, one of the reasons that least squares has found widespread use in surveying is that it tends to produce true values from residuals even though residuals are not true observational errors.

If a single unknown such as a distance x was measured with a steel tape n times, it can be shown that the least squares method gives the arithmetic mean $\bar{x}$ as the most probable value for the distance provided that each measurement is independent and taken under similar conditions.

For n readings $x_1, x_2, \ldots x_n$, the most probable value or sample mean is given by

$$\bar{x} = \frac{x_1 + x_2 + \ldots + x_n}{n} = \frac{\sum_{i=1}^{n} x_i}{n} \tag{6.1}$$

This could have been deduced using one of the general laws of probability, that positive and negative errors of the same magnitude occur with equal frequency.

For the simple case of a single taped distance given above the application of least squares is a straightforward matter, but for more complicated cases involving several related quantities, some of which may be indirectly measured, calculating most probable values by least squares demands specialist knowledge and techniques beyond the scope of this book. However, further examples of least squares estimation are given in the text but only involving small numbers of independent and directly observed variables, and further reading is suggested at the end of this chapter.

6.3 Standard Deviation and Standard Error

As stated in section 6.1, surveying observations with gross and systematic errors removed are subject to random variations which follow the general laws of probability. In order to be able to compare one set of similar observations with another, the spread of a set of residuals must be assessed and to do this, observations and measurements (and therefore residuals) are assumed to follow a normal distribution function, as already shown in figure 6.1. A normal distribution function is defined by two parameters, its expectation or most probable value and its standard deviation. The standard deviation (σ) is a measure of the spread or dispersion of measurements and, in figure 6.2, a small standard deviation (σ_1) indicates a small spread among results whereas a large standard deviation (σ_2) indicates a large spread. The equation for the standard deviation of a variable x measured n times is given by

$$\sigma_x = \pm \sqrt{\frac{\sum\limits_{i=1}^{n} v_i^2}{n}} = \pm \sqrt{\frac{\sum\limits_{i=1}^{n} (\overline{x} - x_i)^2}{n}} \qquad (6.2)$$

where

x_i is an individual measurement
$\overline{x}$ the most probable value (mean value)
v is a residual such that $v_i = (\overline{x} - x_i)$.

The square of the standard deviation (σ^2) is called the *variance*.

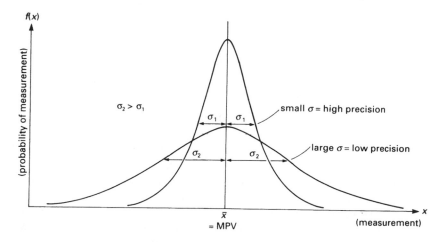

Figure 6.2 *Standard deviation and precision*

The use of the term precision in conjunction with standard deviation and variance is common and this gives an indication of the repeatability of a measurement. In other words, the set of measurements with the smaller standard deviation in figure 6.2 (σ_1) has a higher precision than the other set with the larger standard deviation (σ_2).

For statistical reasons, a standard deviation should be derived from a large number of observations and since surveying measurements are usually taken in small sets, any standard deviations derived from them may be biased. For this reason, a better measure of precision is obtained for surveying observations by using an unbiased estimator for the standard deviation known as the *standard error* (s). This is obtained by replacing n with $(n - 1)$ in equation (6.2) to give

$$\text{standard error in } x = s_x = \pm \sqrt{\frac{\sum\limits_{i=1}^{n} v_i^2}{(n-1)}} = \pm \sqrt{\frac{\sum\limits_{i=1}^{n} (\overline{x} - x_i)^2}{(n-1)}} \qquad (6.3)$$

The standard error of the most probable value (mean value) is given by

$$s_{\bar{x}} = \pm \; \frac{s_x}{\sqrt{n}} \tag{6.4}$$

This demonstrates an important aspect of surveying fieldwork. Suppose a quantity is to be measured using a field procedure and equipment that has a known standard error from previous work. This could be the measurement of a distance with a 30 m tape with a standard error of ±6 mm. If only one measurement was taken, the standard error of the distance measured will be that assigned to the field procedure and tape, that is ±6 mm. Suppose now that a standard error of ±3 mm was needed for a particular distance: equation (6.4) shows that the distance would have to be measured four times in order to double the precision.

6.4 Significance of the Standard Error

The standard error for a series of measurements indicates the probability or chance that the true value for the measurements lies within a certain range of the sample mean and it can be shown that, for the normal distribution function, there is a 68.3 per cent chance that the true value of a measurement lies within the range $x + s_x$ to $x - s_x$. The limits or ranges within which true values are assumed to occur are called *confidence intervals* and these are shown in table 6.1 and graphically in figure 6.3. A reminder is given at this point that the probabilities given in table 6.1 and figure 6.3 assume that measurements are normally distributed and have had all gross and systematic errors removed from them.

TABLE 6.1

Probabilities Associated with the Normal Distribution

Probability (%)	68.3	90	95	95.4	99	99.7	99.9
Confidence interval	±1s	±1.65s	±1.96s	±2s	±2.58s	±3s	±3.29s

6.5 Worked Example: Mean, Standard Error and Confidence Interval

Question
Using the same tape, a setting-out distance was measured ten times by the same engineer under similar field conditions. After systematic corrections had been applied to each of the measurements, the following results were obtained (in metres)

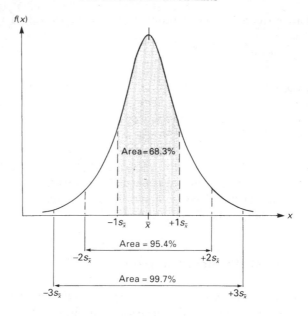

Figure 6.3 *Confidence intervals for normal distribution*

23.287, 23.293, 23.290, 23.289, 23.294,
23.286, 23.283, 23.288, 23.291, 23.289

Calculate the most probable value for this distance, its standard error and confidence interval for a 95 per cent probability.

Solution
Using equation (6.1) the most probable value (mean) for the distance d is given by

$$\bar{d} = \frac{\sum\limits_{i=1}^{n} d_i}{n}$$

The residuals and squares of the residuals for the example are listed along with the original measurements in table 6.2. From table 6.2, $\Sigma d_i = 232.890$ and with $n = 10$

$$\bar{d} = 23.289 \text{ m}$$

Using the square of residuals summation and equation (6.3), the standard error of a single measurement is

TABLE 6.2

i	d_i	$v_i = (\bar{d} - d_i)$ mm	v_i^2 mm^2
1	23.287	2	4
2	23.293	−4	16
3	23.290	−1	1
4	23.289	0	0
5	23.294	−5	25
6	23.286	3	9
7	23.283	6	36
8	23.288	1	1
9	23.291	−2	4
10	23.289	0	0
	$\Sigma d_i = 232.890$	$\Sigma v_i = 0$	$\Sigma v_i^2 = 96$

$$s_d = \pm \sqrt{\frac{\sum_{i=1}^{10} v_i^2}{(n-1)}} = \pm \sqrt{\frac{96}{(10-1)}} = \pm 3.3 \text{ mm}$$

The standard error of the mean distance is obtained from equation (6.4) as follows

$$s_{\bar{d}} = \pm \frac{s_d}{\sqrt{n}} = \pm \frac{3.3}{\sqrt{10}} = \pm 1.0 \text{ mm}$$

This gives, for the 95 per cent probability specified, a confidence interval (see table 6.1) for the true value of

$$\bar{d} \pm 1.96 s_{\bar{d}} = 23.289 \pm 1.96(0.0010) = \textbf{23.287 to 23.291 m}$$

6.6 Redundancy

If a value for a quantity such as a distance is to be found, only one measurement is needed to define the distance assuming there are no gross and systematic errors in the measurement. In the example given in section 6.5, since the distance was measured ten times there are nine redundant observations and without these, it would not have been possible to evaluate standard errors and establish probabilities. Redundant observations are also used to detect mistakes in fieldwork, the classic case being to measure the three angles of a triangle when only two are needed to define it uniquely: the third angle is used to check that the measured angles add up to 180°.

6.7 Precision, Accuracy and Reliability

These terms are used frequently in engineering surveying both by manufacturers when quoting specifications for their equipment and on site by surveyors and engineers to describe results obtained from fieldwork.

Precision represents the repeatability of a measurement and is concerned only with random errors. A set of observations that are closely grouped together and have small deviations from the sample mean will have a small standard error and are said to be precise as shown previously in figure 6.2.

On the other hand, *accuracy* is considered to be an overall estimate of the errors present in measurements including systematic effects. For a set of measurements to be considered accurate, the most probable value or sample mean must have a value close to the true value as shown in figure 6.4a. It is quite possible for a set of results to be precise but inaccurate as in figure 6.4b where the difference between the true value and the mean value is caused by one or more systematic errors. Since accuracy and precision are the same if all systematic errors are removed, precision is sometimes referred to as internal accuracy.

For many surveying measurements, the term *relative precision* is sometimes used and this is the ratio of the precision of a measurement to the measurement itself. For example, if the standard error of the measurement of a distance d is s_d the relative precision is expressed as 1 in d/s_d (say, 1 in 5000). An alternative to this is to quote relative precision in ppm or parts per million (that is, 1 in 1 000 000). This was used for EDM instruments and total stations in chapter 5 and it is also a term used in high precision surveying. The relative precision of a measurement should always be calculated as soon as its precision is known or it may be specified before starting a survey so that the proper equipment and methods can be selected to achieve the relative precision. This is discussed in a number of chapters in the book.

In all types of surveying, attempts are always made to detect and eliminate mistakes in fieldwork and computations and the degree to which a survey is able to do this is a measure of its *reliability*. Unreliable observations are those which may contain gross errors without the observer knowing, whereas reliable observations are unlikely to contain undetected mistakes.

6.8 Propagation of Standard Errors

Basic surveying measurements such as angles and distances are often used to derive other quantities using mathematical relationships. For instance, reduced levels (see chapter 2) are obtained from differences of level staff readings, horizontal distances are obtained from slope distances by a calculation involving vertical angles (see chapter 4) and coordinates are obtained from a combination of horizontal angles and distances (see chapter 7). In each of

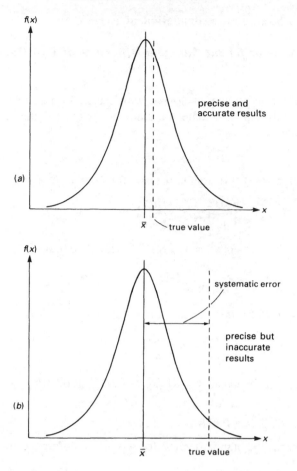

Figure 6.4 *Accuracy and precision*

these cases, the original measurements will be randomly distributed and will have errors, and it follows that any quantities derived from them will also have errors (or residuals).

The law of propagation of standard errors for a quantity U which is a function of independent measurements $x_1, x_2, \ldots, x_n$ where $U = f(x_1, x_2, \ldots, x_n)$ is given by

$$s_U^2 = \left(\frac{\partial U}{\partial x_1}\right)^2 s_{x_1}^2 + \left(\frac{\partial U}{\partial x_2}\right)^2 s_{x_2}^2 + \ldots + \left(\frac{\partial U}{\partial x_n}\right)^2 s_{x_n}^2 \qquad (6.5)$$

where

s_U = standard error of U

$s_{x_1}, s_{x_2}, \ldots, s_{x_n}$ are the standard errors of $x_1, x_2, \ldots, x_n$.

6.9 Worked Examples: Propagation of Errors

(1) Standard Error for the Sum and Difference of Two Quantities

Question
Two distances a and b were measured with standard errors of $s_a = \pm 0.015$ m and $s_b = \pm 0.010$ m. Calculate the standard errors for the sum and difference of a and b.

Solution
Using equation (6.5), the standard error for $D = a + b$ is

$$s_D^2 = \left(\frac{\partial D}{\partial a}\right)^2 s_a^2 + \left(\frac{\partial D}{\partial b}\right)^2 s_b^2$$

$$= (1)^2 s_a^2 + (1)^2 s_b^2 = 0.015^2 + 0.010^2$$

hence

$$s_D = \pm \mathbf{0.018} \text{ m}$$

Using equation (6.5), the standard error for $D = a - b$ is

$$s_D^2 = \left(\frac{\partial D}{\partial a}\right)^2 s_a^2 + \left(\frac{\partial D}{\partial b}\right)^2 s_b^2$$

$$= (1)^2 s_a^2 + (-1)^2 s_b^2 = 0.015^2 + 0.010^2$$

hence

$$s_D = \pm \mathbf{0.018} \text{ m}$$

As can be seen, the standard error for a sum or difference is the same.

(2) Standard Error for Repeated Measurements

Question
The internal angles of an n-sided polygon were all measured using the same equipment and methods such that each angle had the same standard error. What is the standard error for the sum of these angles?

Solution
Applying a similar process to that given in the previous example and with the sum of the angles $A = a_1 + a_2 + \ldots + a_n$

$$s_A^2 = \left(\frac{\partial A}{\partial a_1}\right)^2 s_{a_1}^2 + \left(\frac{\partial A}{\partial a_2}\right)^2 s_{a_2}^2 + \ldots + \left(\frac{\partial A}{\partial a_n}\right)^2 s_{a_n}^2$$

$$= s_{a_1}^2 + s_{a_2}^2 + \ldots + s_{a_n}^2$$

If

$$s_{a_1} = s_{a_2} = \ldots = s_{a_n} = s_a$$

then

$$s_A^2 = ns_a^2 \quad \text{or,} \quad s_A = \pm s_a\sqrt{n}$$

This example shows that the propagated error for the sum of a set of measurements with the same standard error is proportional to the square root of the number of measurements (or observations). In elementary surveying, this is used to determine the allowable misclosures for some types of fieldwork as shown, for example, in sections 2.16 for levelling and section 7.9 for traversing.

(3) Standard Error for the Area of a Rectangle

Question
The two sides of a rectangle were measured as $x = 32.543 \pm 0.010$ m and $y = 17.298 \pm 0.020$ m. Calculate the area of the rectangle and its standard error.

Solution
The area of the rectangle A is given by (noting there are 5 significant figures)

$$A = xy = (32.543)(17.298) = 562.93 \text{ m}^2$$

Using the law for the propagation of standard errors (equation 6.5)

$$s_A^2 = \left(\frac{\partial A}{\partial x}\right)^2 s_x^2 + \left(\frac{\partial A}{\partial y}\right)^2 s_y^2 = y^2 s_x^2 + x^2 s_y^2$$

$$= (17.298)^2(0.010)^2 + (32.543)^2(0.020)^2$$

from which

$$s_A = \pm 0.67 \text{ m}^2$$

(4) Confidence Interval for the Area of a Triangle

Question
Two side lengths of triangle EGH and their included angle (see figure 6.5) have been measured as $e = 256.805 \pm 0.025$ m, $g = 301.465 \pm 0.030$ m and $H = 61°48'20'' \pm 30''$. Calculate the area of the triangle and its 99.9 per cent confidence interval.

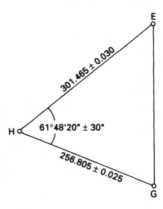

Figure 6.5

Solution

The area of the triangle A is given by

$$A = \tfrac{1}{2}eg\sin H = \tfrac{1}{2}(256.805)(301.465)\sin(61°48'20'')$$

$$= 34\ 116.0\ \text{m}^2$$

The variance of A is given by equation (6.5) as

$$s_A^2 = \left(\frac{\partial A}{\partial e}\right)^2 s_e^2 + \left(\frac{\partial A}{\partial g}\right)^2 s_g^2 + \left(\frac{\partial A}{\partial H}\right)^2 s_H^2$$

$$= (\tfrac{1}{2}\ g\ \sin\ H)^2 s_e^2 + (\tfrac{1}{2}\ e\ \sin\ H)^2 s_g^2 + (\tfrac{1}{2}\ eg\ \cos\ H)^2 s_H^2$$

$$= \left(\frac{A}{e}\right)^2 s_e^2 + \left(\frac{A}{g}\right)^2 s_g^2 + (A\cot H)^2 s_H^2$$

$$= A^2\left[\left(\frac{s_e}{e}\right)^2 + \left(\frac{s_g}{g}\right)^2 + (s_H \cot H)^2\right]$$

In this case, the standard error for H must be converted from an angular error to a linear error by multiplying it by sin 1″ (= 1/206 265). This gives

$$s_A^2 = A^2\left[\left(\frac{0.025}{e}\right)^2 + \left(\frac{0.030}{g}\right)^2 + \left(\frac{30}{206\ 265}\cot H\right)^2\right]$$

from which

$$s_A = \pm 5.4\ \text{m}^2$$

The confidence interval corresponding to a 99.9 per cent probability is $\pm 3.29 s_A$ (see table 6.1) and the true value of A will be in the range

$$A \pm 3.29s_A = 34\ 116.0 \pm 3.29(5.4)$$
$$= \textbf{34 098.2 to 34 133.8 m}^2$$

6.10 Propagation of Errors in Survey Methods

Throughout this chapter, a number of examples have been given showing methods of assigning standard errors to measured quantities so that they can be compared with one another and how these individual errors propagate. By using similar methods, it is possible to study the effect of accumulating errors in survey procedures.

Propagation of Errors in Levelling

Levelling was discussed in chapter 2 where much of the following terminology is defined.

If the standard error for reading a levelling staff is s_s mm m^{-1} of the sight length, and it is assumed that equal sight lengths are used at every instrument set-up, the standard error $s_{\Delta H}$ in the height difference ΔH obtained by levelling through a distance D can be derived.

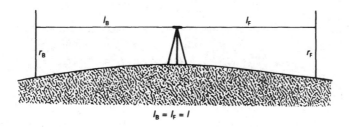

Figure 6.6 *Propagation of errors in levelling*

Figure 6.6 shows a single instrument set-up where a backsight staff reading r_B and a foresight r_F have been taken. With backsight length l_B equal to foresight length l_F, $l_B = l_F = l$ (the sight length). The height difference between A and B, Δh, is given by

$$\Delta h = r_B - r_F$$

The standard error for this height difference, $s_{\Delta h}$, can be obtained from

$$s_{\Delta h}^2 = s_{r_B}^2 + s_{r_F}^2$$

The standard errors in the two staff readings s_{r_B} and s_{r_F} will be equal since the sight lengths are the same. For a sight length l, the standard error in a single staff reading s_r will be

$$s_r = l s_s$$

This gives

$$s_{\Delta h}^2 = 2s_r^2 = 2l^2 s_s^2$$

If n set-ups are required to level through the distance D, the height difference between the ends of the line is

$$\Delta H = \Delta h_1 + \Delta h_2 + \ldots + \Delta h_n$$

The standard error for this sum, $s_{\Delta H}$, is given by

$$s_{\Delta H}^2 = s_{\Delta h_1}^2 + s_{\Delta h_2}^2 + \ldots + s_{\Delta h_n}^2$$

Since equal sight lengths are used

$$s_{\Delta h_1} = s_{\Delta h_2} = \ldots = s_{\Delta h_n} = s_{\Delta h}$$

hence

$$s_{\Delta H}^2 = n s_{\Delta h}^2$$

In a distance D with n set-ups of sight length l

$$D = 2nl$$

and

$$s_{\Delta H}^2 = \frac{D}{2l} s_{\Delta h}^2 = \frac{D}{2l} (2l^2 s_s^2)$$

from which

$$s_{\Delta H} = s_s \sqrt{Dl} \qquad\qquad (6.6)$$

Typical values for a levelling scheme might be $s_s = \pm 0.05$ mm m^{-1}, $l = 40$ m and $D = 1$ km. Substituting these into equation (6.6) gives

$$s_{\Delta H} = \pm 0.05 \sqrt{1000(40)} = \pm 10.0 \text{ mm}$$

Propagation of Errors in Angle Measurement

In section 3.8, it was shown that angles are calculated from directions measured using a theodolite. Random errors in directions are usually caused by an observer not setting and reading a micrometer scale exactly the same each time (applies to optical theodolites only), by uncertainties in digital reading systems (applies to electronic theodolites only) and by not bisecting a target properly (applies to optical and electronic theodolites). Figure 6.7 shows an angle α as the difference between two directions, d_1 and d_2 or

$$\alpha = d_2 - d_1$$

where d_1 is the mean of a face left reading $(d_1)_L$ and a face right reading $(d_1)_R$ such that

$$d_1 = \frac{(d_1)_L + (d_1)_R}{2}$$

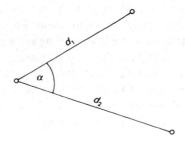

Figure 6.7 *Propagation of errors in angle measurement*

If the standard error in measuring and setting a micrometer scale or the uncertainty in a digital reading system is s_m and that for bisecting a target is s_b, the standard errors in $(d_1)_L$ and $(d_1)_R$ are

$$s^2_{(d_1)L} = s^2_{(d_1)R} = (s^2_m + s^2_b)$$

This gives the standard error in d_1 as

$$s^2_{d_1} = \tfrac{1}{2}(s^2_m + s^2_b)$$

Similarly

$$s^2_{d_2} = \tfrac{1}{2}(s^2_m + s^2_b)$$

and the standard error in α is

$$s^2_\alpha = (s^2_m + s^2_b)$$

If α is the mean of n rounds of angles, then

$$s^2_\alpha = \frac{(s^2_m + s^2_b)}{n} \tag{6.7}$$

When using an electronic theodolite, a standard error of $s_m = \pm 3''$ is typical and if s_b is $\pm 5''$ for a particular target, the propagated error for two rounds of an angle would be

$$s^2_\alpha = \frac{(9 + 25)}{2} = 17 \ \text{sec}^2$$

from which

$$s_\alpha = \pm 4.1''$$

For four rounds using the same equipment

$$s_\alpha = \pm 2.9''$$

Propagation of Errors in Trigonometrical Heighting

Very often in engineering surveying, EDM instruments and total stations are used for trigonometrical height measurement. Figure 6.8 shows the situation where a theodolite-mounted EDM instrument has been centred over a control station A of height H_A and the slope distance L and vertical angle θ have been measured to a pole-mounted reflector held at B. B could be a spot height required for contouring or it could be a point to be set out on site. Whatever the case, the precision of the height obtained for B is of interest. The height of point B, H_B, is given by (see section 3.11)

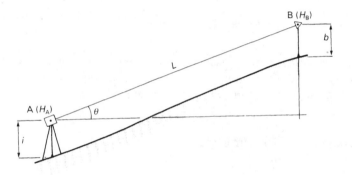

Figure 6.8   *Propagation of errors in trigonometrical heighting*

$$H_B = H_A + i - b + L \sin (\theta + \delta\theta) \qquad (3.4)$$

where

i = height of the theodolite above control station A
b = height of the EDM reflector above point B.

The term $\delta\theta$ represents the effects of curvature and refraction and for this analysis it can be ignored as a systematic error. The height of B is therefore given by

$$H_B = H_A + i - b + L \sin\theta$$

Applying equation (6.5) to this gives

$$s_{H_B}^2 = \left(\frac{\partial H_B}{\partial H_A}\right)^2 s_{H_A}^2 + \left(\frac{\partial H_B}{\partial i}\right)^2 s_i^2 + \left(\frac{\partial H_B}{\partial b}\right)^2 s_b^2 +$$

$$\left(\frac{\partial H_B}{\partial L}\right)^2 s_L^2 + \left(\frac{\partial H_B}{\partial \theta}\right)^2 s_\theta^2$$

$$= s_{H_A}^2 + s_i^2 + s_b^2 + \sin^2\theta \ s_L^2 + L^2 \cos^2\theta \ s_\theta^2$$

Under normal circumstances, H_A, i, b and L will have errors of the order of ± 0.01 m and θ will rarely exceed $\pm 20°$. Assuming the error in the measurement of θ to be $\pm 30''$, the error in H_B will be for $L = 100$ m

$$s_{H_B}^2 = (0.01)^2 + (0.01)^2 + (0.01)^2 + \sin^2 20° \ (0.01)^2 +$$

$$100^2 \cos^2 20° \left(\frac{30}{206\ 265}\right)^2$$

giving

$$s_{H_B} = \pm 0.022 \text{ m}$$

If L is increased to 1 km and all the other parameters remain the same, the error in H_B will be

$$s_{H_B} = \pm 0.138 \text{ m}$$

As the slope distance increases, the effects of s_{H_A}, s_i, s_b and s_L decrease and an approximate value for the error in H_B is given by

$$s_{H_B} = L \cos\theta \ s_\theta = Ds_\theta$$

where D is the horizontal distance between instrument and reflector.

Propagation of Errors in Stadia Tacheometry

With reference to section 4.9, the horizontal distance D between two points is given by vertical staff stadia tacheometry as

$$D = Ks \cos^2\theta + C \cos\theta$$

where

K = the multiplying constant
s = the staff intercept
θ = the vertical angle of the measurement
C = the additive constant.

If $C = 0$, this equation can be written

$$D = Ks \cos^2\theta$$

If the staff is held vertically, the angle between the staff and the normal to the line of sight is θ. However, if the staff is not held vertically, this angle becomes ϕ and D is given by

$$D = Ks \cos\phi \cos\theta$$

Equation (6.5) applied to this gives

$$s_D^2 = \left(\frac{\partial D}{\partial K}\right)^2 s_K^2 + \left(\frac{\partial D}{\partial s}\right)^2 s_s^2 + \left(\frac{\partial D}{\partial \phi}\right)^2 s_\phi^2 + \left(\frac{\partial D}{\partial \theta}\right)^2 s_\theta^2$$

$$= D^2 \left[\left(\frac{s_K}{K}\right)^2 + \left(\frac{s_s}{s}\right)^2 + (s_\phi \tan\phi)^2 + (s_\theta \tan\theta)^2\right]$$

If $K = 100 \pm 0.1$, $s = 0.500 \pm 0.0025$ m, $\phi = 5° \pm 30'$ and $\theta = 5° \pm 30''$, the propagated error in D is

$$s_D^2 = D^2 \left[\left(\frac{0.1}{100}\right)^2 + \left(\frac{0.0025}{0.500}\right)^2 + \left(\frac{1800}{206\ 265} \tan 5°\right)^2 + \right.$$

$$\left.\left(\frac{30}{206\ 265} \tan 5°\right)^2\right]$$

from which

$$s_D = \pm 0.26 \text{ m}$$

The most dominant term in the error of D is s_s/s and if all the others are ignored but are of a similar magnitude to those already given, an approximate value for s_D is given by

$$s_D = D\frac{s_s}{s} = 100s_s$$

The vertical component V of the horizontal distance obtained in stadia tacheometry is given in section 4.9 as

$$V = \tfrac{1}{2} Ks \sin2\theta + C \sin\theta$$

If $C = 0$

$$V = \tfrac{1}{2} Ks \sin2\theta = \tfrac{1}{2} Ks \cos\theta \sin\theta$$

and replacing θ with ϕ

$$V = \tfrac{1}{2} Ks \cos\phi \sin\theta$$

This gives the error in V as

$$s_V^2 = V^2 \left[\left(\frac{s_K}{K}\right)^2 + \left(\frac{s_s}{s}\right)^2 + (s_\phi \tan\phi)^2 + (s_\theta \cot\theta)^2\right]$$

and substituting the values given for the analysis of D

$$s_V = \pm 0.024 \text{ m}$$

As with the horizontal distance D, the predominant term in deriving the error for V is s_s/s and an approximate value for s_V is given by

$$s_V = V \frac{s_s}{s}$$

6.11 Survey Specifications

In surveying, it is usual to specify the precision required for measured and calculated quantities before fieldwork commences. It is then up to the surveyor or engineer on site to decide what type of equipment would be most suitable and what methods should be used in order to achieve these stated precisions. This is discussed further in chapter 14. Another useful application of error propagation techniques is that they can be used to derive the precision required in the individual parts of a measurement when the precision of a derived quantity is known (or specified).

Some examples to demonstrate the application of error theory to survey specifications are given in the following sections.

Precision of Levelling

Suppose a TBM is to be established for a construction site from an existing TBM and that the height difference ΔH between the two bench marks is to have a specified standard error of $s_{\Delta H}$. Equation (6.6) gives

$$s_{\Delta H} = s_s \sqrt{Dl}$$

and if the length between the bench marks D is known and a value for s_s is assumed, the maximum sighting distance l can be calculated for the levelling. Rearranging equation (6.6)

$$l = \frac{s_{\Delta H}^2}{D s_s^2}$$

If the staff readings are taken with a precision of $s_s = \pm 0.05 \text{ mm m}^{-1}$, $D = 250$ m and the height difference between the bench marks is to have an error of ± 0.005 m, the maximum sighting distance allowed in order to achieve this precision is

$$l = \frac{(0.005)^2}{250(0.05 \times 10^{-3})^2} = 40 \text{ m}$$

Precision of Angle Measurement

In section 6.10 it was shown that the precision of an angle is given by (see equation 6.7)

$$s_\alpha^2 = \frac{(s_m^2 + s_b^2)}{n}$$

The number of rounds n to be observed that will achieve a stated precision in the final angle α can be derived from this equation as

$$n = \frac{(s_m^2 + s_b^2)}{s_\alpha^2}$$

If the theodolite to be used for angle observations has a precision of $s_m = \pm 10''$ and targets and observers are to be used such that $s_b = \pm 5''$, the number of rounds that must be observed to achieve a precision of $\pm 5''$ in α is given by

$$n = \frac{(10^2 + 5^2)}{5^2} = 5$$

Precision of Trigonometrical Heighting

The standard error s_{H_B} for the height of an unknown point B was derived as $s_{H_B} = L \cos\theta \, s_\theta$ in section 6.10. If the error in the height of B is not to exceed a specified amount, it is useful to know the precision to which θ must be measured so that this precision is achieved. The equation for s_{H_B} can be rearranged to give

$$s_\theta = \frac{s_{H_B}}{L \cos \theta}$$

With $L = 1$ km and $\theta = 20°$, the precision to which θ must be measured in order to obtain a precision in the height of B of ± 0.10 m is

$$s = \frac{0.10}{1000 \cos 20°} \text{ radians} = \frac{206\ 265\ (0.10)}{1000 \cos 20°} \text{ seconds} = \pm 22''$$

Precision of Slope Corrections

In chapter 4 it was shown that a slope distance L can be reduced to a horizontal distance D by measuring the slope angle θ and by applying the equation $D = L \cos\theta$. The propagated error in D for this reduction is given by

$$s_D^2 = \cos^2\theta \, s_L^2 + (L \sin\theta)^2 s_\theta^2 = s_L^2 + (L \sin\theta)^2 s_\theta^2$$

since θ is a small angle.

The effect of the precision of θ on the precision of D can be obtained by rearranging the equation for s_D to give s_θ as

$$s_\theta^2 = \frac{s_D^2 - s_L^2}{(L \sin\theta)^2}$$

On site, this type of slope correction is usually applied to distances that are measured with EDM instruments. All of these have similar specifications and the value of s_L, for this analysis, can be treated as a constant which has an approximate value of ± 5 mm (see section 5.10). If the precision of D is specified as $s_D = \pm 10$ mm, $L = 100$ m and $\theta = 5°$, the precision required in the measurement of θ is

$$s_\theta^2 = \frac{0.010^2 - 0.005^2}{(100 \ \sin 5°)^2} = 9.873 \times 10^{-7} \text{ radians}^2$$

or

$$s_\theta = \pm 3'25''$$

Another method of applying slope corrections is to obtain D from equation (4.2), where

$$D = L - \frac{\Delta h^2}{2L}$$

This correction is usually applied to taped distances where the height difference between the ends of a line are known. Applying equation (6.5) to this and assuming $D \simeq L$ gives

$$s_D^2 = s_L^2 + \left(\frac{\Delta h}{L}\right)^2 s_{\Delta h}^2$$

Assuming the precision to which the slope distance is measured is known and that the precision of D is specified, the precision required in the height difference is

$$s_{\Delta h}^2 = \left(\frac{L}{\Delta h}\right)^2 (s_D^2 - s_L^2)$$

For a slope distance of $D = 100$ m, a height difference of $\Delta h = 5$ m, a taping precision of $s_L = \pm 5$ mm and a specified precision of $s_D = \pm 10$ mm, the height difference must be measured with a precision of

$$s_{\Delta h}^2 = \left(\frac{100}{5}\right)^2 (0.010^2 - 0.005^2)$$

or

$$s_{\Delta h} = \pm 0.17 \text{ m}$$

Further Reading

M.A.R. Cooper, *Control Surveys in Civil Engineering* (Collins, London, 1987).

B.D.F. Methley, *Computational Models in Surveying and Photogrammetry* (Blackie, Glasgow, 1986).

E.M. Mikhail, *Observations and Least Squares* (Dun-Donnelly, New York, 1976).

E.M. Mikhail and G. Gracie, *Analysis and Adjustment of Survey Measurements* (Van Nostrand Rheinhold, New York, 1981).

P.R. Wolf, *Adjustment Computations* (Landmark Enterprises, Rancho Cordova, 1980).

7

Control Surveys

As outlined in section 1.4, engineering surveys are usually based on horizontal and vertical control networks which consist of fixed points called control stations. A series of control stations forming a network can be used for the production of site plans, for establishing the positions of design points during setting out work and for monitoring (see chapters 9, 14 and 15).

The usual method for determining the vertical positions or heights of control points in engineering surveys is by levelling, which is discussed in chapter 2, and trigonometrical heighting, which is discussed in sections 3.11 and 5.24.

Methods of determining the horizontal positions or rectangular coordinates of control points include traversing, triangulation and trilateration. In addition to these, horizontal control can be extended using intersection and resection. All of these methods are dealt with in this chapter.

In recent years, satellite position fixing systems have been used in engineering surveying to obtain three-dimensional coordinates: these are described in chapter 8.

7.1 Types of Traverse

A traverse is a means of providing horizontal control in which rectangular coordinates are determined from a combination of angle and distance measurements along lines joining adjacent stations.

Closed Traverses

Two cases have to be distinguished with this type of traverse. In figure 7.1, a traverse has been run from station X (of known position) to stations 1, 2, 3 and another known point Y. Traverse X123Y is, therefore, closed at Y. This type of traverse is called a *link, connecting* or *closed-route traverse.*

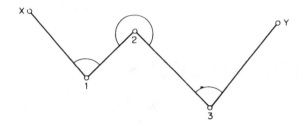

Figure 7.1   *Link traverse*

In figure 7.2, a traverse starts at station X and returns to the same point X via stations 1, 2 and 3. Station X can be of known position or can have an assumed position. In this case the traverse is called a *polygon, loop* or *closed-ring traverse* since it closes back on itself.

In both types of closed traverse there is an external check on the observations since the traverses start and finish on known or assumed points.

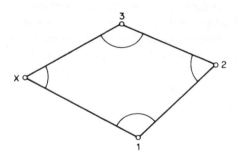

Figure 7.2   *Polygon traverse*

Open Traverses

These commence at a known point and finish at an unknown point and, therefore, are not closed. They are used only in exceptional circumstances since there is no external check on the measurements.

7.2 Traverse Specifications and Accuracy

The accuracy of a traverse is governed largely by the type of equipment used and the observing and measuring techniques employed. These are dictated by the purpose of the survey.

Many types of traverse are possible but three broad groups can be defined and are given in table 7.1.

The most common type of traverse for general engineering work and site surveys would be of typical accuracy 1 in 10 000. This chapter is concerned mainly with an expected accuracy range of about 1 in 5000 to 1 in 20 000.

An important factor when selecting traversing equipment is that the various instruments should produce roughly the same order of precision, that is, it is pointless using a 1″ theodolite to measure traverse angles if the lengths are being measured with a synthetic tape. Table 7.1 gives a general indication of the grouping of suitable equipment.

TABLE 7.1
General Traverse Specifications

Type	Typical accuracy	Purpose	Angular measurement	Distance measurement
Geodetic or Precise	1 in 50 000 or better	(1) Major control for mapping large areas (2) Provision of very accurate reference points for engineering surveys	0.1″ theodolite	EDM
General	1 in 5000 to 1 in 50 000	(1) General engineering surveys, that is, setting out and site surveys (2) Secondary control for mapping large areas	1″ or 20″ theodolite	EDM, steel tapes
Low accuracy	1 in 500 to 1 in 5000	(1) Small-scale detail surveys (2) Rough large-scale detail surveys (3) Preliminary or reconnaissance surveys	20″ or 1′ theodolite	Synthetic tapes, stadia tacheometry

7.3   Traversing Fieldwork: Reconnaissance

This is one of the most important aspects of any survey and must always be undertaken *before* any angles or lengths are measured. The main aim of the reconnaissance is to locate suitable positions for traverse stations and a poorly executed reconnaissance can result in difficulties at later stages in a survey, leading to wasted time and inaccurate work.

To start a reconnaissance, an overall picture of the area is obtained by walking all over the site keeping in mind the requirements of the survey. If an existing map or plan of the area is available, this is a useful aid at this stage.

When siting stations, an attempt should be made to keep the number of stations to a minimum and the lengths of traverse legs should be kept as long as possible to minimise the effect of any centring errors.

If the traverse is being run for a detail survey then the method which is to be used for this subsequent operation must be considered. For most sites a polygon traverse is usually sited around the area at points of maximum visibility. It should be possible to observe cross checks or lines across the area to enable other points inside the area to be fixed and also to assist in the location of angular errors. Traverses for roadworks and pipelines generally require a link traverse since these sites tend to be long and narrow. The shape of the road or pipeline dictates the shape of the traverse.

If distance measurements are to be carried out using tapes, the ground conditions between stations should be suitable for this purpose. Steep slopes or badly broken ground along the traverse lines should be avoided and it is better if there are as few changes of slope as possible. Roads and paths that have been surfaced are usually good for ground measurements.

Stations should be located such that they are clearly intervisible, preferably at ground level, that is, with a theodolite set up at one point, it should be possible to see the ground marks at adjacent stations and as many others as possible. This eases the angular measurement process and enhances its accuracy.

Stations should be placed in firm, level ground so that the theodolite and tripod are supported adequately when observing angles at the stations. Very often stations are used for a site survey and at a later stage for setting out. Since some time may elapse between the site survey and the start of the construction the choice of firm ground in order to prevent the stations moving in any way becomes even more important. It is sometimes necessary to install semi-permanent stations.

Owing to the effects of lateral refraction and shimmer, traverse lines of sight should be well above ground level (greater than 1 m) for most of their length to avoid any possible angular errors due to rays passing close to ground level (*grazing rays*). These effects are serious in hot weather.

When the stations have been sited, a sketch of the traverse should be prepared *approximately to scale*. The stations are given reference letters or numbers. This greatly assists in the planning and checking of fieldwork.

7.4 Station Marking

When a reconnaissance is completed, the stations have to be marked for the duration, or longer, of the survey. Station markers must be permanent, not easily disturbed and they should be clearly visible. The construction and type of station depends on the requirements of the survey.

For general purpose traverses, wooden pegs are used which are hammered into the ground until the top of the peg is almost flush with ground level. If it is not possible to drive the whole length of the peg into hard ground the excess above the ground should be sawn off. This is necessary since a long length of peg left above the ground is liable to be knocked. A nail should be tapped into the top of the peg to define the exact position of the station. Figure 7.3 shows such a station. Several months use is possible with this type of marker.

Stations in roadways can be marked with 75 mm pipe nails driven flush with the surface. The nail surround should be painted for easy identification. These marks are fairly permanent, but it is usually prudent to enquire if the road is to be resurfaced in the near future.

A more permanent station would normally require marks set in concrete; common station designs are shown in figure 14.3. These have to be placed with the permission of land owners as subsurface concrete blocks placed in a field could do considerable damage to farm machinery.

A *reference* or *witnessing sketch* of the features surrounding each station should be prepared, especially if the stations are to be left for any time before being used, or if they will be required again at a much later stage. Measurements are taken from the station to nearby permanent features to enable it to be relocated. A typical sketch is shown in figure 7.4.

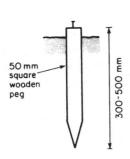

Figure 7.3   *Station peg*

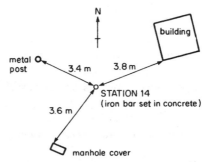

Figure 7.4   *Witnessing sketch*

7.5 Traversing Fieldwork: Angular Measurement

Once the traverse stations have been placed in the ground the next stage in the field procedure is to use a theodolite to measure the included angles between the lines.

This requires two operations: setting the theodolite over each station, and observing the directions to other stations.

In most cases it will be necessary to provide a target at the observed stations since the station marks may not be directly visible. Suitable types of target are described in sections 3.7 and 5.6.

Centring Errors

The measurement of traverse angles requires that the theodolite and targets be located in succession at each station. If this operation is not carried out accurately, centring errors are introduced, the effect of which depends on the length of the traverse leg, as discussed in section 3.10.

If a target displacement of 10 mm occurs on a 300 m traverse leg, the resulting angular error is 7″. The same displacement on a 30 m leg will produce an angular error of 70″. If this occurred during a traverse, the error would be carried through the rest of the traverse, and all subsequent bearings would be incorrect.

Hence, the effect of relatively small centring errors can be serious on short traverse legs. If the theodolite is also displaced a further source of error arises.

The conclusion is that with both theodolites and targets care in centring is vital, especially when traverse legs are short.

Field Procedure and Booking

The method given in chapter 3 for the reading and booking of angles should be adhered to whenever possible.

In the case where no standard booking forms are available, the angles can be entered in a field book, as in figure 7.5, in which *two complete rounds* of angles have been observed and the *zero changed* between rounds. The reasons for this are discussed in section 3.7.

Errors in Angular Measurements

The various sources of error that may arise when measuring traverse angles are summarised as follows.

AT STATION C

STATION	FL	FR	MEAN	ANGLE
① B	00° 07' 20"	180° 07' 10"	00° 07' 20"	
D	192° 23' 40"	12° 23' 20"	192° 23' 30"	192° 16' 10"
② B	87° 32' 40"	267° 32' 20"	87° 32' 30"	
D	279° 49' 20"	99° 49' 00"	279° 49' 10"	192° 16' 40"

MEAN ANGLE = 192° 16' 30"

DIAGRAM

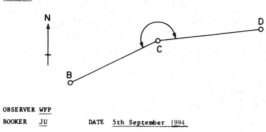

OBSERVER WFP

BOOKER JU   DATE 5th September 1994

Figure 7.5 *Booking traverse angles*

- Inaccurate centring of the theodolite or target.
- Non-verticality of the target.
- Inaccurate bisection of the target.
- Parallax not eliminated.
- Lateral refraction, wind and atmospheric effects.
- Theodolite not level and not in adjustment.
- Incorrect use of the theodolite.
- Mistakes in reading and booking.

7.6 Traversing Fieldwork: Distance Measurement

For the purposes of traversing, measurement of the lengths of the traverse legs is undertaken using steel taping or EDM which are discussed in chapters 4 and 5. If EDM is being used, both angular and distance measurements are usually combined at each traverse station.

7.7 Three-tripod Traversing

Very often, short traverse lines are unavoidable, for example, in surveys in mines, tunnels and on congested sites. One way of reducing the effects of centring errors in such cases is to use three or more tripods during a sur-

vey and to use theodolites (or total stations) that can be detached from their tribrachs and interchanged with a target or prism set (see figure 3.3).
    The system operates as follows, with reference to figure 7.6.

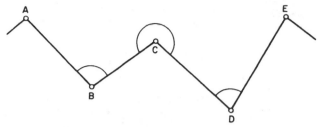

Figure 7.6   *Three-tripod traversing*

When angle ABC is measured

(1) At A a tripod is set up and a tribrach attached to the tripod head. A target or prism set is placed into the tribrach and clamped in position. The target or tribrach will have a tube or circular bubble attached so that the target can be set vertical by levelling using the tribrach footscrews. In order to be able to centre the target, the tribrach usually has an optical plummet.
(2) At B the theodolite (or total station) is set up in the normal manner.
(3) At C a tripod and target is set up as at A.

This enables the horizontal angle at B to be observed and, if a total station or theodolite-mounted EDM unit is being used, enables distances BA and BC to be measured.

When angle BCD is measured

(1) At A the tripod and target are moved to D, where the target is again centred and set vertical.
(2) At B the theodolite is unclamped, removed from its tribrach and interchanged with the target at C. Hence, at B and C, the tripods and tribrachs remain undisturbed and there is no need for recentring.

With the equipment set in this position, the horizontal angle at C and distances CB and CD are measured. A check can be made on the horizontal distances BC and CB at this stage.

When angle CDE is measured

(1) At B the tripod and target are moved to E.

(2) The theodolite and target at C and D are interchanged, the tribrachs (and centring) remaining undisturbed.

The process is repeated for the whole traverse. If *four* tripods (or more) are used this speeds up the fieldwork considerably as tripods can be moved and positioned while angles are being measured.

Nowadays, three-tripod traversing is normal practice, especially when using EDM to measure distances.

7.8 Abstract of Fieldwork

When all the traverse fieldwork has been completed, a single sheet or record containing the mean angles observed and mean horizontal (corrected) lengths measured should be prepared. It is preferable to show all the data on a sketch of the traverse as this helps in the subsequent calculations and can minimise the chance of a mistake.

Such an abstraction of field data is shown in figure 7.7, the angles and lengths being entered on to a traverse diagram. The example shown in figure 7.7 will be referred to in the following sections.

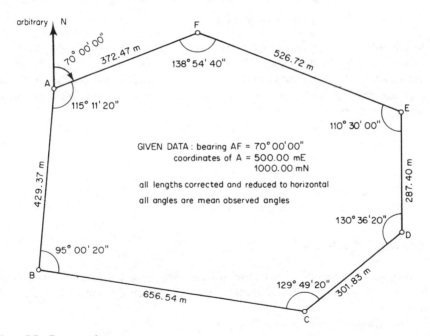

Figure 7.7   *Traverse abstract*

7.9 Angular Misclosure

Determination of Misclosure

For a closed traverse, before any coordinate calculations can commence, the whole circle bearings of all the lines have to be calculated. The first stage in the calculation process is to check that the observed angles sum to their required value.

The observed angles of a *polygon traverse* can be either the *internal* or *external* angles, and angular misclosures are found by comparing the sum of the observed angles with one of the following theoretical values

$$\text{Sum of } internal \text{ angles} = (2n - 4) \times 90° \qquad (7.1)$$

$$\text{or sum of } external \text{ angles} = (2n + 4) \times 90° \qquad (7.2)$$

where n is the number of angles measured.

When the bearings in a *link traverse* are calculated, an *initial back bearing* can usually be determined from known points at the start of the traverse and, to check the observed angles, a *final forward bearing* is computed from known points at the end of the traverse. The method for obtaining a bearing from coordinates is given in section 1.5. The angular misclosure in a link traverse is found using

$$\text{Sum of angles} = (\text{final forward bearing } - \text{ initial back}$$

$$\text{bearing}) + m \times 180° \qquad (7.3)$$

In equation (7.3), m is an integer the value of which depends on the shape of the traverse. In most cases, m will be $(n - 1)$, n or $(n + 1)$ where n is the number of angles measured between the initial back bearing and final forward bearing. The link traverse worked example in section 7.16 shows how m is determined.

For both types of traverse, care must be taken to ensure that the correct angles have been abstracted and summed, that is, the internal *or* external angles in a polygon traverse and the angles on the same side of a link traverse (as in figure 7.1). When the angles have been summed and checked, a very large misclosure probably means that an incorrect angle has been included, or one of the angles has been excluded.

Allowable Misclosure

Owing to the effects of occasional miscentring, slight misreading and small bisection errors, a small misclosure will result when the summation check is made.

The allowable misclosure E is given by

$$E'' = \pm KSn^{\frac{1}{2}} \qquad (7.4)$$

where

> K is a multiplication factor of 1 to 3 depending on weather conditions, number of rounds taken, and so on
>
> S is the smallest reading interval on the theodolite in seconds, for example, 60", 20", 1"
>
> n is the number of angles measured.

The allowable misclosure for the traverse shown in figure 7.7 varies from 50" to 150" assuming a 20" theodolite was used.

Adjustment

When the actual misclosure is known and is compared to its allowable value, two cases may arise.

(1) *If the misclosure is acceptable* (less than the allowable) it is divided *equally* between the observed angles. An equal distribution is the only acceptable method since each angle is measured in the same way and there is an equal chance of the misclosure having occurred in any of the angles.

No attempt should be made to distribute the misclosure in proportion to the size of an angle.

(2) *If the misclosure is not acceptable* (greater than the allowable) the angles should be remeasured if no gross error can be located in the angle bookings or summation.

It may be possible to isolate a gross error in a small section of the traverse if check lines have been observed across it.

Example of Angular Misclosure and Adjustment

The determination of the misclosure and adjustment of the angles of the polygon traverse given in figure 7.7 is shown in table 7.2.

An example in section 7.16 shows how the angles in a link traverse are adjusted.

7.10 Calculation of Whole-circle Bearings

Types and Determination of Bearings

Consider figure 7.8, which shows two legs of a traverse. The decision has been made to calculate the traverse in the direction ... X to Y to Z ... This defines the bearings as follows.

TABLE 7.2

Station	Observed angle	Adjustment	Adjusted angle
A	115°11'20"	−20"	115°11'00"
B	95°00'20"	−20"	95°00'00"
C	129°49'20"	−20"	129°49'00"
D	130°36'20"	−20"	130°36'00"
E	110°30'00"	−20"	110°29'40"
F	138°54'40"	−20"	138°54'20"
Sums	720°02'00"	−02'00"	720°00'00"

Required sum = $((2 \times 6) - 4) \times 90$ Adjustment per angle
            = 720°00'00"   = −(02'00")/6
Misclosure = +02'00"   = −20"

Bearings XY and YZ are *forward bearings* since they are in the same direction in which calculations are proceeding.

Bearings YX and ZY are *back bearings* since they are opposite to the direction in which the traverse calculation is proceeding.

Directions ZY and YZ differ by $\pm$ 180°, as do those of YX and XY. Therefore, the forward bearing of a line differs from the back bearing by $\pm$ 180°.

For the direction of computation shown in figure 7.8, γ_Y is known as the *left-hand angle* at Y since it lies to the left at station Y relative to the direction X to Y to Z.

If γ_Y is added to the back bearing YX it can be seen from figure 7.8 that the resulting angle will be the forward bearing YZ. Thus

forward bearing YZ = back bearing YX + γ_Y

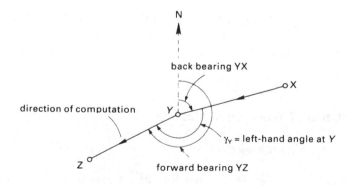

Figure 7.8   *Whole-circle bearing calculation*

Therefore, in general, for any particular traverse station

$$forward\ bearing = back\ bearing + left\text{-}hand\ angle \qquad (7.5)$$

For polygon traverses when working in an *anticlockwise* direction around the traverses, the left-hand angles will be the *internal* angles of the traverse and when working in a *clockwise* direction, the left-hand angles will be the *external* angles.

Example of Bearing Calculation

Some of the bearings of the lines of the traverse shown in figure 7.7 will now be computed using adjusted left-hand angles. Figures 7.9 and 7.10 show sections of this traverse.

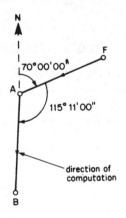

Figure 7.9   Figure 7.10

At station A in figure 7.9

   forward bearing AB = back bearing AF + left-hand angle at A

                = 70°00′00″ (given) + 115°11′00″

                = 185°11′00″

At station B in figure 7.10

   forward bearing BC = back bearing BA + left-hand angle at B

But   back bearing BA = forward bearing AB ± 180°

                = 185°11′00″ ± 180°

                = 365°11′00″ *or* 05°11′00″

                = 05°11′00″ (to keep bearing in range 0

                to 360°)

Hence   forward bearing BC = 05°11′00″   + 95°00′00″

                                         = 100°11′00″

The bearings of all the lines can be computed in a similar manner; the complete calculation is given in table 7.3.

If, at some stage in a bearing calculation, the result for a forward bearing is computed to be greater than 360°, then 360° must be subtracted from the computed bearing to give a bearing in the range 0 to 360°. For example, the forward bearing CD in figure 7.7 is given by (see also table 7.3)

    forward bearing CD = back bearing CB + left-hand angle at C

                       = 280°11′00″ + 129°49′00″

                       = 410°00′00″

Since this is greater 360°

    forward bearing CD = 410°00′00″ − 360° = 50°00′00″

Every bearing calculation finishes by recalculating the initial (given) bearing. This final computed bearing must be in agreement with the initial bearing and, if any difference occurs, an *arithmetic mistake* has been made, and the bearing calculation must be checked before proceeding to the next stage in the calculation.

7.11 Computation of Coordinate Differences

The next stage in the traverse computation is the determination of the co-ordinate differences of the traverse lines.

The information available at this point will be the bearings and horizontal lengths of all the lines.

Examples of Coordinate Difference Calculation

The traverse data of figure 7.7 is again used in the following examples. The bearings are the whole-circle bearings given in table 7.3. With reference to section 1.5, consider line AB in both figure 7.7 and figure 7.11.

From equation (1.1)

$$\Delta E_{AB} = D_{AB} \sin \theta_{AB}$$

$$= 429.37 \sin 185°11′00″ = 429.37 \ (-0.09034)$$

$$= -38.79 \text{ m}$$

TABLE 7.3

LINE	BACK BEARING / ADJUSTED LEFT HAND ANGLE / FORWARD BEARING (° ′ ″)	WHOLE CIRCLE BEARING θ (° ′ ″)	HORIZONTAL DISTANCE D	CALCULATED ΔE	CALCULATED ΔN	ADJUSTMENT δE	ADJUSTMENT δN	ADJUSTED ΔE	ADJUSTED ΔN	COORD E	COORD N	STATION
AF	70 00 00									500.00	1000.00	A
A	115 11 00											
AB	185 11 00	185 11 00	429.37	−38.79	−427.61	−0.04	+0.02	−38.83	−427.59	461.17	572.41	B
BA	05 11 00											
B	95 00 00											
BC	100 11 00	100 11 00	656.54	+646.20	−116.08	−0.05	+0.03	+646.15	−116.05	1107.32	456.36	C
CB	280 11 00											
C	129 49 00											
CD	50 00 00	50 00 00	301.83	+231.22	+194.01	−0.03	+0.01	+231.19	+194.02	1338.51	650.38	D
DC	230 00 00											
D	130 36 00											
DE	00 36 00	00 36 00	287.40	+3.01	+287.38	−0.02	+0.01	+2.99	+287.39	1341.50	937.77	E
ED	180 36 00											
E	110 29 40											
EF	291 05 40	291 05 40	526.72	−491.42	+189.57	−0.04	+0.03	−491.46	+189.60	850.04	1127.37	F
FE	111 05 40											
F	138 54 20											
FA	250 00 00	250 00 00	372.47	−350.01	−127.39	−0.03	+0.02	−350.04	−127.37	500.00	1000.00	A
			Σ 2574	Σ +0.21	Σ −0.12	Σ −0.21	Σ +0.12	Σ 0.00	Σ 0.00			

ACTUAL SUM OF LEFT HAND ANGLES = 720° 02' 00"

REQUIRED SUM OF LEFT HAND ANGLES = (2 × 6 − 4) × 90° = 720°

MISCLOSURE = + 02' 00"

ADJUSTMENT TO EACH OBSERVED ANGLE = − 20"

$e_E = + 0.21$

$e_N = - 0.12$

$e = ((+0.21)^2 + (-0.12)^2)^{\frac{1}{2}} = 0.24$

FRACTIONAL LINEAR MISCLOSURE = 1 in 10 700

ADJUSTMENT TO ΔE/ΔN BY BOWDITCH

Similarly, from equation (1.2)

$$\Delta N_{AB} = D_{AB} \cos \theta_{AB}$$

$$= 429.37 \cos 185°11'00'' = 429.37 \ (-0.99591)$$

$$= -427.61 \text{ m}$$

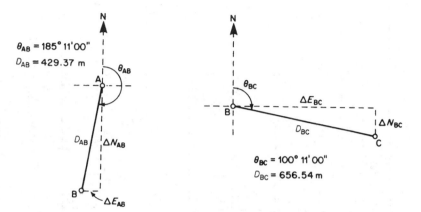

Figure 7.11   Figure 7.12

For line BC shown in figures 7.7 and 7.12, the coordinate differences are given by

$$\Delta E_{BC} = D_{BC} \sin \theta_{BC}$$

$$= 656.54 \sin 100°11'00'' = 656.54 \ (+0.98425)$$

$$= +646.20 \text{ m}$$

$$\Delta N_{BC} = D_{BC} \cos \theta_{BC}$$

$$= 656.54 \cos 100°11'00'' = 656.54 \ (-0.17680)$$

$$= -116.08 \text{ m}$$

As with the bearing calculations, the coordinate difference results are always presented in tabular form since errors are easier to detect.

For the traverse ABCDEFA (figure 7.7) all the calculations for coordinate differences are given in table 7.3.

7.12 Misclosure

When the ΔE and ΔN values have been computed for the whole traverse as in table 7.3, checks can be applied to the computation.

For *polygon* traverses these are

$$\Sigma\Delta E = 0 \text{ and } \Sigma\Delta N = 0 \tag{7.6}$$

since the traverse starts and finishes at the same point.

For link traverses (figure 7.1) these are

$$\Sigma\Delta E = E_Y - E_X \text{ and } \Sigma\Delta N = N_Y - N_X \tag{7.7}$$

where station X is the starting point and station Y the final point of the traverse. Since stations X and Y are of known position, the values of $E_Y - E_X$ and $N_Y - N_X$ can be calculated.

In both cases, owing to field errors in measuring the angles and lengths, there will normally be a misclosure on returning to the starting point on a polygon traverse or on arrival at the final known station in a link traverse.

This *linear* misclosure is computed and any adjustment is allocated appropriately.

Therefore, before the station coordinates are calculated, the ΔE and ΔN values found for the traverse are summed and the misclosures, e_E and e_N, are found by comparing the summations with those expected.

These misclosures form a measure of the linear misclosure of the traverse and can be used to determine the accuracy of the survey. Consider figure 7.13 which shows the starting point A of a polygon traverse.

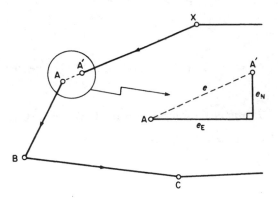

Figure 7.13 *Traverse misclosure*

Owing to field errors, the traverse ends at A′ instead of A. The *linear misclosure, e*, is given by

$$e = (e_E^2 + e_N^2)^{\frac{1}{2}} \tag{7.8}$$

To obtain a measure of the accuracy of the traverse, this misclosure is compared with the total length of the traverse legs, ΣD, to give the *fractional linear misclosure*, where

$$fractional\ linear\ misclosure = 1 \text{ in } (\Sigma D/e) \tag{7.9}$$

This fractional misclosure is *always* computed for a traverse and is compared with the value required for the type of survey being undertaken. For appropriate values of the fractional linear misclosure see table 7.1.

If, on comparison, the fractional linear misclosure is better than the required value, the traverse fieldwork is satisfactory and the misclosures, e_E and e_N, are distributed throughout the traverse.

If, on comparison, the fractional linear misclosure is worse than that required, there is most likely an error in the measured lengths of one or more of the legs. The calculation should, however, be thoroughly checked before remeasuring any lengths.

An example determination of the fractional linear misclosure can be obtained from table 7.3, remembering that the traverse is a polygon. From the table

$$(1) \ \Sigma\Delta E = -38.79 + 646.20 + 231.22 + 3.01 - 491.42 - 350.01$$

$$= +0.21 \text{ m}$$

Hence

$$e_E = +0.21 \text{ m since } \Sigma\Delta E \text{ should be zero}$$

$$(2) \ \Sigma\Delta N = -427.61 - 116.08 + 194.01 + 278.38 + 189.57$$

$$-127.39$$

$$= -0.12 \text{ m}$$

Hence

$$e_N = -0.12 \text{ m since } \Sigma\Delta N \text{ should also be zero}$$

Therefore

$$linear \ misclosure = e = [(+0.21)^2 + (-0.12)^2]^{\frac{1}{2}} = 0.24 \text{ m}$$

From figure 7.7 and table 7.3

$$\Sigma D = 2574 \text{ m}$$

Therefore

$$fractional \ linear \ misclosure = 1 \text{ in } (2574/0.24) \simeq 1 \text{ in } 10\ 700$$

This calculation is also shown at the bottom of table 7.3.

7.13 Distribution of the Misclosure

Many methods of adjusting the linear misclosure of a traverse are possible but, for everyday engineering traverses of accuracy up to 1 in 20 000, one of three methods is normally used.

Bowditch Method

The values of the adjustment found by this method are directly proportional to the length of the individual traverse lines.

Adjustment to ΔE (or ΔN) for one particular traverse leg

$$= \delta E \text{ (or } \delta N) = -e_E \text{ (or } - e_N) \times \frac{\text{length of traverse leg concerned}}{\text{total length of the traverse}}$$

$$(7.10)$$

Transit Method

In this method, adjustments are proportional to the values of ΔE and ΔN for the various lines.

Adjustment to ΔE (or ΔN) for one particular traverse leg

$$= \delta E \text{ (or } \delta N) = -e_E \text{ (or } - e_N) \times$$

$$\frac{\Delta E \text{ (or } \Delta N) \text{ of the traverse leg concerned}}{\text{absolute } \Sigma \Delta E \text{ (or } \Sigma \Delta N) \text{ for the traverse}}$$

$$(7.11)$$

Equal Adjustment

For traverses measured by EDM, the likely error in each distance will be independent of the distance measured for normal work (that is, traverses with lines not much greater than 100–200 m). This can be verified by noting the specifications for distance measurement quoted in tables 5.1 and 5.2. Therefore, for EDM traverses, the error in each measured distance will be of the same order of magnitude and an equal distribution of the misclosure is acceptable. In such cases

$$\delta E \text{ (or } \delta N) \text{ for each line} = \frac{-e_E \text{ (or } -e_N)}{n} \qquad (7.12)$$

where n is the number of traverse lines.

For all methods, the negative signs are necessary since if e_E (or e_N) is positive, the *adjustments* will be negative, and if e_E (or e_N) is negative the *adjustments* will be positive.

For the *Bowditch* method, the adjustment of the values of ΔE and ΔN given in table 7.3 is as follows.

The misclosures have already been determined as $e_E = +0.21$ m and $e_N = -0.12$ m, and the total length of the traverse is 2574 m.

For line AB

$$\delta E_{AB} = -0.21 \times (429/2574) = -0.04 \text{ m}$$

$$\delta N_{AB} = +0.12 \times (429/2574) = +0.02 \text{ m}$$

For line BC

$$\delta E_{BC} = -0.21 \times (657/2574) = -0.05 \text{ m}$$

$$\delta N_{BC} = +0.12 \times (657/2574) = +0.03 \text{ m}$$

This process is repeated for the whole traverse. These adjustments, applied to the ΔE and ΔN values, would normally be tabulated as shown in table 7.3.

Applying the *transit* method to the same example gives

absolute $\Sigma \Delta E = 1761$ m and absolute $\Sigma \Delta N = 1342$ m

Hence *for line AB*

$$\delta E_{AB} = -0.21 \times (39/1761) = 0.00 \text{ m}$$

$$\delta N_{AB} = +0.12 \times (428/1342) = +0.04 \text{ m}$$

and *for line BC*

$$\delta E_{BC} = -0.21 \times (646/1761) = -0.08 \text{ m}$$

$$\delta N_{BC} = +0.12 \times (116/1342) = +0.01 \text{ m}$$

Again, the computation is repeated for each line of the traverse.

An equal adjustment to the example gives the following for all lines

$$\delta E = \frac{-0.21}{6} = -0.03(5) \text{ m}$$

$$\delta N = \frac{+0.12}{6} = +0.02 \text{ m}$$

All of these methods will alter the original bearings by a very small amount. It is *not* necessary to recalculate these bearings unless the traverse is to be used for subsequent control work such as setting out.

Checks on all methods of adjustment should be undertaken as follows. If the adjustment has been carried out successfully

$$\Sigma \delta E \text{ should} = -e_E$$

$$\Sigma \delta N \text{ should} = -e_N$$

These checks must be carried out *before* calculating the adjusted ΔE and ΔN values.

7.14 Calculation of the Final Coordinates

For *polygon* traverses, in order to compute the coordinates of the stations, the coordinates of the starting point have to be known. These starting coordinates may either be assumed for an area to give positive coordinates for the whole survey or may be given if a previously coordinated station is used to start the traverse.

For *link* traverses, the coordinates of the starting and finishing points will be known from a previous survey and the coordinates will be determined relative to these known values.

The coordinates of each point are obtained by adding or subtracting the adjusted ΔE and ΔN values as necessary, working around the traverse.

When all the coordinates have been calculated, there is a final check to be applied.

For a *polygon* traverse, the final and initial coordinates should be equal as these represent the same station.

For a *link* traverse, the final coordinates should equal those of the second known point.

If this check does not hold, there is an *arithmetical mistake* and the calculations should be investigated until it is found.

At this stage for the polygon traverse which has been referred to throughout this discussion (that shown in figure 7.7), the adjusted ΔE and ΔN values have now been determined and, since the coordinates of the starting point, station A, have been given as 500.00 m E and 1000.00 m N, the coordinates of the other traverse stations can be obtained from these initial coordinates and the adjusted ΔE and ΔN values found by the Bowditch method. For example

(1) $E_B = E_A \pm \Delta E_{AB} = 500.00 - 38.83 = 461.17$ m

$N_B = N_A \pm \Delta N_{AB} = 1000.00 - 427.59 = 572.41$ m

(2) $E_C = E_B \pm \Delta E_{BC} = 461.17 + 646.15 = 1107.32$ m

$N_C = N_B \pm \Delta N_{BC} = 572.41 - 116.05 = 456.36$ m

This process is repeated until station A is recoordinated as a check. The complete calculation is shown in table 7.3.

7.15 The Traverse Table

For each particular step in the traverse computation every calculation should be tabulated.

There are many variations of the layout that can be adopted but the format given in table 7.3 is recommended.

Table 7.3 shows the calculation for the polygon traverse ABCDEFA of figure 7.7. This table should be thoroughly studied, referring to the relevant preceding sections of this chapter to enable a complete understanding of how the table is compiled to be gained.

7.16 Worked Examples: Traversing

Polygon Traverse

Question
The traverse diagram of figure 7.14 is a field abstract for a polygon traverse ABCDEA.

Calculate the adjusted coordinates of stations B, C, D and E, adjusting any misclosure by the Bowditch method.

The coordinates of station A are 500.00 m E, 500.00 m N and the line AB has an assumed whole circle bearing of 90°00'00".

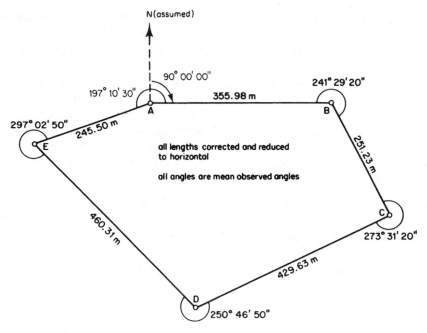

Figure 7.14   *Worked example: polygon traverse*

Solution
The complete solution is given in the traverse table shown in table 7.4.

(1) Since the *external* angles are given, these will be the left-hand angles if the solution follows the *clockwise* direction. For this traverse, no attempt should be made to compute in an anticlockwise direction as this would involve subtraction of angles and errors may result.

(2) The bearing calculation always starts with the assumed or given bearing. In the example, bearing AB is given as 90° and is, for the clockwise direction, a forward bearing and is entered as such in the traverse table.

TABLE 7.4

Angle / bearing computation

LINE / STATION	FORWARD BEARING			WHOLE CIRCLE BEARING θ		
AB	90	00	00	90	00	00
BA	270	00	00			
B	241	29	10			
BC	151	29	10	151	29	10
CB	331	29	10			
C	273	31	10			
CD	245	00	20	245	00	20
DC	65	00	20			
D	250	46	40			
DE	315	47	00	315	47	00
ED	135	47	00			
E	297	02	40			
EA	72	49	40	72	49	40
AE	252	49	40			
A	197	10	20			
AB	90	00	00			

Coordinate differences and coordinates

LINE	HORIZONTAL DISTANCE D	CALCULATED ΔE	CALCULATED ΔN	ADJUSTMENTS δE	ADJUSTMENTS δN	ADJUSTED ΔE	ADJUSTED ΔN	COORD E	COORD N	STATION
								500.00	500.00	A
AB	355.98	+355.98	0.00	-0.01	-0.02	+355.97	-0.02	855.97	499.98	B
BC	251.23	+119.93	-220.76	-0.01	-0.01	+119.92	-220.77	975.89	279.21	C
CD	429.63	-389.39	-181.53	-0.02	-0.03	-389.41	-181.56	586.48	97.65	D
DE	460.31	-321.01	+329.91	-0.02	-0.03	-321.03	+329.88	265.45	427.53	E
EA	245.50	+234.56	+72.48	-0.01	-0.01	+234.55	+72.47	500.00	500.00	A
	Σ 1743	Σ+0.07	Σ0.10	Σ-0.07	Σ-0.10	Σ 0.00	Σ 0.00			

ACTUAL SUM OF LEFT HAND ANGLES = 1260° 00' 50"

REQUIRED SUM OF LEFT HAND ANGLES = (2 X 5 + 4) X 90° = 1260°

MISCLOSURE = + 00' 50"

ADJUSTMENT TO EACH OBSERVED ANGLE = -10"

$e_E = +0.07$

$e_N = +0.10$

$e = ((+0.07)^2 + (+0.10)^2)^{\frac{1}{2}} = 0.12$

ADJUSTMENT TO ΔE/ΔN BY BOWDITCH

FRACTIONAL LINEAR MISCLOSURE = 1 in 14 500

(3) The sum of the left-hand angles gives a misclosure of $+00'50''$ and since there are five angles, each has an adjustment of $-10''$.
(4) The fractional linear misclosure is rounded off to 1 in 14 500. It is not necessary to quote this to better than three significant figures. 1 in 14 500 would be acceptable for most engineering work.
(5) Adjustment of the ΔE and ΔN values by the Bowditch method gives the adjustments as shown. For example calculations, consider line CD.

$$\delta E_{CD} = -e_E \times (\text{length CD/}\Sigma D)$$

$$= -0.07 \times (430/1743) = -0.02\text{m}$$

$$\delta N_{CD} = -e_N \times (\text{length CD/}\Sigma D)$$

$$= -0.10 \times (430/1743) = -0.03 \text{ m}$$

Note that lengths of each line and ΣD need only be used to three significant figures for required adjustments of two significant figures and that the total of the individual adjustments for the ΔE and ΔN values *must equal* $-e_E$ and $-e_N$ respectively.
(6) The coordinate computation starts and ends with the station of known position A. The final check is to ensure that the derived coordinates of A agree with the start coordinates of A.

Link Traverse

Question
A link traverse was run between stations A and X as shown in the traverse diagram of figure 7.15.

The coordinates of the controlling stations at the ends of the traverse are as follows

	E (m)	N (m)
A	1769.15	2094.72
B	1057.28	2492.39
X	2334.71	1747.32
Y	2995.85	1616.18

Calculate the coordinates of stations 1, 2, 3 and 4, adjusting any misclosure by the Transit method.

Solution
The complete solution is given in the traverse table shown in table 7.5.

(1) The solution follows the direction A to X as this will give the left-hand angles, as shown in figure 7.15.
(2) When link traversing, the starting and closing bearings may either be given directly or implied by the coordinates of the stations used to

TABLE 7.5

LINE / STATION	BACK BEARING — Adjusted Left Hand Angle / Forward Bearing	WHOLE CIRCLE BEARING θ	HORIZONTAL DISTANCE D	CALCULATED ΔE	CALCULATED ΔN	ADJ. δE	ADJ. δN	ADJUSTED ΔE	ADJUSTED ΔN	COORD E	COORD N	STATION
AB	299 11 20											
A	115 37 20									1769.15	2094.72	A
A1	54 48 40	54 48 40	208.26	+170.20	+120.01	+0.02	+0.01	+170.22	+120.02	1939.37	2214.74	1
1A	234 48 40											
1	168 19 30											
12	43 08 10	43 08 10	193.47	+132.28	+141.18	+0.02	+0.01	+132.30	+141.19	2071.67	2355.93	2
21	223 08 10											
2	281 13 00											
23	144 21 10	144 21 10	326.71	+190.40	-265.49	+0.02	+0.02	+190.42	-265.47	2262.09	2090.46	3
32	324 21 10											
3	242 54 00											
34	207 15 10	207 15 10	309.15	-141.57	-274.83	+0.02	+0.02	-141.55	-274.81	2120.54	1815.65	4
43	27 15 10											
4	80 26 40											
4X	107 41 50	107 41 50	224.79	+214.15	-68.33	+0.02	+0.00	+214.17	-68.33	2334.71	1747.32	X
X4	287 41 50											
X	173 31 20											
XY	101 13 10	101 13 10										
			Σ 1262	Σ+565.46	Σ-347.46	Σ+0.10	Σ+0.06	Σ+565.56	Σ-347.40			

ACTUAL SUM OF LEFT HAND ANGLES   = 1061°59'50"
REQUIRED SUM OF LEFT HAND ANGLES   = 1062°01'50"
MISCLOSURE   =   -02'00"
ADJUSTMENT TO EACH OBSERVED ANGLE   =   +20"

$\Sigma\Delta E$ = +565.46 $\Sigma\Delta N$ = -347.46   abs $\Sigma\Delta E$ = 849

$E_X - E_A$ = +565.56 $N_X - N_A$ = -347.40   abs $\Sigma\Delta N$ = 870

e_E =   -0.10 e_N =   -0.06   ADJUSTMENT TO $\Delta E / \Delta N$ BY TRANSIT

$e = ((-0.10)^2 + (-0.06)^2)^{\frac{1}{2}} = 0.12$

FRACTIONAL LINEAR MISCLOSURE = 1 in 10 500

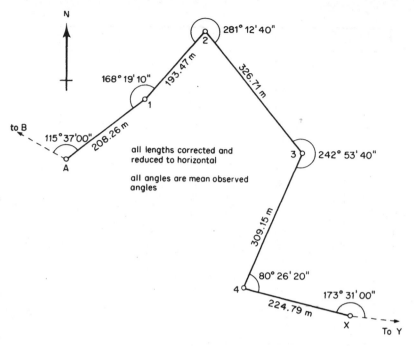

Figure 7.15   *Worked example: link traverse*

start and end the traverse. In this case, coordinates are given and it is necessary to compute the initial and final bearings.

(a) *Initial back bearing AB*

Figure 7.15 is a sketch of the traverse, approximately to scale, and, therefore, shows that the bearing AB is in the *fourth* quadrant. Hence, the whole-circle bearing, θ_{AB}, is given by (see section 1.5)

$$\theta_{AB} = \tan^{-1} (\Delta E_{AB}/\Delta N_{AB}) + 360°$$

$$= \tan^{-1} [(1057.28 - 1769.15)/(2492.39 - 2094.72)] + 360°$$

$$= \tan^{-1} (-711.87/397.67) + 360°$$

$$= \tan^{-1} (1.790\ 10) + 360°$$

Hence

$$\theta_{AB} = -60°48'40'' + 360°$$

Therefore

$$\theta_{AB} = 299°11'20''$$

Alternatively, a rectangular/polar conversion can be used to obtain θ_{AB}.

(b) *Final forward bearing XY*

From figure 7.15, the bearing XY lies in the *second* quadrant hence

$$\theta_{XY} = \tan^{-1} (\Delta E_{XY}/\Delta N_{XY}) + 180°$$

$$= \tan^{-1} [(2995.85 - 2334.71)/(1616.18 - 1747.32)] + 180°$$

$$= \tan^{-1} (661.14/-131.14) + 180°$$

$$= \tan^{-1} (-5.041\ 48) + 180°$$

Hence

$$\theta_{XY} = -78°46'50'' + 180°$$

Therefore

$$\theta_{XY} = 101°13'10''$$

Again, a rectangular/polar conversion can also be used.
(3) The angular misclosure is found as follows

$$\text{Actual sum of left-hand angles} = 1061°59'50''$$

From equation (7.3), Required sum of left-hand angles
$$= (\text{final forward bearing} - \text{initial back bearing}) + m \times 180°$$
$$= (101°13'10'' - 299°11'20'') + m \times 180°$$
$$= -(197°58'10'') + m \times 180°$$
The value of m is obtained by assuming that no very large error was made when measuring the left-hand angles and that their actual sum is approximately correct. In such a case, the value of m needed to give a required sum close to the actual sum of $1061°59'50''$ is $m = 7 = (n + 1)$ in this case, where n is the number of left-hand angles measured. Therefore, the required sum of the left-hand angles
$$= -(197°58'10'') + (7 \times 180°)$$
$$= 1062°01'50''$$
The misclosure is, therefore, $-02'00''$ and each left-hand angle is adjusted by adding $20''$ to it.
(4) To evaluate the misclosures e_E and e_N for the link traverse the following formulae are used

$$\Sigma \Delta E - (E_X - E_A) = e_E$$

$$\Sigma \Delta N - (N_X - N_A) = e_N$$

These are evaluated as shown in table 7.5.
(5) Adjustments to the ΔE and ΔN values are by the Transit method. Example derivations are given for the line joining stations 1 and 2 as follows

$$\delta E_{12} = -e_E \times (\Delta E/\text{abs } \Sigma \Delta E) = +0.10 \times (132/849)$$

$$= +0.02 \text{ m}$$

$$\delta N_{12} = -e_N \times (\Delta N/\text{abs } \Sigma \Delta N) = +0.06 \times (141/870)$$

$$= +0.01 \text{ m}$$

The terms abs $\Sigma\Delta E$ and abs $\Sigma\Delta N$ are the summations of the ΔE and ΔN values regardless of sign.

The ΔE, ΔN, abs $\Sigma\Delta E$ and abs $\Sigma\Delta N$ values are required only to three significant figures.

(6) The check on the final coordinates is satisfactory since the derived coordinates for station X agree with those given.

7.17 Triangulation and Trilateration

In common with traversing, triangulation and trilateration are used to locate control points or stations which form a network.

A *triangulation network* consists of a series of single or overlapping triangles as shown in figure 7.16, the points (or vertices) of each triangle forming control stations. Position is determined by measuring all the angles in the network and by measuring the length of one or more baselines such as AB or HJ in figure 7.16. Starting at a baseline, application of the Sine Rule in each triangle throughout the network enables the lengths of all triangle sides to be calculated. These lengths, when combined with the measured angles, enable the coordinates of the stations to be computed.

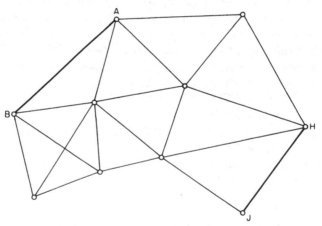

Figure 7.16 *Triangulation network*

A *trilateration network* also takes the form of a series of single or overlapping triangles but in this case position is determined by measuring all the distances in the network instead of all the angles. To enable station coordinates to be calculated, the measured distances are combined with angle values derived from the side lengths of each triangle.

Until the advent of EDM, the measurement of distances in a trilateration scheme with sufficient accuracy was a very difficult and time-consuming process and because of this trilateration techniques were seldom used for establishing horizontal control. Traversing techniques were also limited since it was not possible to maintain a uniformly high accuracy when traversing over long distances. As a result, triangulation was used extensively in the past to provide control for surveys covering very large areas. For example, the triangulation network throughout Great Britain that provides control for mapping was first established by the Ordnance Survey (see section 1.9) between 1783 and 1853, and was subsequently resurveyed from 1935 to 1962.

Nowadays, because of the high precision and accuracy of modern EDM equipment, traversing, triangulation and trilateration can all be used as methods of establishing horizontal control. However, although traversing is the most popular method for providing control on site, combined triangulation and trilateration is often used; this involves the measurement of angles and distances throughout a network rather than between selected stations as in traversing. On construction sites, combined networks are used where horizontal control is required to be spread over large areas and they are also used to provide reference points for control extension, for monitoring (see chapter 15) and for precise engineering work.

In the following sections, combined triangulation and trilateration schemes are simply referred to as networks.

7.18 Network Configurations

Although combined networks could be made up entirely from single triangles as in figure 7.17a, it is often better to use a more complicated scheme involving such figures as braced quadrilaterals (figure 7.17b) and centre-point polygons (figures 7.17c and 7.17d). Compared with a network consisting of simple triangles, these figures usually require more fieldwork and the subsequent computations are often more complicated. However, the advantage of incorporating figures more elaborate than simple triangles in a control scheme *strengthens the network* by increasing the number of redundant measurements taken. These redundant measurements are used in an adjustment of the network, they enable errors to be detected and they can be used to estimate the accuracy of a network.

An example of a control network is shown in figure 7.18. This was set up for the construction of Munich Airport and shows the control point layout with an average distance of about 1 km between stations. A combination of angles and distances was observed for this network in order to obtain as many redundant measurements as possible. Using this as a reference, a site network was established for setting-out purposes, the layout of which depended on the work in progress. The average point-to-point distance for this network was about 60 m.

The primary control network used for the Channel Tunnel project is shown

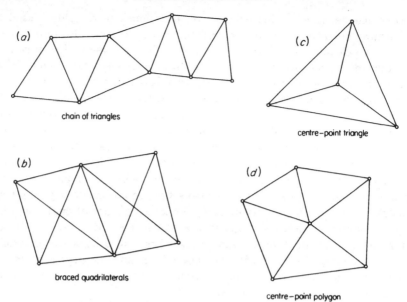

Figure 7.17 *Network figures*

in figure 7.19. This network is clearly much larger than that used at Munich Airport and many special techniques were used to measure all the angles and distances.

Despite the difference in size, both of these networks incorporated many braced quadrilaterals, centre-point polygons and overlapping figures to strengthen them. As well as this, small angles of less than about 20–25° were seldom used so that any errors associated with poor intersections were reduced. A network is said to be *well-conditioned* if small angles are avoided.

7.19 Triangulation and Trilateration: Fieldwork

The methods that can be used to establish and observe a combined network vary considerably with its size and it is emphasised that the following sections are concerned solely with civil engineering and construction sites where distances between control stations seldom exceed 1 km.

Reconnaissance

The reconnaissance for a network is the most important part of the survey and is carried out to determine the positions of the control stations. Since this is linked to the size and shape of the figures to be used in the scheme

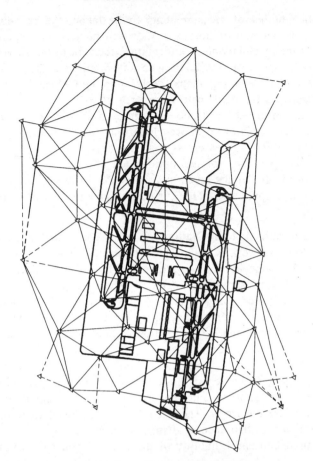

Figure 7.18   *Control network for Munich Airport (courtesy Leica UK Ltd)*

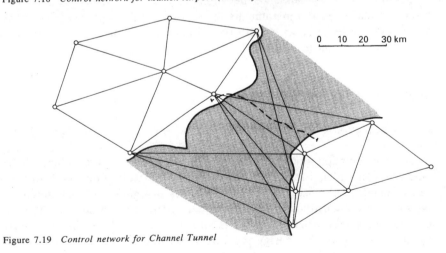

Figure 7.19   *Control network for Channel Tunnel*

and to the number of measurements to be taken, the reconnaissance will determine the amount of fieldwork that will have to be undertaken.

To start the reconnaissance, information relevant to the survey area should be gathered, especially that relating to any previous surveys. Such information may include existing maps, aerial photographs and any site surveys already prepared for the construction project.

From this information, a network diagram should be prepared, approximately to scale, showing proposed locations for the stations.

Following this, it is essential that the survey area is visited, at which time the final positions for the stations are chosen.

Many of the guidelines given in section 7.3 for reconnaissance when traversing are also applicable here, but particular attention must be paid to the following.

(1) When establishing the stations, it is essential that a well-conditioned and strong network is obtained. In general, it is advisable not to include angles less than 25° in the scheme and braced quadrilaterals and centre-point polygons should be included wherever possible. A reliable diagram is required for determining the strength of the network and, to construct this, it may be necessary to take approximate measurements of some angles and distances in the field to supplement the network diagram already prepared.

(2) As with all control surveys, the layout of stations in relation to the survey work for which they are intended must be carefully planned.

(3) At least two stations must be established such that they are very unlikely to be disturbed. These could then be used to reinstate the network if it was damaged or disturbed in some way.

(4) The precision and reliability of the network must be assessed.

Based on the reconnaissance, decisions regarding the measurements to be taken are made and the instruments to be used for the survey are specified. More importantly, a check should be made to ensure that the survey meets its specification and to ensure that the costs are acceptable.

Station Marks and Signals (Targets)

Upon completion of the reconnaissance, the survey stations are marked in some way.

The triangulation stations set up by the OS consist of an elaborate arrangement in which a metal plate is set into a concrete pillar, both of these being centred over an underground marker as in figure 1.17. Although this type of station construction could be used in engineering surveys, the cost is high and a less expensive pillar is shown in figure 14.3. Also shown in figure 14.3 is a suitable design for a mark set into a buried concrete block.

For surveys of a temporary nature (a few months only) wooden pegs can be used.

To enable angles and distances to be observed in a network, each station must have some form of target erected vertically above the station mark and reflecting prisms have to be set at each station. The type of target/prism used depends on the length of each line: suitable targets and prism sets are described in sections 3.7 and 5.6.

Distance Measurement

During the observation of a network, the lengths of as many of the triangle sides as possible are measured using some form of EDM equipment. When using the EDM equipment, the meteorological conditions at the time of measurement must be monitored carefully and suitable corrections made; also any systematic instrumental errors present in the equipment must be allowed for by careful calibration of the equipment. For surveys that are to be based on the National Grid, the scale factor is applied to each measured distance and, if the distance has been measured at an appreciable elevation, a height correction must be applied since mean sea level is the datum height for the National Grid.

All of the above corrections to EDM measurements are discussed in sections 5.19 to 5.23.

Angle Measurement

The instrument normally required for measurement of the angles in networks is a 0.1″/0.2″ or 1″ double reading optical micrometer theodolite as described in section 3.3 or an electronic theodolite of similar precision as described in section 3.4. The theodolite is set up and the angles are observed and booked in rounds using the methods given in sections 3.6 and 3.7.

Very often, a total station or theodolite-mounted EDM system is used to observe a network and distances and angles are measured simultaneously at each station.

Orientation

As in traversing, the North axis of the rectangular grid on which a network is based must be orientated to a specified direction. In engineering work, one of a number of north directions may be selected as described in section 1.5.

Generally, it is usual to set the scheme to align with one of the following.

(1) The National Grid, by using a baseline defined by two existing OS pillars. The coordinates of the points can be used to calculate the bearing of the baseline.
(2) Any other grid, by using existing points defined by another survey.
(3) Any other north direction to suit site conditions such as a structural or site grid (see section 14.6).

7.20 Network Computations: Least Squares

Nowadays, network coordinates are often calculated using methods based on least squares (see section 6.2). As already stated in section 7.18, adding more data than is needed to a survey makes the computations more complex but gives a stronger network through redundant observations. When dealing with these redundant observations, a step-by-step approach in which each figure throughout a network is adjusted and solved in turn can be used to obtain coordinates, but it is possible to obtain slightly different values depending on which route of computation is chosen.

A least squares adjustment, however, accounts for all angles and distances measured in a network and, making full use of all the redundancy in a network, performs a simultaneous adjustment of field data and calculation of coordinates. In other words, least squares will produce a single solution no matter how the original data is collected and processed. In addition to computing the best adjustment, least squares is also capable of providing a complete analysis of a survey including details of the positional accuracy of each coordinated station. This information can be used to detect errors and can be used at the planning stage to ensure that a survey meets its specification.

In summary, the advantages of using least squares to compute a survey are as follows.

(1) A mathematically correct solution is obtained for all types of network.
(2) A single solution is computed, no matter how complex the survey.
(3) Standard errors can be applied to all the observations and the effects of these can be included in the adjustment.
(4) It allows flexibility during data collection.
(5) Details of the accuracy of each point surveyed are obtained.
(6) The detection of gross errors in field data is made easier.
(7) Survey planning is possible.

For many years, least squares could only be implemented on a mainframe computer because, even for a relatively small network, the calculations are quite complicated. As a result, it was difficult to use least squares until personal computers of sufficient speed and storage capacity were developed. Using modern software techniques, least squares is now much easier to use and most networks can be designed, calculated and analysed using a desk-top or laptop computer with an adjustment program of some sort. Although it is possible to develop 'in-house' software for network analysis and computation, several commercial packages are available for this purpose and, for many civil engineering sites, the use of commercial software for network computations and analysis is increasing.

The application of least squares, sufficient for a thorough understanding of the subject is beyond the scope of this book and it is not included here. However, further reading is suggested at the end of chapter 6 for those requiring a specialist knowledge of the subject.

7.21 Network Computations: Equal Shifts

The semi-rigorous adjustment and computation of simple networks can be carried out by the method of equal shifts. Although superseded to some extent by the increasing use of least squares adjustment software, the method is included here as it produces perfectly acceptable results for general site work. It is stressed, however, that a full least squares adjustment must be carried out for complex, overlapping figures and where a first-order precision is required.

The computation of coordinates by equal shifts of two of the most commonly used figures in control surveys, the braced quadrilateral and centre-point polygon, are given in the following sections.

7.22 Worked Example: Adjustment and Computation of a Braced Quadrilateral

Question
The field abstract of figure 7.20 shows the observed angles for a braced quadrilateral PQRS. Using this data, calculate the coordinates of station R and S.

Solution
Four geometric conditions must be satisfied when adjusting the observed angles of a braced quadrilateral: three *angle conditions* and a *side condition*. The angle conditions are, referring to the numbering system of figure 7.20

(*a*) Σ angles 1 to 8 = 360°
(*b*) Angles 1 + 2 = Angles 5 + 6
(*c*) Angles 3 + 4 = Angles 7 + 8.

In this example

$$\Sigma \text{ angles 1 to 8} = 359°59'54''$$

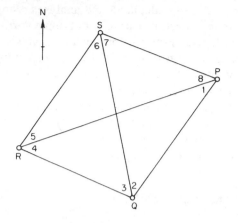

Angle	Observed value
1	30° 20' 50"
2	54 10 45
3	55 44 38
4	39 43 39
5	41 53 49
6	42 37 47
7	54 54 56
8	40 33 30

Station	Coordinates	
	mE	mN
P	1885.82	1632.47
Q	1401.00	1045.76

Figure 7.20   *Abstract for braced quadrilateral*

and 0.75″ is added to each angle to adjust them to 360°. This is shown in columns 1 and 2 of table 7.6 which shows the full adjustment.

Angle conditions (b) and (c) are known as *adjustments to opposites* and for figure 7.20

$$\text{Angles } 1 + 2 = 84°31'35''$$
$$\text{Angles } 5 + 6 = \underline{84°31'36''}$$
$$\text{Difference } \Delta \quad = \quad \underline{\quad 01''}$$

In order to satisfy the condition (angles $1 + 2 =$ angles $5 + 6$), each angle must be changed by an amount $\Delta/4 = 0.25''$. Since $(5 + 6) > (1 + 2)$ in this case, 0.25″ is subtracted from 5 and 6 and added to 1 and 2. These adjustments, including those for $3 + 4$ and $7 + 8$ are shown in column 3 and in the lower part of table 7.6.

Application of the adjustment to 360° and the adjustment to opposites gives the first adjusted angles of column 4.

The fourth geometric condition to be satisfied in a braced quadrilateral is a side condition of the form

$$|v''| = \frac{a - c}{(ab + cd) \sin 1''} \tag{7.13}$$

where

$|v''| =$ *magnitude* of a side adjustment to be applied to each first adjusted angle
$a = \sin 1 \times \sin 3 \times \sin 5 \times \sin 7$
$b = \cot 1 + \cot 3 + \cot 5 + \cot 7$
$c = \sin 2 \times \sin 4 \times \sin 6 \times \sin 8$
$d = \cot 2 + \cot 4 + \cot 6 + \cot 8$

TABLE 7.6
Equal Shifts Adjustment of Braced Quadrilateral

Angle	Observed value	Adjustment to 360°	Adjustment to opposites	First adjusted angles	Side adjustment	Final adjusted angles	Final angles
	(1)	(2)	(3)	(4)	(5)	(6)	(7)
1	30°20'50"	+0.75"	+0.25"	30°20'51.00"	+2.13"	30°20'53.13"	30°20'53"
2	54°10'45"	+0.75"	+0.25"	54°10'46.00"	-2.13"	54°10'43.87"	54°10'44"
3	55°44'38"	+0.75"	+2.25"	55°44'41.00"	+2.13"	55°44'43.13"	55°44'43"
4	39°43'39"	+0.75"	+2.25"	39°43'42.00"	-2.13"	39°43'39.87"	39°43'40"
5	41°53'49"	+0.75"	-0.25"	41°53'49.50"	+2.13"	41°53'51.63"	41°53'52"
6	42°37'47"	+0.75"	-0.25"	42°37'47.50"	-2.13"	42°37'45.37"	42°37'45"
7	54°54'56"	+0.75"	-2.25"	54°54'54.50"	+2.J3"	54°54'56.63"	54°54'57"
8	40°33'30"	+0.75"	-2.25"	40°33'28.50"	-2.13"	40°33'26.37"	40°33'26"
	359°59'54"			360°00'00.00"		360°00'00.00"	360°00'00"

Angles 1 + 2 = 84°31'35"   Angles 3 + 4 = 95°28'17" a =0.228 202 099

Angles 5 + 6 = 84°31'36"   Angles 7 + 8 = 95°28'26" b =4.206 102 699

Δ =   1" Δ =   9" c =0.228 221 840

$\frac{\Delta}{4}$ =   0.25" $\frac{\Delta}{4}$ =   2.25" d =4.179 873 602

$|v"|$ =2.13"

Final a = 0.228 212 011

c = 0.228 211 989 (check)

a, b, c, and d are all computed using the first adjusted angles.
    For PQRS of figure 7.20

$$a = 0.228\ 202\ 099 \qquad\qquad c = 0.228\ 221\ 840$$
$$b = 4.206\ 102\ 699 \qquad\qquad d = 4.179\ 873\ 602$$

which gives

$$|v''| = 2.1''$$

Since $c > a$ for this quadrilateral, the side adjustment of 2.1″ is subtracted from the angles used to compute c, that is, angles 2, 4, 6 and 8, and added to the angles used to compute a, that is, angles 1, 3, 5 and 7. The application of the side adjustment is shown in columns 5 and 6 of table 7.6.
    In the case where $a > c$ for a quadrilateral, the side adjustment is added to angles 2, 4, 6 and 8 and subtracted from angles 1, 3, 5 and 7.
    It is important to note that the application of the side adjustment given here (and the adjustments to opposites) refers only to the numbering sequence adopted in figure 7.20.
    The side adjustment is checked by computing further a and c values using the final adjusted angles of column 6. In *all* adjustments, a should equal c or very nearly so to give $|v''| = 0$ which indicates a properly satisfied side condition. Since the final a and c values in table 7.6 agree, when rounded, to the seventh decimal place, the side adjustment has been applied correctly. If the final a and c values in any side adjustment do not agree to at least the sixth decimal place, the adjustment has not been applied correctly and should be checked. A common mistake is to allocate the incorrect $+$ or $-$ sign to the adjustment such that it is added when it should have been subtracted and vice versa.
    Although not always necessary, column 7 shows the final adjusted angles of column 6 rounded to the same precision as the original observations.

    The procedures involved in the calculation of the coordinates of R and S are as follows.
    The baseline length D_{PQ} and bearing θ_{PQ} for PQRS are given by (see section 1.5)

$$D_{PQ} = [(E_Q - E_P)^2 + (N_Q - N_P)^2]^{\frac{1}{2}} = 761.104 \text{ m}$$

$$\theta_{PQ} = \tan^{-1}\left[\frac{E_Q - E_P}{N_Q - N_P}\right] + 180° = 219°34'05''$$

By application of the Sine Rule to triangle PQR, the following can be written

$$\frac{\sin 1}{D_{QR}} = \frac{\sin(2+3)}{D_{PR}} = \frac{\sin 4}{D_{PQ}}$$

or

$$D_{PR} = \frac{\sin(2 + 3)}{\sin 4} D_{PQ}$$

and

$$D_{QR} = \frac{\sin 1}{\sin 4} D_{PQ}$$

Since all the angles have been observed in the quadrilateral, these equations give the unknown side lengths D_{PR} and D_{QR}. A similar set of calculations in triangle PQS will give the unknown side lengths in that triangle.

The other triangles QRS and PSR in the quadrilateral should also be solved since these triangles provide a check on the side length RS and show the consistency in calculating the side lengths QS and PR.

Table 7.7 shows the complete calculation of the side lengths in PQRS using the final adjusted angles already calculated.

The bearings (θ) of PQRS can be evaluated as follows.
For triangle PQR

$$\begin{array}{ll}
\theta_{PQ} = 219°34'05'' & \theta_{QP} = 39°34'05'' \\
+ \; 1 \; = 30°20'53'' & -(2 + 3) = 109°55'27'' \\
\hline
\theta_{PR} = 249°54'58'' & \theta_{QR} = 289°38'38''
\end{array}$$

$$\theta_{RQ} - \theta_{RP} = 109°38'38'' - 69°54'58'' = 39°43'40'' = \text{angle } 4$$
$$\text{(check)}$$

For triangle PQS

$$\begin{array}{ll}
\theta_{PQ} \; = 219°34'05'' & \theta_{QP} = 39°34'05'' \\
+(1 + 8) \; = 70°54'19'' & - \; 2 = 54°10'44'' \\
\hline
\theta_{PS} \; = 290°28'24'' & \theta_{QS} = 345°23'21''
\end{array}$$

$$\theta_{SQ} - \theta_{SP} = 165°23'21'' - 110°28'24'' = 54°54'57'' = \text{angle } 7$$
$$\text{(check)}$$

When all the lengths and bearings of the quadrilateral sides have been computed, the coordinates of R and S are evaluated by computing traverses to include these unknown stations. The method is identical to that for a link traverse (see section 7.16) and the coordinates of R and S are derived in traverses QRP and QSP as shown in table 7.8.

7.23 Worked Example: Adjustment and Computation of a Centre-Point Triangle

Question
The field abstract for a triangulation scheme established for a small con-

SURVEYING FOR ENGINEERS

TABLE 7.7

Side Length Calculation for a Braced Quadrilateral

Triangle	Side	Formula	Station	Angle	Horizontal Length (m)
PQR	PQ	Baseline			761.104
			P	1 = 30°20'53"	
			Q	2 + 3 = 109°55'27"	
			R	4 = 39°43'40"	
				180°00'00"	
	PR	D_{PQ} sin (2 + 3)/sin 4			1119.546
	QR	D_{PQ} sin 1/sin 4			601.666
PQS	PQ	Baseline			761.104
			P	1 + 8 = 70°54'19"	
			Q	2 = 54°10'44"	
			S	7 = 54°54'57"	
				180°00'00"	
	PS	D_{PQ} sin 2/sin 7			754.165
	QS	D_{PQ} sin (1 + 8)/sin 7			878.919
QRS	QR	Baseline			601.666
			Q	3 = 55°44'43"	
			R	4 + 5 = 81°37'32"	
			S	6 = 42°37'45"	
				180°00'00"	
	QS	D_{QR} sin (4 + 5)/sin 6			878.922 (check)
	RS	D_{QR} sin 3/sin 6			734.296
PSR	PS	Baseline			754.165
			P	8 = 40°33'26"	
			S	6 + 7 = 97°32'42"	
			R	5 = 41°53'52"	
				180°00'00"	
	PR	D_{PS} sin (6 + 7)/sin 5			1119.544 (check)
	SR	D_{PS} sin 8/sin 5			734.293 (check)

struction site is shown in figure 7.21. Using this data, calculate the coordinates of stations S and W.

Solution

The procedure for solving the centre-point triangle is as follows.

(1) The observed angles are first adjusted using the equal shifts method as shown in table 7.9. The geometric conditions to be satisfied in a centre-point triangle are

  (a) the angles in any triangle must sum to 180° (refer to columns 1 and 2 in table 7.9)

TABLE 7.8
Coordinate Calculations in a Braced Quadrilateral

Triangle	Side	Bearing	Horizontal Length (m)	ΔE(m)	ΔN(m)	Co-ordinates mE	Co-ordinates mN	Station
PQR						1401.00	1045.76	Q
	QR	289°38'38"	601.666	-566.649	202.264			
						834.35	1248.02	R
	RP	69°54'58"	1119.545	1051.466	384.447			
						1885.82	1632.47	P
						(check)		
PQS						1401.00	1045.76	Q
	QS	345°23'21"	878.921	-221.710	850.498			
						1179.29	1896.26	S
	SP	110°28'24"	754.165	706.528	-263.785			
						1885.82	1632.47	P
						(check)		

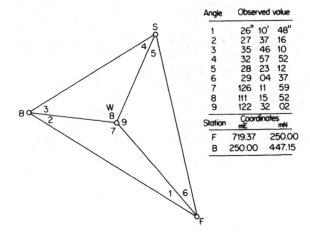

Angle	Observed value
1	26° 10' 48"
2	27 37 16
3	35 46 10
4	32 57 52
5	28 23 12
6	29 04 37
7	126 11 59
8	111 15 52
9	122 32 02

Station	Coordinates mE	mN
F	719.37	250.00
B	250.00	447.15

Figure 7.21 *Abstract for centre-point triangle*

(b) the angles at the centre station must sum to 360° without altering any previous adjustment (see columns 3 and 4)

(c) the side condition $|v''| = \dfrac{a - c}{(ab + cd) \sin 1''}$

where

$$a = \sin 1 \times \sin 3 \times \sin 5 \qquad b = \cot 1 + \cot 3 + \cot 5$$
$$c = \sin 2 \times \sin 4 \times \sin 6 \qquad d = \cot 2 + \cot 4 + \cot 6$$

The application of the side condition is shown in columns 5, 6 and 7 of table 7.9.

(2) The coordinates of the baseline FB are used to give $\theta_{FB} = 292°47'02''$ and $D_{FB} = 509.09(4)$ m. Since the coordinates of F and B are given for the network, they must NOT be altered.

(3) The lengths of the sides of all the triangles are calculated as shown in table 7.10.

(4) The bearings of the sides of all the triangles are calculated in the sequences given below.

For triangle FBW

$$
\begin{array}{ll}
\theta_{FB} = 292°47'02'' & \theta_{BF} = 112°47'02'' \\
+1 \;\;= \;\;26°10'45'' & -2 \;\;= \;\;27°37'16'' \\
\hline
\theta_{FW} = 318°57'47'' & \theta_{BW} = \;\;85°09'46''
\end{array}
$$

$$\theta_{WB} - \theta_{WF} = 126°11'59'' = \text{angle } 7 \qquad \text{(check)}$$

For triangle WBS

$$
\begin{array}{ll}
\theta_{WB} = 265°09'46'' & \theta_{BW} = \;\;85°09'46'' \\
+8 \;\;= 111°15'55'' & -3 \;\;= \;\;35°46'10'' \\
\hline
\theta_{WS} = \;\;16°25'41'' & \theta_{BS} = \;\;49°23'36''
\end{array}
$$

$$\theta_{SB} - \theta_{SW} = 32°57'55'' = \text{angle } 4 \qquad \text{(check)}$$

For triangle FWS

$$
\begin{array}{ll}
\theta_{FW} = 318°57'47'' & \theta_{WF} = 138°57'47'' \\
+6 \;\;= \;\;29°04'41'' & -9 \;\;= 122°32'06'' \\
\hline
\theta_{FS} = 348°02'28'' & \theta_{WS} = \;\;16°25'41''
\end{array}
$$

$$\theta_{SW} - \theta_{SF} = 28°23'13'' = \text{angle } 5 \qquad \text{(check)}$$

(5) The coordinates of W and S are evaluated in link traverses BWF and BSF as shown in table 7.11.

7.24 Intersection and Resection

Two techniques commonly employed in extending horizontal control surveys and in setting out are intersection and resection.

Intersection is a method of locating a point without actually occupying it. In figure 7.22 points A and B are stations in a control network already

TABLE 7.9

Equal Shifts Adjustment of a Centre-point Triangle

Angle	Observed Value (1)	Adjustment to 180° (2)	Adjusted Centre Angles (3)	Adjustment to 360° (4)	First Adjusted Angles (5)	Side Adjustment (6)	Final Adjusted Angles (7)	Final Angles (8)
1	26°10'48"	-1.0"		-0.5"	26°10'46.5"	-1.3"	26°10'45.2"	26°10'45"
2	27°37'16"	-1.0"		-0.5"	27°37'14.5"	+1.3"	27°37'15.8"	27°37'16"
7	126°11'59"	-1.0"	126°11'58"	+1.0"	126°11'59"		126°11'59"	126°11'59"
	180°00'03"				180°00'00.0"			
3	35°46'10"	+2.0"		-0.5"	35°46'11.5"	-1.3"	35°46'10.2"	35°46'10"
4	32°57'52"	+2.0"		-0.5"	32°57'53.5"	+1.3"	32°57'54.8"	32°57'55"
8	111°15'52"	+2.0"	111°15'54"	+1.0"	111°15'55"		111°15'55"	111°15'55"
	179°59'54"				180°00'00.0"			
5	28°23'12"	+3.0"		-0.5"	28°23'14.5"	-1.3"	28°23'13.2"	28°23'13"
6	29°04'37"	+3.0"		-0.5"	29°04'39.5"	+1.3"	29°04'40.8"	29°04'41"
9	122°32'02"	+3.0"	122°32'05"	+1.0"	122°32'06"		122°32'06"	122°32'06"
	179°59'51"		359°59'57"		180°00'00.0"			

$a = \sin 1 \sin 3 \sin 5$ $= 0.122\ 607\ 224$

$b = \cot 1 + \cot 3 + \cot 5$ $= 5.272\ 607\ 166$

$c = \sin 2 \sin 4 \sin 6$ $= 0.122\ 599\ 318$

$d = \cot 2 + \cot 4 + \cot 6$ $= 5.251\ 373\ 055$

$$|v''| = \frac{a - c}{(ab + cd)\sin 1''} = 1.3'' \quad a > c$$

Final $a = 0.122\ 603\ 150$

$c = 0.122\ 603\ 375$ (check)

TABLE 7.10

Calculation of Side Lengths in a Centre-point Triangle

Triangle	Side	Formula	Station	Angle	Horizontal Length (m)
FBW	FB	Baseline			509.094
			F	1 = 26°10'45"	
			B	2 = 27°37'16"	
			W	7 =126°11'59"	
				180°00'00"	
	FW	D_{FB} sin 2/sin 7			292.489
	WB	D_{FB} sin 1/sin 7			278.330
WBS	WB	Baseline			278.330
			W	8 = 111°15'55"	
			B	3 = 35°46'10"	
			S	4 = 32°57'55"	
				180°00'00"	
	BS	D_{WB} sin 8/sin 4			476.685
	WS	D_{WB} sin 3/sin 4			298.992
WFS	WF	Baseline			292.489
			W	9 = 122°32'06"	
			F	6 = 29°04'41"	
			S	5 = 28°23'13"	
				180°00'00"	
	FS	D_{WF} sin 9/sin 5			518.667
	WS	D_{WF} sin 6/sin 5			298.996 (check)

TABLE 7.11

Coordinate Calculations in a Centre-point Triangle

Triangle	Side	Bearing	Horizontal Length (m)	ΔE(m)	ΔN(m)	Co-ordinates m_E	m_N	Station
BWF						250.00	447.15	B
	BW	85°09'46"	278.330	277.339	23.470			
						527.34	470.62	W
	WF	138°57'47"	292.489	192.032	-220.620			
						719.37	250.00	F
BSF						250.00	447.15	B
	BS	49°23'36"	476.685	361.897	310.256			
						611.90	757.41	S
	SF	168°02'28"	518.667	107.473	-507.410			
						719.37	250.00	F

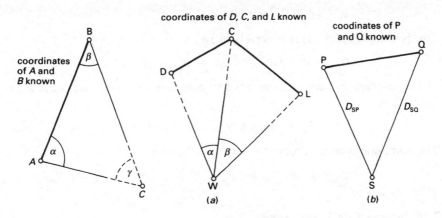

Figure 7.22 *Intersection*   Figure 7.23 *(a) Angular resection; (b) distance resection*

surveyed and, in order to coordinate unknown point C which lies at the
intersection of the lines from A and B, angles α and β are observed.

Resection is a method of locating a point by taking observations from it
to other known stations in a network. In figure 7.23a, point W can be fixed
by observing angles α and β subtended at resection point W by control
stations D, C and L. This type of resection could be used if stations D, C
and L were inaccessible. If EDM equipment is used and two accessible control stations are available, a distance resection can be performed by measuring the distances to each station from the unknown point (D_{SP} and D_{SQ} in
figure 7.23b).

As with other types of network surveying, well-conditioned figures must
be used for intersection and resection if the best results are to be obtained.

7.25 Intersection by Solution of Triangle

In triangle ABC of figure 7.22, the length and bearing of baseline AB are
given by (see section 1.5)

$$D_{AB} = [(E_B - E_A)^2 + (N_B - N_A)^2]^{\frac{1}{2}}$$

$$\theta_{AB} = \tan^{-1}\left[\frac{E_B - E_A}{N_B - N_A}\right]$$

The Sine Rule gives

$$D_{BC} = \frac{\sin \alpha}{\sin \gamma} D_{AB} \qquad D_{AC} = \frac{\sin \beta}{\sin \gamma} D_{AB}$$

where

$$\gamma = 180° - (\alpha + \beta)$$

The bearings in the triangle are given by

$$\theta_{AC} = \theta_{AB} + \alpha \qquad \theta_{BC} = \theta_{BA} - \beta$$

These bearings and distances are used to compute the coordinates of A along line AC as

$$E_C = E_A + D_{AC} \sin \theta_{AC} \qquad N_C = N_A + D_{AC} \cos \theta_{AC}$$

The computations are checked along line BC using

$$E_C = E_B + D_{BC} \sin \theta_{BC} \qquad N_C = N_B + D_{BC} \cos \theta_{BC}$$

7.26 Intersection Using Angles

Adopting the clockwise lettering sequence used in figure 7.22, the coordinates of C can be obtained directly from

$$E_C = \frac{(N_B - N_A) + E_A \cot \beta + E_B \cot \alpha}{\cot \alpha + \cot \beta} \qquad (7.14)$$

$$N_C = \frac{(E_A - E_B) + N_A \cot \beta + N_B \cot \alpha}{\cot \alpha + \cot \beta} \qquad (7.15)$$

A disadvantage of this method compared with solving the triangle is that no check is possible on the calculations.

7.27 Intersection Using Bearings

If the bearings of lines AC (θ_{AC}) and BC (θ_{BC}) in figure 7.22 are known, the coordinates of C are given by

$$E_C = \frac{E_A \cot \theta_{AC} - E_B \cot \theta_{BC} - N_A + N_B}{\cot \theta_{AC} - \cot \theta_{BC}} \qquad (7.16)$$

$$N_C = \frac{N_A \tan \theta_{AC} - N_B \tan \theta_{BC} - E_A + E_B}{\tan \theta_{AC} - \tan \theta_{BC}} \qquad (7.17)$$

As with intersection using angles, no check on the computations is possible.

7.28 Intersection from Two Baselines

When solving intersections using the formulae given in the previous sections, two quantities are observed in each case (α and β or θ_{AC} and θ_{BC}) to define two unknowns E_C and N_C. Consequently, no redundancy exists in the fixation and it is not possible to check the observations. It is, of course, possible to check the computations when solving the triangle but this method does not enable the angles α and β to be checked.

One method of detecting gross errors in the observations is to observe additional angles from a second baseline. This is shown in figure 7.24 where the angles δ and ϕ have been added to those already observed in figure 7.22.

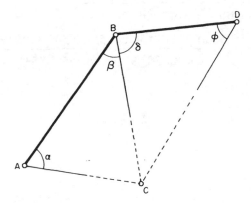

Figure 7.24   *Intersection from two baselines*

The coordinates of point C in figure 7.24 are found by solving the intersections formed by the triangles ABC and BDC, the two sets of coordinates obtained being compared. If the differences between the two intersections are small, it is assumed that the observations contain no gross errors and the average coordinates from the two sets are taken as the final values.

7.29 Worked Example: Intersection

Question
The coordinates of stations S, A and L are $E_s = 1309.12$ m E, $N_s = 1170.50$ m N, $E_A = 1525.43$ m E, $N_A = 958.87$ m N, $E_L = 1231.08$ m E and $N_L = 565.81$ m N. Calculate the coordinates of point B which has been located by intersection from stations S, A and L by observing the following angles: $B\hat{S}A = 85°38'49''$, $S\hat{A}B = 55°50'53''$, $B\hat{A}L = 41°41'48''$ and $A\hat{L}B = 68°09'32''$.

Solution
Referring to figure 7.25 and clockwise triangle SAB, the coordinates of B are given by the angles method (see section 7.26) as

$$E_B = \frac{(N_A - N_S) + E_S \cot S\hat{A}B + E_A \cot B\hat{S}A}{\cot B\hat{S}A + \cot S\hat{A}B}$$

$$= \frac{(958.87 - 1170.50) + 1309.12 \cot 55°50'53'' + 1525.43 \cot 85°38'49''}{\cot 85°38'49'' + \cot 55°50'53''}$$

$$= 1050.45 \text{ m}$$

$$N_B = \frac{(E_S - E_A) + N_S \cot S\hat{A}B + N_A \cot B\hat{S}A}{\cot B\hat{S}A + \cot S\hat{A}B}$$

$$= \frac{(1309.12 - 1525.43) + 1170.50 \cot 55°50'53'' + 958.87 \cot 85°38'49''}{\cot 85°38'49'' + \cot 55°50'53''}$$

$$= 862.45 \text{ m}$$

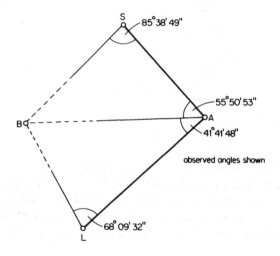

Figure 7.25 *Worked example: intersection*

To check the fieldwork and computations, the intersection of B in triangle BAL must also be computed as follows

$$E_B = \frac{(N_L - N_A) + E_A \cot A\hat{L}B + E_L \cot B\hat{A}L}{\cot B\hat{A}L + \cot A\hat{L}B} = 1050.50 \text{ m}$$

$$N_B = \frac{(E_A - E_L) + N_A \cot A\hat{L}B + N_L \cot B\hat{A}L}{\cot B\hat{A}L + \cot A\hat{L}B} = 862.46 \text{ m}$$

Since the two results for E_B and N_B agree within 0.05 m, no gross error has occurred in the observations and the final coordinates are the mean values from the two sets, that is

$$E_B = 1050.48 \text{ m} \quad N_B = 862.46 \text{ m}$$

7.30 Angular Resection

This resection is carried out in the field by observing the angles subtended at the unknown point by at least three known stations and in the *three-point resections* shown in figure 7.26, P is located in each case by measurement of angles α and β.

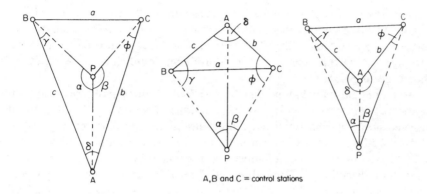

Figure 7.26 *Possible configurations for a three-point angular resection*

A three-point resection can be solved in a number of ways. However, no matter which method is used, it must be noted that if points A, B, C and P in figure 7.26 all lie on the circumference of the same circle then the resection is indeterminate. This condition is present when $\delta + \alpha + \beta = 180°$.

One method of solving a three-point resection is as follows. In each quadrilateral ABPC of figure 7.26

$$\alpha + \beta + \gamma + \phi + \delta = 360°$$

or

$$\gamma = [360° - (\alpha + \beta + \delta)] - \phi = R - \phi$$

where R can be deduced.

In triangles ABP and APC

$$D_{AP} = \frac{\sin \gamma}{\sin \alpha} c = \frac{\sin \phi}{\sin \beta} b$$

From which

$$K = \frac{\sin \gamma}{\sin \phi} = \frac{b \sin \alpha}{c \sin \beta}$$

which can be evaluated.

Substituting $\gamma = R - \phi$ gives a further expression for K

$$K = \frac{\sin(R - \phi)}{\sin \phi} = \frac{\sin R \cos \phi - \cos R \sin \phi}{\sin \phi} = \sin R \cot \phi - \cos R$$

Therefore

$$\cot \phi = \frac{K + \cos R}{\sin R}$$

This expression enables ϕ and all the angles in ABPC to be found, which in turn enables the coordinates of P to be calculated by solving triangles ABP and APC (see section 7.25). Both triangles are solved in order to provide a check on the calculations since the coordinates found for P in each triangle should be identical.

Although the calculations can be checked in three-point resections, the fieldwork cannot be checked since a unique position is obtained for the resected point by observing only three directions and deriving two angles.

To introduce some redundant data into a resection requires further directions to be observed and for most engineering surveys it is normal to observe four directions (giving three angles), the extra angle being used to check the fieldwork. The method of applying this check in a *four-point resection* is as follows.

(1) Choose three directions out of the four observed and compute a three-point resection. The observed angles (or combinations of these) that give the two resection angles nearest to 90° should be used in this calculation.
(2) Using the coordinates of P found in (1), *calculate* the value of one of the angles not used in the three-point resection.
(3) Compare the angle calculated in (2) with its observed value. If the two are in close agreement, it is assumed that no gross error has occurred in the observations and the resection coordinates obtained in (1) are accepted for further work.

7.31 Worked Example: Four-point Resection

Question
Using the resection data given in table 7.12, calculate the coordinates of point A.

Solution
The layout of the four control stations and point A are shown in figure 7.27. Using this, observed directions AM, AT and AW are selected for the coordinate calculation since the geometry of these directions gives resection angles at A closest to the optimum of 90°. The remaining direction AP will be used to check the fieldwork.

TABLE 7.12

Station	m_E	m_N	Angle	Observed value
M	845.11	1952.50	MÂP	30°40′11″
P	1312.59	2205.90	PÂT	26°47′52″
T	1621.29	1835.07	TÂW	56°47′08″
W	1729.04	1158.60		

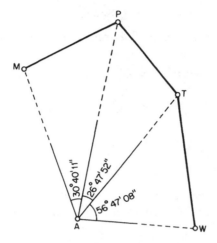

Figure 7.27 *Worked example: four-point resection*

For the resection formed at A by M, T and W (see figure 7.28), angle γ is given by

$$\gamma = \cot^{-1}\left[\frac{K + \cos R}{\sin R}\right]$$

where

$$K = \frac{D_{MT}\sin\beta}{D_{WT}\sin\alpha} = \frac{785.01\ \sin 56°47′08″}{685.00\ \sin 57°28′03″} = 1.137\ 219$$

and

$$R = 360° - (\alpha + \beta + \delta) = 138°05′37″$$

which gives

$$\gamma = \cot^{-1}\left[\frac{1.137\ 219 + \cos 138°05′37″}{\sin 138°05′37″}\right] = \cot^{-1}\ (0.588\ 372)$$

$$= 59°31′43″$$

The coordinates of A are found by solving triangle AMT as follows

$$M\hat{T}A = 180° - (\gamma + \alpha) = 180° - (59°31'43'' + 57°28'03'') = 63°00'14''$$

$$D_{MA} = \frac{\sin M\hat{T}A}{\sin \alpha} D_{MT} = \frac{\sin 63°00'14''}{\sin 57°28'03''} (785.01) = 829.66 \text{ m}$$

$$\theta_{MA} = \theta_{MT} + \gamma = 98°36'11'' + 59°31'43'' = 158°07'54''$$

$$E_A = E_M + D_{MA} \sin \theta_{MA} = 1154.14 \text{ m}$$

$$N_A = N_M + D_{MA} \cos \theta_{MA} = 1182.54 \text{ m}$$

A check on the *computations only* is provided by solving triangle ATW in which

$$\phi = R - \gamma = 78°33'54'' \quad A\hat{T}W = 180° - (\phi + \beta) = 44°38'58''$$

$$D_{WA} = \frac{\sin A\hat{T}W}{\sin \beta} D_{WT} = 575.40 \text{ m} \quad \theta_{WA} = \theta_{WT} - \phi = 272°23'05''$$

$$E_A = E_W + D_{WA} \sin \theta_{WA} = 1154.14 \text{ m}$$

$$N_A = N_W + D_{WA} \cos \theta_{WA} = 1182.54 \text{ m}$$

The *observations* for the resection are checked, in this example, by comparing the observed and calculated values for angle $M\hat{A}P$.

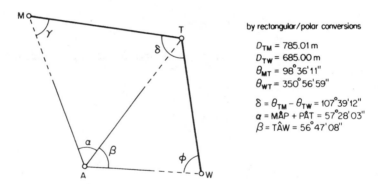

by rectangular/polar conversions

$D_{TM} = 785.01 \text{ m}$
$D_{TW} = 685.00 \text{ m}$
$\theta_{MT} = 98°36'11''$
$\theta_{WT} = 350°56'59''$

$\delta = \theta_{TM} - \theta_{TW} = 107°39'12''$
$\alpha = M\hat{A}P + P\hat{A}T = 57°28'03''$
$\beta = T\hat{A}W = 56°47'08''$

Figure 7.28

The coordinates found above for A give, by calculation

$$\theta_{AP} = 08°48'05'' \qquad \theta_{AM} = 338°07'53''$$

from which

$$M\hat{A}P = \theta_{AP} - \theta_{AM} = 30°40'12''$$

Since MÂP (observed) = 30°40′11″, a difference of only 1″ exists between the two values and therefore no gross errors have occurred. Hence, the co-ordinates of A are

$$E_A = 1154.14 \text{ m}, \qquad N_A = 1182.54 \text{ m}$$

7.32 Distance Resection

This type of resection is usually carried out using EDM equipment or a total station. Referring to figure 7.29, point P is fixed by measurement of distances D_{PW} and D_{PF}. To solve for the coordinates of P, the angles in triangle PWF are calculated using the cosine rule, remembering that D_{WF} is obtained from the coordinates of W and F. All three angles should summate to 180° and, if α is measured during the resection, this can also be used to check the angle calculations. If the angles and lengths of triangle PWF are known, the coordinates of P can be calculated in the same way as that de-scribed in section 7.25 for an intersection by solution of triangle.

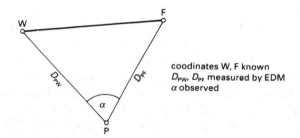

Figure 7.29 *Distance resection*

In order to check fieldwork, a second resection can be observed and cal-culated using different control stations. In practice, a third control station is usually introduced and a resection carried out with this and either station W or F.

The majority of total stations currently available include a software func-tion for performing a distance resection (see section 5.11).

8

Satellite Position Fixing Systems

Another method of determining horizontal and three-dimensional position for engineering surveys is by processing measurements from artificial Earth satellites. Although a number of different systems can be used for satellite positioning, the TRANSIT system and *Global Positioning System* (GPS) are the two that have had most engineering applications.

The TRANSIT system, also known as NAVSAT or NNSS, was developed in the 1960s by the United States Navy for updating the positions of submarines. Although intended as a military system, civilian use of TRANSIT was permitted in 1967 and various methods for locating the positions of fixed points have been developed since then. At best, TRANSIT produces standard errors of about 0.3 m in coordinate differences, but this is possible over several hundred kilometres. Consequently, its main applications in surveying have been in strengthening existing national triangulation networks and in positioning offshore structures.

The development of GPS (also known as NAVSTAR for NAVigation System using Timing And Ranging) began in 1973. Designed primarily for military users, GPS is managed and is under the control of the United States Department of Defense (US DoD). Compared with TRANSIT, which only gives intermittent position fixes as a satellite passes overhead, GPS has been developed so that a user at any point on or near the Earth can obtain three-dimensional coordinates instantaneously. These fixes can be taken at any time of the day or night and in any weather conditions.

The accuracy of GPS equipment and methods continues to improve and its possible applications in engineering surveying are far greater than those of the TRANSIT system. For this reason, GPS is described in detail in the following sections.

8.1 GPS Space Segment

GPS consists of three segments called the space segment, control segment and user segment.

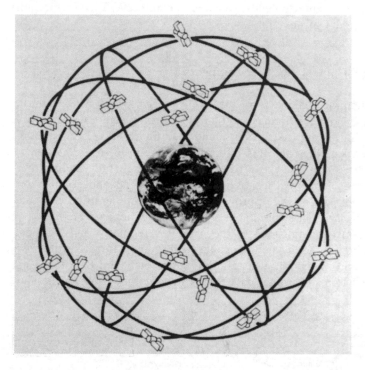

Figure 8.1   *GPS Space segment (courtesy Leica UK Ltd)*

When fully operational, the *space segment* (figure 8.1) will consist of 24 satellites all of which will be in orbits at an altitude of 20 200 km. At this altitude, each satellite will orbit the Earth every 12 hours and this, together with a suitable choice of orbital plane for each satellite, ensures that at least four (the minimum needed for a position fix) will be in view at any time.

All satellites in the constellation transmit two L-band signals known as L1 and L2 of frequencies 1575.42 MHz and 1227.60 MHz respectively. L1 is modulated (see section 5.4) with two binary codes referred to as the C/A (coarse acquisition) and P (precise) codes and a data message. The data message consists of an almanac giving the approximate position of all the satellites in the system, the satellite ephemeris which contains precise information about the position of the host satellite and subsidiary information such as clock corrections and the status of the system. The L2 carrier is modulated with the P-code and data message only.

The C/A-code allows access to the *Standard Positioning Service* which has an intended accuracy for single-point positioning of the order of 150 m and the P-code allows access to the *Precise Positioning Service* which has an intended accuracy of about 15 m. By giving an approximate position, the C/A-code is able to help a receiver acquire the P-code for more precise measurement of position.

8.2 GPS Control and User Segments

Five monitor stations form the *control segment* of GPS: a master control station at Colorado Springs and four other control stations at Hawaii, Kwajalein, Diego Garcia and Ascension Island (see figure 8.2). Each station tracks all the GPS satellites and this information is relayed to the master control station where it is used to predict future orbits for all satellites. In addition, the clock on board each satellite is monitored and comparison with the GPS clock at Colorado Springs enables corrections to be computed to keep satellite clocks in step with GPS time. The ephemeris predictions and clock corrections are uploaded to the satellites regularly and the data message transmitted by each satellite changes every hour. In case problems arise with the tracking network, each satellite stores sufficient data to be able to predict and transmit orbital data for 14 days without any update.

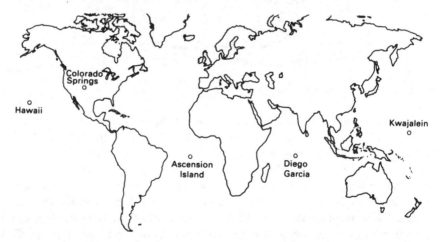

Figure 8.2 *GPS control segment*

The GPS *user segment* consists of all civil and military users of the system. In addition to land surveyors, the number of other civilian users of GPS is considerable since it is capable of dynamic positioning and has applications in hydrographic surveying, vehicle navigation and all forms of aviation.

8.3 GPS Positioning Methods: Pseudo-ranging

In its simplest form, GPS positioning is carried out with a single receiver determining pseudo-ranges.

Both the C/A and P-codes transmitted by each GPS satellite are digital pseudo-random timing codes. The C/A-code has a frequency of 1.023 MHz and repeats every 1 ms, whereas the P-code has a frequency of 10.23 MHz but is 38 weeks long. At a frequency of 1.023 MHz, the spacing between binary digits on the C/A-code is about 300 m and for the P-code, the spacing is about 30 m. These spacings (or the frequency of the codes) dictate what accuracy is possible from pseudo-range measurements with GPS. Since the C/A-code repeats every 1 ms, it is easy for a ground receiver to acquire without knowing the pseudo-random sequence. However, unless a user has prior knowledge of the P-code, it is impossible to decode because of the long pseudo-random sequence involved. This helps the US DoD to deny unauthorised users access to the P-code and hence to the precise positioning service.

When a receiver locks onto a satellite, the incoming signal triggers the receiver to generate a C/A-code identical to that produced by the satellite. The replica code generated by the receiver is then compared with the satellite code in a process known as cross-correlation in which the receiver code is shifted until it is in phase with the satellite code. The amount by which the receiver-generated code is shifted is equal to the transit time of the signal between satellite and receiver. Multiplied by the speed of light, this gives the distance between satellite and receiver. If measurements are to be taken using the P-code, the receiver repeats the cross-correlation process and the so-called 'hand-over word' instructs the receiver which portion of the P-code to generate.

All GPS satellites are fitted with very accurate caesium clocks which are all kept synchronised in GPS time by applying frequent corrections. On the other hand, receiver clocks are of poorer quality and are not usually synchronised with GPS time. Consequently, the receiver generated codes contain a clock error which affects transit times. Because of this, all ranges measured by a receiver will be biased and are called *pseudo-ranges*.

As well as pseudo-range measurement, a receiver will also decode the data message and from this computes the position of the satellite at the time of measurement.

8.4 Calculation of Position

If pseudo-range measurements are taken to other satellites, computation of receiver position in a GPS survey is similar to a distance resection (see section 7.32). In this case, the satellites are control stations of known coordinates and distances have been measured to these from a receiver located at a point whose position is to be fixed as shown in figure 8.3. In figure 8.3, each range measured from point P to satellite S can be defined as

$$R_{sp} = c(\Delta t - e) \tag{8.1}$$

where

$$\begin{aligned}
R_{sp} &= \text{true range from satellite S} \\
c &= \text{speed of light} \\
\Delta t &= \text{transit time} \\
e &= \text{clock error.}
\end{aligned}$$

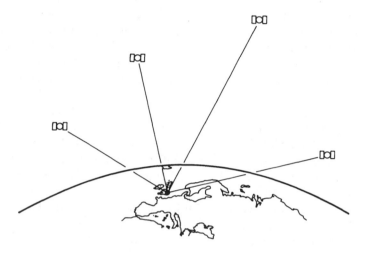

Figure 8.3   *Point positioning with GPS*

In terms of satellite coordinates X_s, Y_s and Z_s this becomes

$$[(X_s - X_p)^2 + (Y_s - Y_p)^2 + (Z_s - Z_p)^2]^{\frac{1}{2}} = c(\Delta t - e) \tag{8.2}$$

in which the transit time Δt is called the observable and unknowns X_p, Y_p, Z_p (the position of P) and e exist. Observations to four satellites will give the four equations in four unknowns to give the receiver position and clock error.

8.5 Ionospheric and Atmospheric Effects

So far, it has been assumed that the value of c, the speed of light, remains constant. As described for EDM in section 5.19, it is well known that the speed of an electromagnetic wave is affected by the medium through which it is passing. GPS signals, which are electromagnetic, propagate through the ionosphere and then the atmosphere both of which affect the speed of the signal. The effect of this is to introduce a timing error into the pseudo-range measurements which can produce the worst source of error in GPS point positioning. Propagation errors are reduced by taking measurements using both the L1 and L2 signals with a dual frequency receiver or by application of differential GPS which is described in section 8.7.

8.6 Accuracy Denial

When designing GPS, the US DoD intended that pseudo-range measurements with the C/A-code would only give an accuracy of about 150 m for absolute single-point positioning and that more precise positioning to about 15 m would only be possible using the P-code. This would enable the DoD to restrict the use of the P-code to the US military and would stop civilian or other users obtaining a high accuracy from GPS.

In practice, however, it has been possible to obtain accuracies as good as 10 m even with the C/A-code. This is looked upon as a security risk by the DoD because GPS could be used for precise positioning by those hostile towards the US. This has led to the implementation of an accuracy denial system by the DoD known as *Selective Availability* (SA). The effect of this is twofold: firstly, the accuracy of pseudo-ranges is made worse by applying a *dither* to satellite clocks so that transit times cannot be measured precisely and, secondly, the data message is altered so that satellite positions are computed incorrectly (this is known as *epsilon*). With SA, the accuracy of single-point positioning with GPS downgrades to about 100 m from 10 m.

As well as accuracy denial, it is also current DoD policy to deny the P-code to all users other than those authorised by the US military when GPS is fully operational. This will be carried out by implementing *anti-spoofing* (AS) in which the P-code is encrypted and changed to a secret Y-code. At present, the P-code is unrestricted.

For both accuracy denial and anti-spoofing, users of GPS approved by the US DoD will have receivers capable of removing dither and epsilon and will have access to the P- and Y-codes.

8.7 Differential GPS

At even the 10 m level of accuracy, GPS is of little use to the surveyor and methods have been developed to improve this. The most important of these is differential GPS in which two or more receivers work together (figure 8.4). One of the receivers is located at a precisely known point and, as it processes information from the satellites, it is able to compute a position

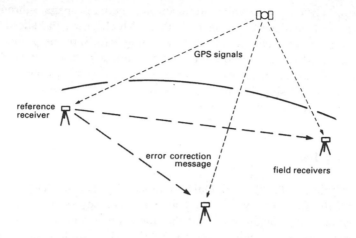

Figure 8.4 *Differential GPS*

based entirely on satellite data. This is compared with the known position for that point and any discrepancies are assumed to be due to atmospheric and ionospheric errors, incorrect satellite orbits and downgrading due to SA. Since any other receiver operating in the vicinity of the reference receiver will be using signals that propagated along similar paths from the same satellites, they will be affected by the same errors. So, the reference receiver calculates corrections based on its known and computed positions and these are transmitted to all the other receivers working in the area. With suitable post-processing software, the accuracy of differential GPS can approach 1 metre. Even this, however, is not good enough for most engineering surveying activities and, for a better accuracy, different techniques have to be adopted.

8.8 GPS Positioning Methods: Carrier Phase Measurements

In section 5.2, the method of distance measurement by phase comparison is described in which distance is given by $n\lambda + \Delta\lambda$. EDM instruments measure the fractional component of the distance $\Delta\lambda$ by comparing the phases of signals transmitted and received by the same instrument. Although the equipment and methods used in EDM are very different to GPS, GPS receivers also use phase comparison to measure distances to satellites but this is carried out by comparing the phase of an incoming satellite signal with a similar signal generated by the receiver. These phase measurements are taken on the L1 and L2 carrier waves with a resolution of about 1° and because the carriers have very short wavelengths of 0.19 m (L1 signal) and 0.24 m (L2 signal) it is possible to observe a range with millimetre precision.

At the start of a measurement, a GPS receiver must first remove the codes from the L1 and L2 signals so that it can access the carrier wave. This is usually done by either reconstructing the original carrier or by using a signal-squaring technique. In order to be able to *reconstruct the original carrier*, an exact knowledge of pseudo-random binary codes (usually the P-codes) is required. *Squaring techniques*, on the other hand, require no knowledge of codes (this is known as the codeless approach) and give a carrier, with codes eliminated, at twice the original frequency. Because of this, the squaring technique is capable of being more accurate since phase measurements are taken at half the original wavelength. Unfortunately, this method suffers the disadvantage that the squaring process destroys the data message and an external ephemeris must be used to obtain satellite positions.

The biggest problem with GPS phase measurements is determining the integer number of carrier wavelengths between satellite and receiver. In other words, the problem is how to find n in $n\lambda + \Delta\lambda$ in order to resolve the *carrier phase ambiguity*. Many techniques have been developed for estimating integer phase ambiguities both rapidly and reliably, the most successful of which have used strategies based on dual-frequency phase and P-code measurements. Unfortunately, the P-code will be encrypted when GPS is fully operational and will be restricted to authorised users only. As a result, some receivers are capable of switching to a signal squaring technique when P-code encryption is active.

When resolving GPS ambiguities, the satellites are always moving and phase differences change with time. However, most receivers measure phase continuously and will count the number of complete wavelengths due to relative motion between satellite and receiver from the time a receiver has locked on to a satellite.

8.9 Precise Differential Positioning and Surveying

This section describes how geodetic GPS receivers taking carrier phase measurements can achieve accuracies at the centimetre level.

One of the problems with carrier phase measurements is that, like pseudo-range measurements, they are subject to clock errors and propagation delays, the effect of which must be removed from any observations. Different methods can be adopted when post-processing satellite ranges so that clock and atmospheric errors cancel, ambiguities can be resolved and positional computations simplified. All of these are based on differential techniques which rely on at least two receivers collecting data for a period of time at different stations.

Differential methods enable the *double difference* equations to be formed by differencing range measurements with respect to satellites and ground stations (figure 8.5*a*) and the *triple difference* equations to be formed by differencing range measurements with respect to the satellites, ground stations and time (figure 8.5*b*).

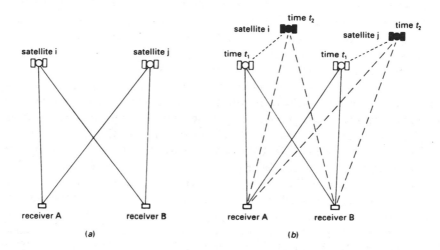

Figure 8.5   *Differential positioning: (a) double difference; (b) triple difference*

Both of these data sets are extensively used in GPS algorithms and for each point in a survey, a considerable amount of positional information is generated even in a relatively short observation period. Consequently, a least squares solution is necessary to deal with all the data collected and all GPS surveying relies entirely on sophisticated computer software to determine position. This software also performs other functions such as mission planning (which gives information as to when satellites are overhead), orbit improvement calculations, network adjustments, datum transformations and so on.

Static Differential Positioning

This was the first high-precision differential method developed for GPS. It requires at least two receivers at different locations to collect data for extended periods from 30 minutes to several hours. The reason a relatively long observation period is needed is to allow the satellite geometry to change sufficiently so that enough data is available to resolve integer ambiguities and to allow systematic errors to be removed. All the data collected is simultaneously post-processed to give the relative position between the two (or more) receivers used.

This method is the standard GPS method for determining the length of long baselines and has an accuracy of about 5–10 mm plus 1–2 ppm of the baseline length. Since these accuracies can be achieved over distances of several hundreds of kilometres, the static method is used extensively for establishing control networks that cover large areas.

For surveys that cover much smaller areas where baseline lengths are up to several kilometres only, some of the systematic errors in carrier phase measurements can be ignored. In such cases, the static method can be replaced with one of several other methods, all of which reduce the time of occupancy at each station.

Stop-and-go Surveying (Kinematic or Semi-kinematic Surveying)

In this method, a reference receiver remains at one end of a baseline while the second receiver is moved from the other end of the baseline to points whose coordinates are required. When the second or roving receiver stops at the unknown points, it collects data for periods that can vary from a few seconds to a few minutes. Before the roving receiver can move at the start of a survey, integer ambiguities must be resolved and a number of different techniques can be used to do this including static observations at both ends of the baseline, starting with known coordinates at each end of the baseline or by using a technique known as the antenna-swap method.

At all times during a stop-and-go survey, the roving receiver must maintain phase lock to at least four GPS satellites otherwise the survey will not be successful. This means that the method is of no use in areas where signal shading occurs, such as tunnels, wooded areas and in the vicinity of tall buildings or when the receiver is moved too fast.

The accuracy of coordinates computed from stop-and-go surveys is usually in the 10–30 mm range.

Pseudo-kinematic Surveying (Reoccupation Surveying)

This technique was developed to overcome the problem of maintaining lock on at least four satellites during an entire survey with the stop-and-go method. In the field, a similar procedure to stop-and-go surveying is followed and a reference receiver occupies one end of a baseline and a roving receiver occupies a series of remote, unknown, sites in sequence. The roving receiver collects data at each point surveyed for a few minutes but, as the alternative title suggests, the whole procedure is repeated within 30 minutes to two hours and all the remote sites are re-occupied. The time interval between the first and second runs can be critical and depends on the satellite geometry at the time measurements are taken.

During the survey, there is no need for the roving receiver to maintain phase lock on any satellites and it can even be switched off while moving between sites. This is clearly an advantage over stop-and-go surveying in areas where problems with signal reception occur. However, at each site to be surveyed, the usual requirement that four satellites be in view at all times still applies but this need not be the case between sites.

All of the data collected on the two runs is processed simultaneously using algorithms similar to those used in static positioning.

Pseudo-kinematic surveying works well when a large number of sites have to be visited as this tends to reduce waiting periods between station reoccupations. The accuracy of coordinates obtained using this method is similar to stop-and-go surveying and varies between 10 and 30 mm.

Rapid Static Surveying (Fast Static Surveying)

This is a technique similar to conventional static positioning but has oc-cupation times of minutes rather than hours.

The method relies on a faster ambiguity solution by one of two methods. The first of these combines P-code pseudo-range measurements with carrier phase measurements and involves search routines rather than a least squares solution to solve for ambiguities. Some GPS receivers are able to record data from all visible satellites simultaneously which produces redundant carrier phase measurements. The second rapid static method processes these using sophisticated statistical software to resolve ambiguities.

Rapid static GPS is capable of determining baseline lengths with an accu-racy of 10–30 mm.

Full Kinematic GPS

Sometimes known as On-the-Fly Ambiguity Resolution, this technique is not yet fully realised and is the subject of much research. When imple-mented, truly kinematic millimetric GPS surveying will enable integer am-biguities to be resolved instantly (even when the receiver is moving), will not require the receiver to lock onto the satellites at all times and will not require the operator to stop at unknown points.

8.10 GPS Coordinates and Heights

In engineering surveying, the horizontal positions of control and other points are defined using rectangular coordinates (see section 1.5). For many surveys, local grids are used that have arbitrary origins and north directions but the National Grid (see sections 1.5 and 1.9) can also be used.

GPS software always computes position from pseudo-ranges and carrier phase measurements on the global *WGS84 coordinate system*. A GPS user will, however, want position on a different system to this and WGS84 coordinates must be transformed to whatever local coordinate system is in use. For surveys based on the National Grid of Great Britain, GPS positions are transformed to the *OSGB36 coordinate system* established by the Ordnance Survey. For local surveys, a further transformation is required. These coordinate transformations are carried out using post-processing software.

All heights or reduced levels used for the majority of engineering surveys are referred to mean sea level or a line parallel to this (see section 2.1). Unfortunately, GPS heights are based on a different surface and because mean sea. level is rather irregular, it can be difficult to make corrections from one datum to the other. A full knowledge of geodesy is needed in order to understand how to apply such corrections and it may be necessary to use specialised GPS field procedures if first-order accuracy is required for heights.

8.11 GPS Instrumentation

Figure 8.6 shows the Wild GPS System 200, the 4000 SSE from Trimble Navigation and the Ashtech Z-12, a sample of GPS receivers now suitable for engineering surveys.

The *Wild GPS System 200* (figure 8.6a) from Leica has a 9-channel dual-frequency receiver which means it can track 9 satellites simultaneously and can take measurements on both L1 and L2 signals. It uses a reconstructed carrier in phase measurements but should the P-code become encrypted, it can switch to the signal squaring method. The System 200 supports all the measurement modes used for precise GPS surveying and, with their SKI post-processing software, the accuracy quoted by Leica for baseline measurements is 5 mm + 1 ppm of the baseline length. For single-point positioning with pseudo-ranges, the accuracy is 15 m subject to SA.

The *4000 SSE Geodetic Surveyor* from *Trimble Navigation* (figure 8.6b) is also a dual-frequency 9-channel receiver. Normally, it uses P-code measurements on both the L1 and L2 frequencies for ambiguity resolution but during periods of P-code encryption the receiver measures the cross-correlation of the encrypted P-codes in conjunction with the C/A-code instead. This combination of observables, according to the manufacturer, provides faster ambiguity resolution than squaring techniques. When used for static positioning, the 4000 SSE has a quoted accuracy of 5 mm + 1 ppm times the baseline length and when used in the various kinematic surveying modes,

(a)

(a)

(b)   (c)

Figure 8.6   *GPS instrumentation: (a) Wild GPS System 200 (courtesy Leica UK Ltd); (b) Trimble
Navigation 4000 SSE Geodetic Surveyor (courtesy Trimble Navigation Europe Ltd); (c)
Ashtech Z-12 (courtesy Ashtech)*

it has a quoted accuracy of 20 mm + 1 ppm of the baseline length. Data
processing for the 4000 SSE is carried out with a software package known
as GPSurvey.

The *Ashtech Z-12* (figure 8.6c) is a 12-channel GPS receiver that uses
the P-code on both L1 and L2 frequencies and the C/A-code to obtain carrier phase and pseudo-range measurements. These are all combined to resolve carrier phase ambiguities. When anti-spoofing (AS) is turned on, the

instrument automatically activates its Z-Tracking mode which cancels the effects of AS. The Z-12 has an accuracy quoted in millimetres, the exact figure depending on observation times and operating mode.

8.12 Applications of GPS

As can be seen from the preceding sections, GPS is a rather complex system that can be used in many ways. For basic point positioning and navigation, handheld receivers with an accuracy at the 100 m level have found widespread use while at the other end of the GPS spectrum, geodetic receivers with a computer and post-processing software are now starting to be used for routine survey work at the centimetre level.

Although the accuracy is important, some surveyors feel that the main advantage of GPS compared with conventional surveys is that it can be used in any weather conditions day or night. This enables GPS surveying to be carried out over extended periods at any time of the year without restrictions such as rain, fog and poor visibility delaying work. Another advantage when surveying with GPS is that intervisibility between stations or points surveyed is not necessary. This allows control stations to be placed where convenient and not at locations which may be difficult to get to in order to establish lines of sight.

At the moment, the full potential of GPS has not been realised even though the accuracy required for engineering surveys can be achieved. One of the reasons for this is the cost of GPS surveying which can be uneconomical compared with conventional surveying. These high costs are caused by, firstly, the receivers which are between five and ten times more expensive than total stations and, secondly, the fact that GPS is not fully kinematic and there are problems with satellite coverage, both of which can result in long occupation times. Added to this, there are difficulties in defining heights above survey datums such as mean sea level and with real-time data processing and control.

Despite these drawbacks, GPS has been very successfully used for control surveys, where it has joined traversing, triangulation and trilateration as a method for coordinating stations in a network. The best applications identified so far for GPS have been for improving existing national control networks and for surveys in remote areas. GPS is also used on engineering projects that extend over large areas, especially where a high precision is required. A good example of this is the Channel Tunnel network (see figure 7.19) where GPS observations were combined with theodolite and EDM measurements in order to strengthen the network. Another application where GPS has been successfully used in engineering surveying is in providing control for a number of major route location and highway maintenance schemes. In both these examples, GPS provided what is known as primary control, or points with a high precision spread out over relatively long distances. These were used as reference points for providing further control by, for example, link traversing which was carried out between the GPS reference points using total stations or combined theodolite and EDM systems. This may well be the best use for GPS in the future where it is integrated with other methods of surveying rather than trying to compete with them.

One of the biggest users of GPS in the UK is the OS who use it for the provision of minor control for mapping. Since it is being used daily and over wide areas, the OS claim it is very cost effective. In time, according to the OS, stations on the existing National Grid will be replaced by a GPS network of stations at 25 km intervals. These will be used for relative positioning and instead of being on hill tops, will be located in much more accessible places beside roads and in car parks.

As far as detail surveys and setting out are concerned, GPS is not used extensively in civil engineering and construction as it cannot compete with conventional large-scale surveying at present, particularly regarding costs. However, the possible applications in engineering surveying for a low-cost, small-size GPS 'black box' capable of high-precision, real-time surveys are enormous. Such a surveying system would be integrated with or even replace existing methods for control surveys, detail surveys and setting out and would completely change surveying as it is known today. Much research is being carried out to achieve this, and developments in receiver technology and associated software continue at an ever increasing pace.

A final note of caution is expressed here. However advanced GPS becomes, it will always be a military positioning system and its use could be denied to civilians for some political reason. For any country or organisation to abandon completely all other methods of surveying in favour of GPS would therefore seem ill-advised. Consequently, GPS should be seen as another method for surveying that can be used alongside those already established and not one that should be used to replace all others.

9

Detail Surveying and Plotting

The two previous chapters dealt with the various methods by which networks of control points can be established. In engineering surveying, such a network is required for one of two purposes. Either to use as a base on which to form a *plan of the area* in question or to use as a series of points of known coordinates from which to *set out* a particular engineering construction. Often, these two purposes are linked in a three-stage process. First, using the control points, a contoured plan of the area is produced showing all the existing features. Second, the project is designed and superimposed on the original plan. Third, the designed points are set out on site with the aid of the control points used to prepare the original plan. While the design stage may be undertaken by any of the engineering team, the production of the original plan and the setting out of the project are the responsibility of the engineering surveyor. Consequently, it is essential that correct surveying procedures are adopted to ensure that these two activities are carried out successfully. The purpose of this chapter, therefore, is to describe the various methods by which accurate contoured plans at the common engineering scales of between 1:50 and 1:1000 can be produced from control networks. The techniques by which control networks are used to set out engineering projects are covered in depth in chapter 14.

9.1 Principles of Plan Production

The procedures involved in the production of a contoured plan follow a step-by-step process.

(1) The accuracy of the survey is specified (see section 9.2).
(2) Suitable drawing paper or film is chosen as described in section 9.3.
(3) An accurate coordinate grid is established on the drawing paper or

293

film at the required scale and the control network is plotted. This is
described in section 9.4.

(4) The positions of the features in the area are located on site from the
control network. This information is then brought back into the office
and plotted on the drawing. The term *detail* is used to describe the
various features that are found on the ground surface. This is discussed in section 9.5. The expression *detail surveying* is used to describe the process of locating or *picking up* detail on site from the
control network. Depending on the type of detail and the accuracy
specified in (1), this can be done by one of two techniques, either by
using offsets and ties or by using radiation methods. *Offsets and ties*
can only locate detail in plan position, heights must be added later.
This method is described in section 9.6. *Radiation methods* usually
enable both plan and height information to be obtained simultaneously.
Three radiation techniques are available: *by stadia tacheometry, using
a theodolite and tape* and *using electromagnetic distance measuring
equipment.* These are discussed in sections 9.7, 9.8 and 9.9, respectively.

(5) Once all the detail has been plotted, the plan is completed by adding
a title block containing the location of the survey, a north sign, the
scale, the date, the key and other relevant information. This is discussed in section 9.10.

The steps outlined above should be followed when producing any hand-
drawn plan. Increasingly, however, plans are being produced not by hand
but with the aid of computer software on special multi-pen plotters. A
wide range of such software and plotters is now available which enables
contoured plans to be produced very quickly to a high degree of accuracy.
These have fundamentally changed the ways in which surveying drawings
are produced and, although the end products may look the same, the methods
by which they are obtained differ in a number of respects from those outlined above for hand drawings. For example, they normally store the field
survey observations in a *database* and use this to prepare a three-dimensional representation of the ground surface, known as a *digital terrain
model* (DTM). Using the database and the DTM, plans, contour overlays,
sections and perspective views can be obtained at virtually any scale in a
variety of colours. Some of the currently available software packages and
the techniques involved in the production of computer-aided survey drawings are discussed in sections 9.11 and 9.12.

9.2 Specifications for Detail Surveys

The accuracy required in detail surveying should always be considered
before the survey is started. This is governed by two factors: the scale of
the finished plan, and the accuracy with which it can be plotted.

TABLE 9.1

	Scale				
	1:50	1:100	1:200	1:500	1:1000
Contour vertical interval	0.05 m	0.1 m	0.25 m	0.5 m	1 m
Spot level grid size	2 m	5 m	10 m	20 m	40 m

For *plan positions*, it is usual to assume a plotting accuracy for detail of 0.5 mm and for various scales this will correspond to certain distances on the ground. These lengths are an indication of the accuracy required at the scales in question. However, even if a plan is to be plotted at 1:500, part of it may at a later date be enlarged to 1:50 so it is always better to take measurements in the field to a greater accuracy than that required for the initial plan. A good compromise is obtained by recording all distances, where possible, to the nearest 0.01 m.

For *contours*, the vertical interval depends on the scale of the plan and suitable vertical intervals are listed in table 9.1 for general purpose engineering surveys. Usually, the accuracy of a contour is guaranteed to one-half of the vertical interval.

All *spot levels* taken on soft surfaces (for example, grass) should be recorded to the nearest 0.05 m and those taken on hard surfaces to 0.01 m. In areas where there is insufficient detail at which spot levels can be taken, a grid of spot levels should be surveyed in the field and plotted on the plan, the size of the grid depending on the scale, as shown in table 9.1.

In addition to accuracy considerations, it is necessary to decide on the amount and type of detail that is to be located and the intensity of the spot levels that are to be taken. These decisions depend on a number of factors, for example, the *topography of the site* will influence the number of spot levels (see section 2.24), the *purpose of the survey* will dictate the type of detail to be located and the *time available* for the work may restrict the amount that can be picked up. In addition, one very important factor that can have a considerable bearing on what is shown on the finished drawing is the budget available for its production. Since greater detail requires a greater amount of work and higher accuracy requires more expensive equipment, a survey which requires high accuracy and intensive detail will be more expensive to achieve.

Consequently, it is essential that the accuracy and the aims of the survey are considered and established well before any fieldwork begins. This calls for careful planning and the preparation of a survey specification in which the various requirements are listed. To help with this, the Royal Institution of Chartered Surveyors (RICS) has produced two publications on the preparation of specifications for surveying and mapping. Their titles are

Specification for Mapping at Scales between 1:1000 and 1:10 000 and *Specification for Surveys of Land, Buildings and Utility Surveys at Scales of 1:500 and Larger*. The purpose of the former is to provide a general technical specification for contract mapping worldwide which can be modified as required by commissioning agencies and surveyors to meet the particular needs of individual mapping projects. The purpose of the latter is to provide a standard specification for *large-scale* surveys. Both are relevant to engineering surveys although the latter is more applicable since the majority of engineering plans are produced at scales of 1:500 or larger. Each takes the form of a series of sections under a range of surveying related headings. Within each section, sub-sections are provided itemising the various parameters which may be included if required. They have been set out in such a way that they can be used directly as contract documents by entering appropriate details in the spaces provided, deleting sections and numbered paragraphs which are not required, and adding appendices or additional sections and numbered paragraphs to include special requirements not covered in the documents themselves. As an example of this, figure 9.1 shows Section 3 of the *Specification for Surveys of Land, Buildings and Utility Surveys at Scales of 1:500 and Larger* which has been reproduced by kind permission of the RICS.

Although a fuller description of these specifications is beyond the scope of this book, their use is strongly recommended when planning engineering surveys. They are referenced at the end of this chapter.

9.3 Drawing Paper and Film

Nowadays, there is a variety of drawing media available on which to produce engineering survey plans either by hand or on computer-aided plotters. All, however, fall into one of two categories; either *paper* or *film*. Both are produced in a number of different grades, each being specified by its degree of translucency and either its weight or its thickness. Different surface finishes are also available, for example, matt, semi-matt and gloss. During the course of a survey, several different types of paper or film may be used at various stages of the work. Some of the most popular types are summarised below.

Light-weight paper, 60 grammes per square metre (gsm), is an inexpensive opaque medium. However, it is easily torn, has poor dimensional stability and should be handled with care. It is ideal for pencil work and may be used for preparing an initial check plot in order to verify that the survey data is correct. Ink is possible but is not recommended since absorption can occur resulting in unsightly smudges.

Opaque bond paper, 90 gsm, is ideal for all pencil drawings and will also take ink. It can be used for hand drawings and computer plots. Some-

Section 3 – Detail to be Surveyed for Final Plans
(drawn and/or digital)

(Delete items not required. Complete where applicable)

3.1 Planimetric information

3.1.1   The planimetric features listed below shall be surveyed (specific items not required should be deleted from the list below and from the corresponding lists in Annexe B if used)

- control points
- buildings and structures
- boundary features
- roads, tracks and footpaths
- street furniture and visible service features
- railways
- water, drainage and coastal features
- slopes and earthworks
- woods, trees and recreational areas
- industrial features
- other special requirements (give details under Annexe B, Section 11)

3.1.2   The absolute plan position of any well defined point of detail shall be correct to within ± 0.3mm r.m.s.e. at the plan scale when checked from the nearest control point.

3.1.3   Features which cannot be surveyed to the specified accuracy without extensive clearing shall be

(a) surveyed approximately and annotated accordingly, or

(b) cleared to enable a survey to the specified accuracy to be undertaken.

3.2 Height information

3.2.1   Sufficient spot levels shall be surveyed such that the true ground configuration is accurately represented on the plans.

3.2.2   The maximum distance between adjacent spot levels shall not exceed . . . . . . metres.

3.2.3   Additional spot levels shall be observed on the features listed in Annexe C.

3.2.4   Ground survey spot levels on hard surfaces shall be correct to ± 10mm r.m.s.e. and elsewhere to ± 50mm r.m.s.e. except on ploughed or otherwise broken surfaces. Photogrammetic spot levels shall be correct to ± 100mm r.m.s.e.

3.2.5   Contours shall be shown at vertical intervals of . . . . . . metres.

3.2.6   At least 90% of all contours shall be correct to within one half of the specified contour interval. Any contour which can be brought within this vertical tolerance by moving its plotted position in any direction by an amount equal to 1/10th of the horizontal distance between contours, or 0.5mm at plan scale, whichever is the greater, shall be considered as correct.

3.2.7   Contours which cannot be surveyed to the specified accuracy without extensive clearing shall be:

(a) surveyed approximately and annotated accordingly; or

(b) cleared to enable a survey to the specified accuracy to be undertaken.

3.2.8   In flat areas where the horizontal distance between contours generally exceeds 50mm at plan scale, supplementary spot levels shall be surveyed at intervals not exceeding 50mm at plan scale, parallel to the contours.

Figure 9.1   *Specifications for detail surveys (reproduced by permission of the RICS which owns the copyright)*

times referred to as *cartridge paper*, it usually has good dimensional stability and may be used for the production of a master pencil drawing which is later to be traced. For direct ink work, a quality must be chosen which is sufficiently high to prevent the ink from being absorbed. For both ink and pencil, its quality must be such that repeated erasures do not make the surface fibrous or cause tearing.

Tracing paper, 90 gsm, has an excellent translucency and is ideal for an ink tracing of the original pencil drawing. A high quality should be chosen to give good dimensional stability and to reduce the chances of tearing and creasing. This is particularly important for computer plots. An extra smooth surface will give a high standard of linework with no absorption. Tracing paper provides a good original for dyeline printing. However, with age, it can become brittle and discolour.

Vellum, 80 gsm, has a higher quality than tracing paper and a good translucency. It is ideal for ink and does not become brittle or discolour with age. Vellum is an excellent medium for computer plots from which dyeline copies can be produced and where drawings are to be stored for a long period of time.

Polyester based films are the top quality media available. They are translucent and are referenced by their thickness rather than their weight, for example, 75 microns. With first-class surfaces and excellent dimensional stability they are ideal for all survey drawings. They are particularly good for computer-aided plotters as they are capable of giving sharper plotted ink lines at faster speeds. Pencil and ink can be used on them very easily and both can be erased without leaving unsightly marks. Such films are waterproof, difficult to tear and can be used in any reprographic process.

All paper and film can be purchased either in roll form or as packs of cut sheets. Rolls come in a variety of lengths, typically 50 m and 100 m, and a wide range of roll widths is available. Similarly, cut sheets come in packs of, typically, 100 to 250 and in a wide choice of sheet sizes. Increasingly, however, for both rolls and sheets, there is a trend towards the use of standard widths and sizes based on the dimensions recommended by the International Standards Organisation (ISO).

International Standards Organisation Paper Sizes

ISO sizes for cut sheets are based on rectangular formats which have a constant ratio between their sides of $1:\sqrt{2}$. Succeeding sizes in the range are created by either halving the longer side or doubling the shorter side. This is best illustrated by considering the most important ISO series of paper sizes, the *A-Series*.

The basic unit of the A-series is designated A0 and has an area of 1 m^2 with sides in the ratio $1:\sqrt{2}$. This gives its dimensions as 841 mm $\times$ 1189 mm.

TABLE 9.2
A-series Paper and Film Sizes

Classification	Sheet size	Equivalent roll width
A0	841 mm × 1189 mm	841 mm
A1	594 mm × 841 mm	594 mm
A2	420 mm × 594 mm	420 mm
A3	297 mm × 420 mm	297 mm
A4	210 mm × 297 mm	210 mm

To create the next larger size, the smaller dimension is doubled and an even whole number inserted before the A0 symbol, for example

$$2A0 = 1189 \text{ mm} \times 1682 \text{ mm} \quad 4A0 = 1682 \text{ mm} \times 2378 \text{ mm}$$

In practice, for survey drawings, sizes larger than A0 are rarely used.

To create smaller sizes, the larger dimension is halved and a different number is inserted after the A, that is, A1, A2, A3 and so on. Hence, A1 is half the area of A0, A2 is half the area of A1 (and quarter of the area of A0) and so on. This enables two of each subsequent sheet size to be obtained by folding the previous size in half, for example, an A0 size sheet folded in half provides two A1 size sheets. Table 9.2 lists the sizes of the A-series cut sheets commonly used for survey drawings.

Although the A-series strictly applies to cut sheets, rolls are available having widths based on the dimensions it uses. Those suitable for survey drawings are listed in table 9.2. Each subsequent roll width is dimensioned such that two of the A-series sheet sizes can be cut from it, for example, from a roll width of 841 mm, both A0 sized sheets (841 mm × 1189 mm) and A1 sized sheets (594 mm × 841 mm) can be cut.

9.4 Plotting the Control Network

When a control network has been computed, as described in chapter 7, the end-product is a set of coordinates. This is normally in the form of a list of eastings (E) and northings (N) for each of the points in the network, for example, a control station R may have coordinates (4571.56 m E, 7765.86 m N). Such coordinates have many uses in engineering surveying operations, one being to provide a framework for the production of survey drawings, either by computer-aided plotting or by hand.

For *computer-aided plotting*, the grid on which the control network is based is used by the computer to locate all the control stations and detail points in their correct E, N positions on the drawing. Different line types and symbols are then added to the points (usually in a range of colours using multiple pens) to form the finished plan and the grid itself can be added if required. Such computer-aided plotting techniques are discussed further in section 9.11. The remainder of this section deals with the procedures involved in plotting the control network by hand.

When *plotting by hand*, a common mistake is to plot the stations using the angles (or the bearings) and the lengths between the stations. This is known as plotting by *angle and distance* or *bearing and distance* and involves the use of a 360° protractor and a graduated straight edge. However, the accuracy of such a method is limited and, although it can be used to plot some of the detail (see section 9.7), it is NEVER used to plot the control stations. Instead, the preferred and most accurate method of plotting these is to use the coordinate grid on which the computation of their coordinates was based. The procedure for this involves three steps. First, the grid on which the control network is based is carefully orientated such that the survey area falls centrally on the plotting sheet. Second, using a beam compass and a graduated straight edge, the grid lines are accurately plotted. Third, the control stations are plotted with the help of the grid and their positions carefully checked. The control stations are then used, together with the grid, to help with the plotting of the detail. The three stages involved are discussed in the following sections.

Orientating the Survey and Plot

Before commencing any plot (that is, before constructing the grid), the extent of the survey should be taken into account such that the plotted survey will fall centrally on to the sheet.

In the case where the north direction is stipulated, the north–south and east–west extents of the area should be determined and the stations plotted for the best fit on the sheet.

If an arbitrary north is to be used, the best method of ensuring a good fit is to assign a bearing of 90° or 270° to the longest side. This line is then positioned parallel to the longest side of the sheet so that the survey will fit the paper properly.

Sometimes it may be necessary to set the arbitrary north to a particular direction in order to ensure that the survey will fit a particular sheet size. This will often be the case with long, narrow site surveys where, to save paper and for convenience, the plot of the survey is to go on to a single or a minimum number of sheets or the minimum length of a roll. The boundary of the survey should be roughly sketched and positioned until a suitable fit is obtained. Again, the longest or most convenient line should be assigned a suitable arbitrary bearing.

Where the north point is arbitrary, it has to be established before the coordinate calculation takes place. In order to estimate the extent of the survey it should be sketched, roughly to scale, using the left-hand angles and the lengths between stations.

Plotting the Coordinate Grid

The first stage in plotting the control network is to establish a coordinate grid on the drawing medium.

Coordinated lines are drawn at specific intervals, for example, 10 m, 20 m, 50 m, 100 m in both the east and north directions to form a pattern of squares. The stations are then plotted in relation to the grid.

When drawing the grid, T-squares and set squares should *not* be used since they are not accurate enough. Instead, the grid is constructed in the following manner.

(1) From each corner of the plotting sheet two diagonals are drawn, as shown in figure 9.2*a*.

(2) From the intersection of these diagonals an equal distance is scaled off along each diagonal using a *beam compass*. This scaled distance must be large, see figure 9.2*b*.

(3) The four marked points on the diagonals are joined using a *steel straight edge* to form a rectangle. This rectangle will be perfectly true and is used as the basis for the coordinate grid (see figure 9.2*c*).

On all site plans and maps it is conventional to have the north point (true, magnetic or arbitrary) on the drawing such that the north direction is from the bottom to the top of the sheet and roughly parallel to the sides of the sheet. This will be achieved if the grid framework is constructed as described.

(4) By scaling equal distances along the top and bottom lines of the rectangle and joining the points, the vertical (E) grid lines will be formed. The horizontal (N) grid lines are formed in a similar manner using the other sides of the rectangle (see figure 9.2*d* and *e*).

All lines must be drawn with the aid of a steel straight edge and all measurements must be taken from lines AB and BC and not from one grid line to the next. This avoids accumulating errors.

(5) The grid lines should now be numbered accordingly. The size of a grid square should not be greater than 100 mm by 100 mm. It is not necessary to plot the origin of the survey if it lies outside the area concerned.

Plotting the Stations

Let the station to be plotted have coordinates 283.62 m E, 427.45 m N and let it be plotted on a 100 m grid prepared as described in the previous section.

(1) The grid intersection 200 m E, 400 m N is located on the prepared grid.

(2) Along the 400 m N line, 83.62 m is scaled off from the 200 m E

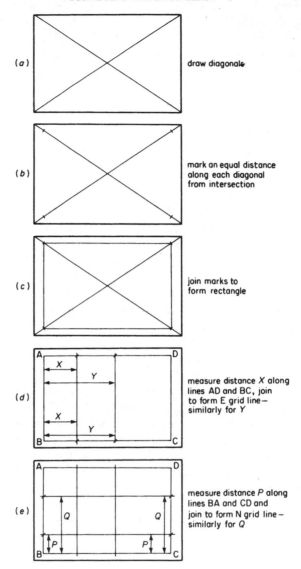

Figure 9.2 *Plotting the coordinate grid*

intersection towards the 300 m E intersection and point a is located
(see figure 9.3). Similarly, point b is located along the 500 m N line.
Points a and b are joined with a pencil line.

(3) Along the 200 m E line, 27.45 m is scaled from the 400 m N intersection towards the 500 m N intersection to locate point c. Point d is
found by scaling 27.45 m along the 300 m E line. Points c and d are
joined.

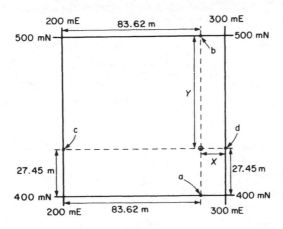

Figure 9.3 *Plotting the stations*

(4) The intersection of lines ab and cd gives the position of the station.

(5) To check the plotted position, dimensions X and Y are measured from the plot and compared with their expected values. In this case, X *should* equal $100.00 - 83.62 = 16.38$ m and Y *should* equal $100.00 - 27.45 = 72.55$ m.

(6) When all the stations have been plotted, the lengths between the plotted stations are measured and compared with their accepted values.

(7) The control lines are added by carefully joining the plotted stations. This is to aid in the location of detail (see sections 9.6 to 9.9).

9.5 Detail

The term 'detail' is a general one that implies features both above and below ground level and at ground level.

Buildings, roads, walls and other constructed features are called *hard detail*, whereas natural features including rivers and vegetation are known as *soft detail*. Other definitions include *overhead detail* (for example, power and telephone lines) and *underground detail* (for example, water pipes and sewer runs).

Many types of symbols are used for representing detail and a standard format has yet to be universally agreed. Those symbols and abbreviations shown in figure 9.4 are fairly comprehensive and their use is recommended. However, it will be noted from figure 9.4 that more abbreviations rather than symbols are given for detail. This is due to the fact that, at the large scales used for engineering surveys, the actual shapes of many features can be plotted to scale and, therefore, do not need to be represented by a symbol.

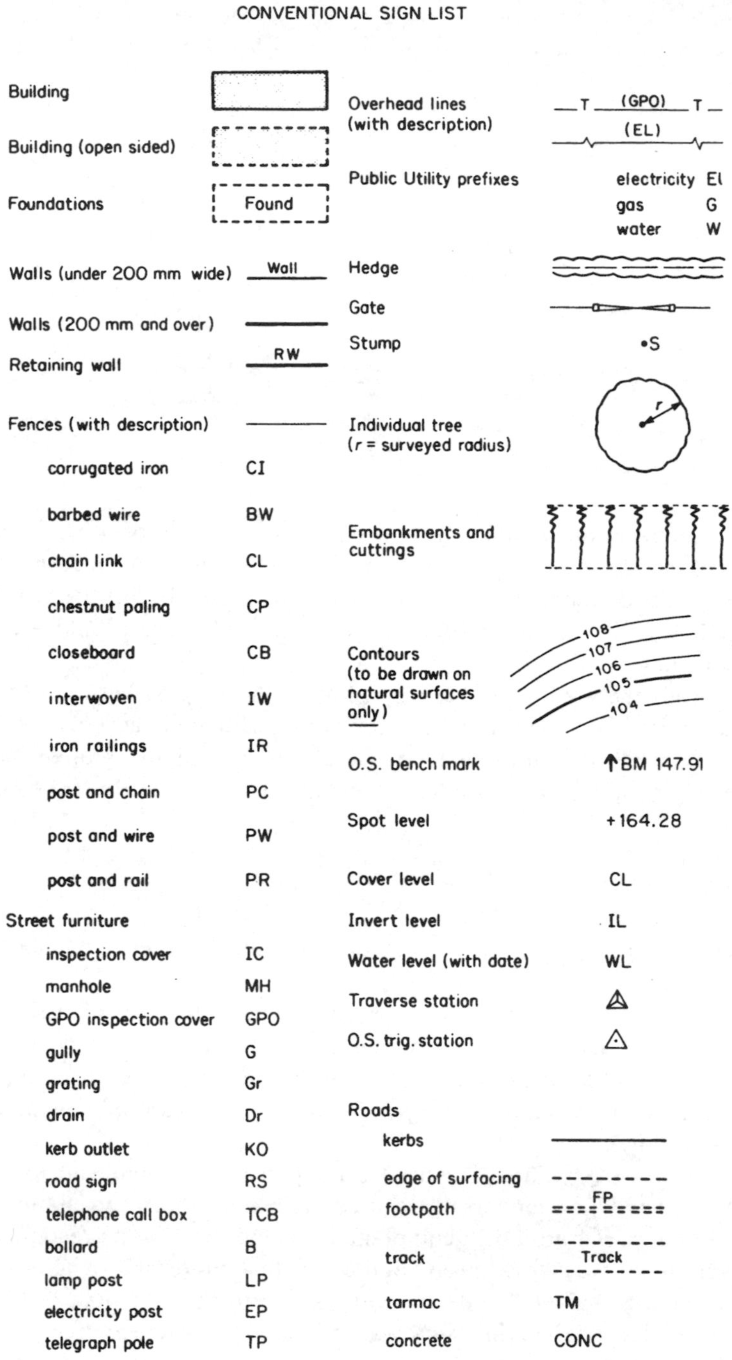

CONVENTIONAL SIGN LIST

Building		Overhead lines (with description)	T __ (GPO) T __ / (EL)
Building (open sided)		Public Utility prefixes	electricity El
Foundations	Found		gas G
			water W
Walls (under 200 mm wide)	Wall	Hedge	
Walls (200 mm and over)		Gate	
Retaining wall	RW	Stump	•S
Fences (with description)		Individual tree (r = surveyed radius)	
corrugated iron	CI		
barbed wire	BW	Embankments and cuttings	
chain link	CL		
chestnut paling	CP		
closeboard	CB	Contours (to be drawn on natural surfaces only)	108 107 106 105 104
interwoven	IW		
iron railings	IR		
post and chain	PC	O.S. bench mark	↑BM 147.91
post and wire	PW	Spot level	+164.28
post and rail	PR	Cover level	CL
Street furniture		Invert level	IL
inspection cover	IC	Water level (with date)	WL
manhole	MH	Traverse station	▲
GPO inspection cover	GPO	O.S. trig. station	△
gully	G		
grating	Gr	Roads	
drain	Dr		
kerb outlet	KO	kerbs	
road sign	RS	edge of surfacing	
telephone call box	TCB	footpath	FP
bollard	B	track	Track
lamp post	LP		
electricity post	EP	tarmac	TM
telegraph pole	TP	concrete	CONC

Figure 9.4 *Conventional sign list*

When detail surveying, the amount and type of detail that is located (or *picked up*) for any particular survey varies enormously with the scale (see section 1.6) and the intended use of the plan. Detail can be located from the control network by one of two methods, either by using offsets and ties or by using radiation methods.

Offsets and *ties* can only locate detail in the plan position. They are discussed in section 9.6. If height information is required, spot levels must be obtained at a later date by levelling at points of detail that have already been located.

Radiation methods usually enable both plan and height information to be obtained. Several techniques are available and these are discussed in sections 9.7, 9.8 and 9.9.

9.6 Offsets and Ties

Figure 9.5a shows the method of *offsets* in which lengths x and y are recorded in the field in order to locate two trees. The offsets are taken at right angles to the lines running between control points. A variation on the offset method is shown in figure 9.5b where *ties* from two (or more) points are used to locate the corner of a building.

In practice, a synthetic tape is usually laid along the control line to measure the x distances shown in figure 9.5, and the offsets and ties (the y distances) are also measured using a synthetic tape.

Since an offset is a line measured at right angles to a survey line to a particular feature, it is necessary to establish a right angle. This is achieved

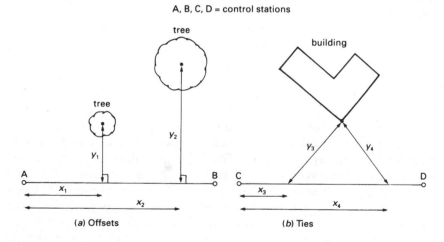

(a) Offsets   (b) Ties

Figure 9.5   *Locating detail*

by holding the zero of the x tape on the point of detail and swinging this over the y tape. The minimum reading obtained on the x tape occurs at the perpendicular and the reading on the y tape at this point indicates the distance along the control line for the offset measurement.

For best results, an offset should never be longer than 10 m owing to possible errors in the tape length and the uncertainty of establishing an accurate right angle. Where detail has to be located beyond 10 m from the survey line, ties should be used. Usually two ties are taken but, occasionally, if the distances involved are long, three should be measured. The maximum length of a tie should not be greater than one tape length.

As well as measuring offsets and ties, the dimensions of certain features may be recorded, for example, the widths of paths or roads with parallel sides, the spread and girth of trees, the lengths around buildings and the radius of circular features. Sometimes it is acceptable to survey rectangular buildings by fixing the two corners of the longest side by ties or offsets and by recording the remaining dimensions and plotting accordingly.

Detail surveying using offsets and ties can only locate detail in the plan position. Since all survey plans must include height information this has to be added at some stage in the survey by taking spot heights at points of detail that have already been located (see section 2.24).

Booking Offsets and Ties

When booking offsets and ties, the field book should always be neat and consistent and, as a general standard, all fieldwork must be capable of being plotted by someone who was not involved in the field survey. An emphasis should, therefore, be placed on clear, legible writing and large diagrams.

Figure 9.6 shows a typical booking. This may be drawn either in a field book or on loose-leaf sheets. Conventionally, a double line is drawn through the centre of the page and this represents the survey line. Entries start at the bottom of the page and, standing at station A, facing station B, detail that is on the right-hand side of the line is booked on the right-hand side of the page and vice versa. Only continuous lengths from station A are recorded in between the double lines, the total length between A and B being written sideways with a line drawn on each side. No attempt is made to draw the sketch to scale and complicated features are exaggerated. The running dimensions around the buildings are also shown in figure 9.6 and are distinguished from other measurements by being inserted in brackets.

Long survey lines should be continued over as many pages as necessary and each new line should be started on a new page. It is important that all necessary information is recorded and explanatory notes should be given where appropriate since nothing should be left to memory. This applies especially to unusual features which do not have a conventional symbol.

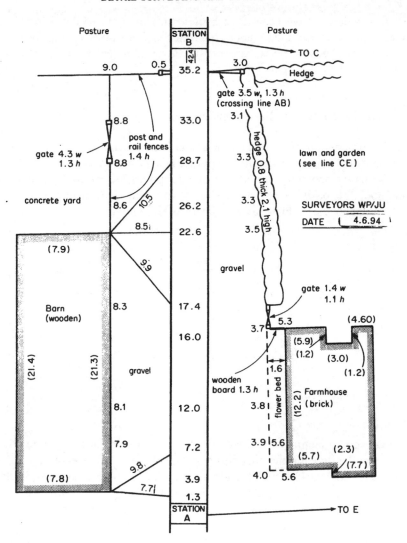

Figure 9.6   *Example booking for offsets and ties*

Plotting Offsets and Ties

When the basic control network has been plotted (see section 9.4), the detail for each control line can be added by marking off the distances along each line corresponding to the points on the tape at which offsets and ties were taken. From these points, the relevant offset and tie lengths can be scaled to fix the points of detail with the aid of a set-square and a pair of compasses.

If spot levels have been taken by levelling at some of the points of detail, these can be added and contours interpolated from them (see section 2.25).

All construction marks are erased after the detail and contours have been added to the plan.

9.7 Radiation by Stadia Tacheometry

A full description of the theory of stadia tacheometry is given in section 4.9 and its application to detail surveying is considered here.

Stadia tacheometry is used to locate points of detail by the radiation method, the basis of which is shown in figure 9.7, where r and θ are measured in the field to locate the tree at point P.

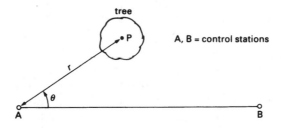

Figure 9.7   *Radiation method*

The component r is measured by tacheometry and θ by reading the horizontal circle of the theodolite or level used. An important difference between stadia tacheometry and the method of offsets and ties is that the height of each point is obtained in addition to its plan position. Tacheometry can, therefore, be used effectively in contouring, particularly in open areas where there are no points of clearly defined detail. It can, also, be used for a complete detail survey but, assuming a plotting accuracy of 0.5 mm, stadia tacheometry can be used only at scales less than 1:200 or for soft detail.

Fieldwork for Stadia Tacheometry

A network of control stations is again used as a base for the survey and, during the reconnaissance, it must be remembered that the length of a single tacheometric observation is limited to 50 m. This implies that, for each station, the maximum coverage on unrestricted sites should be a radius of 50 m.

Several methods of observation are possible and the following description is given as a general purpose approach. It is assumed that the staff is held vertically, that a theodolite is being used and that the reduced levels of the control stations are known.

(1) Set the instrument up over a station mark and centre and level it in the usual way. For a detail survey it is standard practice to measure horizontal and vertical angles on one face only and hence the theodolite should be in good adjustment.

(2) Measure and record the height of the trunnion axis above the station mark.

(3) Select a suitable station as reference object (RO), sight this point and record the horizontal circle reading. It may be necessary to erect a target at the RO if it is not well defined. All the detail in the radiation pattern will now be fixed in relation to this chosen direction. Some engineers prefer to set the horizontal circle to zero along the direction to the RO, although this is not essential.

(4) With the staff in position at a point of detail, rotate the telescope until the staff is aligned along the vertical hair in the field of view. Turn the vertical slow motion screw until the lowest reading stadia hair is set to a convenient mark on the staff such as 1 m, 2 m or the nearest 0.1 m. Read and record the three hairs.

A check can be applied to the stadia readings since the centre or middle reading should be the mean of the other two within ± 2 mm.

(5) Signal the person holding the staff to move to the next staff point.

(6) To save time, while the staff is moving, the vertical circle is read. This is followed by a reading of the horizontal circle. These readings need only be taken with an accuracy of ± 1′.

(7) The procedure is repeated until all the observations have been completed. As far as is practicable, each of the staff points should be selected in a clockwise order to keep the amount of walking done by the person holding the staff to a minimum.

(8) The final sighting should be back to the RO to check that the setting of the horizontal circle has not been altered during observations. If it has, all the readings are unreliable and should be remeasured. Hence, it is advisable that, during a long series of tacheometric readings, a sighting back to the RO should be taken after, say, every 10 points of detail.

Booking and Calculating Stadia Tacheometry Observations

A systematic approach to booking is essential.

Various systems of booking can be used and a sample field sheet, suitable for most types of work, is shown in table 9.3. All the information in columns (1) to (4) and the REMARKS column is recorded in the field, the remainder being computed in the office at a later stage. The vertical circle readings entered in column (2) must be those as read directly on the instrument, reduction being carried out in column (5) where necessary. Particular note should be paid to the accuracy of the computation. The horizontal distance in column (7) is recorded to 0.1 m and the reduced levels in column (10) to 0.01 m.

The booking form should also incorporate a sketch identifying all the staff points (see figure 9.8). In addition, this diagram should indicate miscellaneous information such as types of vegetation, widths of tracks and roads, heights and types of fences and so on.

Plotting Stadia Tacheometry

The network of control stations is first plotted as described in section 9.4. To plot the detail and spot levels, a 360° protractor and a scale rule are required. With reference to figure 9.8 and table 9.3, the procedure is as follows to plot the detail located from station D.

(1) Attach the protractor to the survey plan using masking tape such that its centre is at station D and it is orientated to give the same reading to the RO, station E, as was obtained in the field on the horizontal circle of the theodolite; in this case, 00°00′.

(2) Plot the positions of the horizontal circle readings taken to the detail points around the edge of the protractor. Identify each by its staff position, that is, D1, D2, D3 and so on. Since it is impossible to plot the horizontal circle readings to the same accuracy as that to which they were measured, errors can occur at this stage. These should not have a noticeable effect on *soft* detail. However, if any *hard* detail has been fixed, some adjustments to the initial plotted positions may be necessary before the plan is finalised.

(3) Remove the protractor and very faintly join point D to the plotted horizontal circle positions. Extend these lines.

(4) Using the calculated horizontal distance values from table 9.3, measure from point D along each direction, allowing for the scale of the plan, to fix the plan positions of the points of detail.

(5) Write the appropriate reduced level value taken from table 9.3 next to each point of detail.

(6) Using the field sketches, the detail is now filled in between these points and the contours drawn by interpolation (see section 2.25). All construction marks are erased after the detail and contours have been added.

TABLE 9.3

Example Tacheometric Booking

SURVEY FLAG HOUSING DEVELOPMENT
INSTRUMENT STATION D
NOTES
REDUCED LEVEL STATION D — 47.15
HEIGHT OF INSTRUMENT — 1.73
RL OF INSTRUMENT AXIS — 48.88
OBSERVER JU
BOOKER WFP
DATE 4 July 94

STAFF POINT	VERTICAL CIRCLE	HORIZONTAL CIRCLE	STAFF READINGS CENTRE m	STAFF READINGS UPPER / LOWER	θ	s	D	V	± V−m	REDUCED LEVEL	REMARKS
E (RO)		00-00									Station E
1	355-40	08-05	0.932	1.064 / 0.800	-04-20	0.264	26.2	-1.99	-2.92	45.96	Edge of gravel track
2	357-20	51-20	0.359	0.618 / 0.100	-02-40	0.518	51.7	-2.41	-2.77	46.11	Post and barbed wire fence
3	358-07	124-44	0.709	0.818 / 0.600	-01-53	0.218	21.8	-0.72	-1.43	47.45	Fence meets track
4	02-14	143-15	1.675	1.750 / 1.600	+02-14	0.150	15.0	+0.58	-1.10	47.78	Hedge next to track
5	00-31	297-52	1.974	2.048 / 1.900	+00-31	0.148	14.8	+0.13	-1.84	47.04	Grass meets track
6	00-26	286-10	1.402	1.604 / 1.200	+00-26	0.404	40.4	+0.31	-1.09	47.79	Hedge next to track
7	00-27	293-34	2.098	2.295 / 1.900	+00-27	0.395	39.5	+0.31	-1.79	47.09	Opposite 6, edge of grass
8	358-14	305-38	1.420	1.639 / 1.200	-01-46	0.439	43.9	-1.35	-2.77	46.11	S Corner of store
9	358-12	314-26	1.786	1.971 / 1.600	-01-48	0.371	37.1	-1.16	-2.95	45.93	SE Corner of store
10	358-10	330-53	1.678	1.956 / 1.400	-01-50	0.556	55.5	-1.78	-3.46	45.42	NE Corner of store
E (RO)		00-00									Station E
(1)	(2)	(3)	(4)		(5)	(6)	(7)	(8)	(9)	(10)	

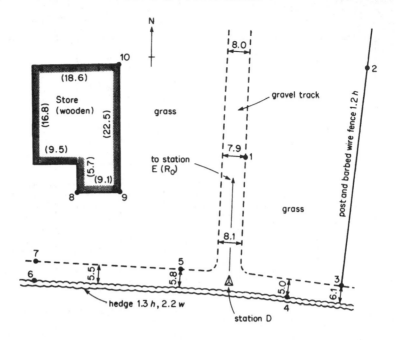

Figure 9.8   *Example sketch for a tacheometric survey*

9.8 Radiation Using a Theodolite and Tape

This technique is very similar to stadia tacheometry, except that no staff is read and no vertical angles recorded. Instead, the distance to each point is measured directly using a steel or synthetic tape with the tape being held horizontally in each case. The disadvantages of this are that no levels are obtained, the range is limited to one tape length unless ranging is introduced and care must be taken to ensure that the tape is horizontal. The best application of this method is in dense detail where stadia tacheometry would become tedious owing to the amount of office work involved, and especially over distances less than 30 m.

9.9 Radiation Using EDM Equipment and Total Stations

This technique utilises some of the latest electronic surveying equipment (as discussed in chapter 5) to locate detail and it is closely linked to computer-aided plotting methods. The instruments that can be used range from combined theodolite/EDM systems in which the observer has to write down all the observations by hand to total stations where the observations are stored directly on plug-in electronic data recorders for subsequent downloading into computer-aided surveying software packages.

Nowadays, most detail surveys are undertaken using EDM radiation techniques with the majority of these being calculated and plotted on computer-aided systems. The accuracy attainable is of the highest order and the cost of the necessary surveying equipment and computer hardware and software has fallen to such an extent that they are now well within the reach of even the smallest engineering surveying firm. The basic technique they employ, however, is one of radiation in which the instrument is set over a control station and the EDM prism mounted on a *detail pole* is held at the point of detail being picked up.

Fieldwork for Radiation by EDM

This is similar to that for radiation by stadia tacheometry as discussed in section 9.7, the main differences being

(1) Instead of a levelling staff, a prism mounted on a detail pole is held at the point being fixed and is observed by the instrument.
(2) The actual observations booked at each pointing will depend on the type of instrument being used. In the simplest case, readings of the horizontal circle, the vertical circle and the slope distance (L) are booked. On instruments which give the horizontal distance (D) and the vertical components (V) of the slope distance directly, these two values should be booked together with the horizontal circle reading. If using an instrument which calculates and displays the coordinates (E, N) and reduced level (RL) of the point on its screen, then these can be recorded directly.
(3) The height of the centre of the prism above the bottom of the detail pole (h_p) must be recorded. Since detail poles are telescopic, this height can be set as required.

If hand calculations are being done (see section 5.24), these are simplified if h_p is set to the same value as the height of the instrument above the control station (h_i) as shown in figure 9.9. In such a case, h_i is cancelled out by h_p and the vertical component (V) of the slope distance (L) is equal to the difference in height between ground points I and P (ΔH).

If the observations are being recorded in a data logger then h_p can be set to any convenient value. However, once it has been set, it is recommended that the height of the prism is not altered unless absolutely necessary since every time h_p is changed, its new value must be keyed into the data logger by hand. As well as being time consuming, this can easily be forgotten causing errors.

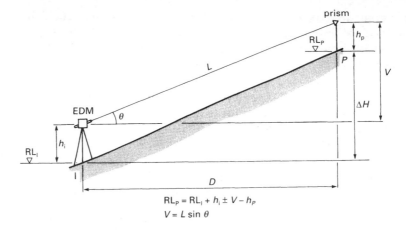

$$RL_P = RL_I + h_i \pm V - h_P$$
$$V = L \sin \theta$$

Figure 9.9   *Measuring heights by EDM*

(4) Since the accuracy of combined theodolite/EDM instruments and total stations is very high, typically ± 5 mm for distances and ± 6″ (and better) for angles, there is no need to round any of the observations taken. The values displayed on the instruments' screens can be booked directly onto the booking sheets.
(5) Its high-accuracy capability enables EDM radiation to be used to pick up any type of detail, hard or soft.

Booking and Calculating Radiation by EDM

If electronic data recorders are being used, no booking sheets are required since the software contained in the logger will prompt the observer for all the necessary information. This is input directly from the built-in keyboards on either the logger and/or the instrument. Section 9.11 gives further details.

For hand booking, however, some type of standard sheet is necessary in order that the observations are recorded accurately and neatly and can be used by others not necessarily involved with the fieldwork. Ideally, space should be provided on such a sheet to record the following for each point: horizontal circle, vertical circle, slope distance (L), horizontal distance (D), vertical component (V), coordinates (E, N) and reduced level (RL). As discussed in the previous section, the type of information actually booked on the form will depend on the instrument being used. In some cases, horizontal circle, vertical circle and L values must be booked while, in others, D and V, or E, N and RL can be recorded directly. The method of plotting to be used can also influence the type of information recorded. This is discussed in the next section. Great care must always be taken when booking, particularly when horizontal circle, vertical circle and L

values are not recorded, since any error in booking, say, D and V directly, cannot later be corrected by calculation. In all cases when booking, a REMARKS column must be included in which to record details of the points being observed.

Although individuals tend to develop their own methods of booking as they gain experience, an example of a suitable form for booking and calculating radiation by EDM is shown in table 9.4. This has similarities with that recommended for radiation by stadia tacheometry as shown in table 9.3. In order that the reader can compare these two booking sheets directly, tables 9.3 and 9.4 both apply to the detail survey shown in figure 9.8. In the case of table 9.4, the survey has been carried out using an EDM theodolite from which horizontal circle, vertical circle and L values have been booked and used to calculate the D, V, E, N and RL values. During the fieldwork, the height of the prism was set equal to the height of the instrument.

Plotting Radiation by EDM

If electronic data loggers have been used, the plotting is carried out using computer-aided methods as discussed in section 9.11. If the booking and calculating have been done by hand, the detail can be plotted by one of the following two methods.

(1) Exactly as described in section 9.7 for radiation by stadia tacheometry using a 360° protractor and a scale rule. In such a case, H, D and RL values are required from the booking forms.
(2) By using the control grid drawn earlier (see section 9.4) to plot the rectangular coordinates of each point of detail. The RL of the point is then written alongside its plan position. In such a case, E, N and RL values are required from the booking forms.

Of the two methods, (1) is quicker but less accurate than (2) which requires additional computations in the form of polar–rectangular (P–R) conversions (see section 1.5) if the instrument cannot provide the E, N values directly.

9.10 The Completed Survey Plan

The end-product of a detail survey is an accurate plan of the area in question at a known scale. This is a very important document which may be used in any contracts that are signed in connection with the construction of an engineering project in the area (see section 14.1). Consequently, it must be accurate and it must look professional. A drawing that is technically correct but has been badly plotted with poor linework and lettering will

TABLE 9.4
Example Radiation by EDM Booking

SURVEY	*Flag Housing Development*						RL OF STATION D	47.15		COORDINATES (E, N)		OBSERVER	JU	
INSTRUMENT STATION	D						HEIGHT OF INSTRUMENT	1.55	D	719.36, 911.72		BOOKER	WFP	
NOTES							RL OF INSTRUMENT AXIS	48.70	E	724.75, 1023.97		DATE	4 JULY 94	

PRISM POINT	VERTICAL CIRCLE (° ' ")	HORIZONTAL CIRCLE (° ' ")	L	h_p	θ (° ' ")	D	V	$\pm V - h_p$	RL	E	N	REMARKS
E(RO)		00 00 00										Station E
1	92 35 40	08 04 27	26.248	1.55	−02 35 40	26.221	−1.188	−2.738	45.96	724.28	937.47	Edge of gravel track
2	91 09 18	51 19 48	51.753	1.55	−01 19 18	51.742	−1.043	−2.593	46.11	761.26	942.08	Post and barbed wire fence
3	89 12 55	124 44 09	21.789	1.55	+00 47 05	21.787	+0.298	−1.252	47.45	736.65	898.46	Fence meets track
4	87 35 06	143 15 17	15.044	1.55	+02 24 54	15.031	+0.634	−0.916	47.78	727.76	899.26	Hedge next to track
5	90 24 42	297 51 52	14.758	1.55	−00 24 42	14.758	−0.106	−1.656	47.04	706.66	919.24	Grass meets track
6	89 05 44	286 09 51	40.356	1.55	+00 54 16	40.351	+0.637	−0.913	47.79	681.19	924.80	Hedge next to track
7	90 05 03	293 34 11	39.471	1.55	−00 05 03	39.471	−0.058	−1.608	47.09	683.98	929.22	Opposite 6, edge of grass
8	91 21 15	305 37 53	43.886	1.55	−01 21 15	43.874	−1.037	−2.587	46.11	684.97	938.96	S Corner of store
9	91 53 08	314 25 51	37.167	1.55	−01 53 08	37.147	−1.223	−2.773	45.93	694.11	938.97	SE Corner of store
10	91 47 09	330 53 17	55.581	1.55	−01 47 09	55.554	−1.732	−3.282	45.42	694.69	961.50	NE Corner of store
E(RO)		00 00 00										Station E

probably be mistrusted and rejected. To ensure that a professional standard is achieved, great care must be taken at all stages of the work. Good quality paper and drawing equipment must be used and a uniform approach should be adopted. The following are examples of this.

(1) Freehand must never be used on survey drawings. Straight lines should be plotted using *straight edges* such as those on steel rulers or high quality set squares. When joining points to form curved lines, *french curves* or *flexicurves* should be used.

(2) All *annotation* (lettering and numbering) should be at such an orientation that it can be read without having to turn the plan upside-down. In addition, annotation of equal importance should be of equal size.

(3) If *spot levels* are to be shown, one method is to plot each as a small cross with the relevant reduced level written alongside. This will look much neater if the size and orientation of all the crosses are the same.

(4) Control stations are often shown in case they are needed for future use. However, the lines joining the control points are not usually shown since they do not actually exist. The only *imaginary* lines normally included on survey drawings are contours (see section 2.22).

(5) If contours are included, they should normally be on *natural* surfaces only and they should not run through embankments and cuttings which have their own symbols.

In practice, if a *hand drawing* is being produced, the original survey plan is usually prepared in pencil by the surveyor who undertook the fieldwork. This is known as the *master survey drawing* and usually shows all the required detail and the control information. When it has been completed, the master drawing is then traced in ink onto plastic film. Since the traced drawing will be used in reprographic processes to obtain copies for use on site, extreme care must be taken with the ink tracing since any errors will be transferred to all the copies taken. Consequently, the job of producing the tracing is usually given to someone who has been specially trained in draughting techniques. During the tracing stage, other relevant information is added to that shown on the master drawing to produce the *completed survey plan*. The information added may have been included in the specifications as a requirement of the survey, for example, a list of the coordinates and reduced levels of the control points, or it may be necessary to help users to interpret the drawing, for example, a *key*.

If a *computer-aided drawing* is being produced then any additional relevant information can be programmed into the finished drawing and viewed on the computer screen before the final plan is plotted. This is discussed further in section 9.11.

However, whether the drawing is done by hand or by computer, the following should normally be included on the finished plan in addition to the actual surveyed area:

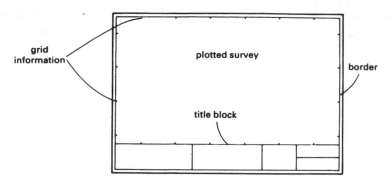

Figure 9.10   *Border, title block and grid layout*

(1) A rectangular or square *border* surrounding the whole of the surveyed area. This provides a neat boundary to the drawing.

(2) A *title block*, within the border and running along one edge of the drawing. This is subdivided into smaller rectangles into which the additional information can be slotted. Figure 9.10 shows an arrangement of the border and title block which would be suitable for most surveys.

(3) The *location* of the survey. Sometimes, a smaller scale *locating map* is included in the title block to show the relationship of the survey to its surrounding area.

(4) The *scale* of the drawing.

(5) The *date* of the survey.

(6) A *north* direction. This may be arbitrary, magnetic or true north depending on the type of survey (see section 1.5).

(7) A *key* (or a *legend*) illustrating any symbols, line-types and abbreviations used.

(8) Details of the *control grid* used. This can either be shown in full, as a series of crosses indicating the grid intersections or as a series of short lines along the sides of the border and title block. This last alternative is shown in figure 9.10 and represents a good compromise in that it does not obscure the drawing but enables the grid to be reconstructed if required.

(9) A list of the *coordinates* and *reduced levels* of the control stations.

(10) The *names* of those who undertook the fieldwork and those who produced the drawing. This is useful if problems arise when using the finished plan.

(11) A separate box within the title block should be allocated to recording any *amendments* that have been made to the drawing. The nature of each amendment should be recorded together with its reference number or letter and the date on which it was included, for example, *Amend-*

ment A: overhead power lines added – 26th August 1994. This is very important in engineering projects where changes often occur as construction proceeds. Engineers and surveyors must be kept informed as drawings are amended in order that they are fully aware of any changes. Care must be taken on site to ensure that the drawings showing the latest amendments are always available.

In addition, other information may be added in the title block depending on the purpose of the survey. For example, a forestry plan could include a numbered schedule of the trees listing their types, girths and spreads, whereas a survey of underground services could include details of pipe diameters, lengths and depths.

Figure 9.11 shows a section of a survey plan and its title block containing some of the information detailed above.

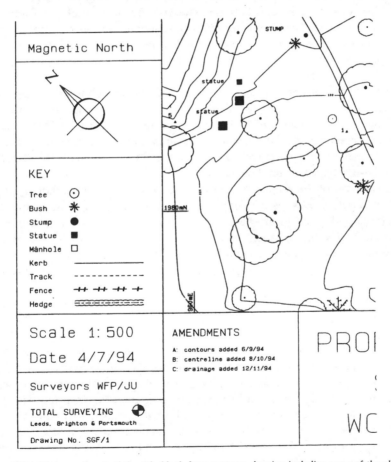

Figure 9.11 *Close-up of part of the title block from a survey drawing including some of the plan detail*

9.11 Computer-Aided Plotting

In the previous sections of this chapter, methods were given for plotting large-scale detail surveys by hand. The process of hand drawing survey plans is extremely time consuming and very often the individual surveyor or engineer does not have the ability to achieve the required high standard of presentation in inked work. This problem is usually overcome by passing the master survey drawing (as discussed in section 9.10) to someone skilled in hand draughting techniques for tracing and annotating. However, this adds to the expense and further increases the time taken to produce the finished survey plan.

For many years, there was no alternative to this process and every engineering and surveying firm employed draughtsmen and draughtswomen to undertake the skilled task of tracing and annotating survey drawings by hand. However, this is no longer the case. The significant technological advances that have been made in both surveying instrumentation and computer hardware and software in recent years have led to a fundamental change in the methods by which surveying calculations are performed and survey drawings are produced.

Surveying programs for undertaking coordinate calculations first appeared in the 1960s for use on the then newly developed mainframe computers although data input was often cumbersome involving as it did the use of punched tape or cards. Nevertheless, the great benefit of computers for performing error free calculations was quickly realised and specialised surveying software began to be developed.

The arrival of the relatively inexpensive *desktop* microcomputers (or personal computers, PCs, as they have become known) in the 1980s led to their widespread adoption in all walks of life, including surveying. Having very large memories and excellent colour display screens, desktop PCs are ideal for high-quality surveying computations and output, particularly when linked to one of the current large range of multi-colour plotters capable of producing drawings up to A0 size.

Although desktop PCs are now widely used for surveying activities, there is an increasing trend towards the use of smaller more portable PCs which can be used directly on site. Initially, *laptop* microcomputers were developed for this purpose. These have similar specifications to desktops (RAM memory, hard disc capacity, ports for peripherals) but are only the size of a small attache case. They generally have a mono rather than a colour screen although colour is possible at a corresponding increase in price. However, laptops have themselves been superseded by *notebook* microcomputers which have exactly the same capabilities but are much smaller, lighter and easier to carry. Typically, they are A4 in size (see table 9.2) and come complete with a small carrying case. As with laptops, notebooks usually have mono

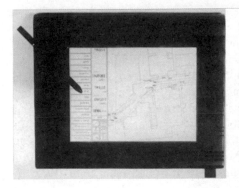

Figure 9.12   *Leica Pen-Map system (courtesy Leica UK Ltd)*

screens as standard but colour is available, if needed. Generally, however, colour is not essential for immediate site work where it is much more important to ensure that the data is correct. Notebooks enable such data verification to be carried out immediately. The data can later be transferred into a desktop if required. The most recent development in computer technology for surveying applications is the emergence of hand-held touch-screen pads which have no keyboard. Instead, all input is done by writing or pressing on the screen directly using an interactive pen. An example of such a system is the Pen-Map marketed by Leica (see figure 9.12) which has many applications including updating existing maps and creating new ones.

In parallel with the evolution of microcomputers, electronic circuitry began to be incorporated into surveying equipment enabling field instruments to be integrated with PCs to produce complete data collection, analysis and design systems. Combined angle and distance measurers (total stations) were developed and linked to hand-held data recorders for the automatic storage of survey observations (see chapter 5). These could then be downloaded directly into a PC through a suitable interface. Once safely inside the PC, the data could be manipulated and analysed as required using specially developed software packages.

As a result of such computing and electronic advances, engineers and surveyors are now presented with a vast array of desktop and notebook PCs and survey mapping and design software from which to choose. These have revolutionised surveying activities with computers and software being increasingly involved at all stages of survey work from the initial data collection on site, through all the analysis and design procedures to the final output of drawings, setting-out data and other numerical information. The following sections discuss the role of software in *survey mapping*. The application of software to *highway design* is discussed in section 11.23.

Survey Mapping Software

As computers evolved, many survey organisations and civil engineering firms developed their own in-house software for the plotting of survey plans. However, it was not long before commercially produced software began to be marketed and there is now a large number of such packages available. The current range includes names such as *IntSURVEYOR* from Applications in CADD, *Landscape* from Blue Moon Systems, *LSS* from Hall & Watts Systems, *Cadsite* from JTC Computer Systems, *Geocomp* from Monostar, *Survpro* from National Survey Software, *STARDUST for Windows* from Softcover International, *SDRMapping and Design* from Sokkia and *CIVILCAD* from Survey Supplies.

Given this wide choice, it is impossible to provide a detailed review of all the various packages in a general textbook such as this. Should the reader wish to know more about the capabilities and costs of these and other packages the two excellent articles by Mike Fort referenced in Further Reading at the end of the chapter are strongly recommended. However, given the increasing use of such software, it is essential that the concepts on which they are based are understood together with the field methods and computing techniques they involve. Fortunately, many of the packages are based on similar principles and utilise similar techniques. Normally, a three-stage process is involved as illustrated in figure 9.13. First, the raw data is *acquired* by taking field measurements. Second, the raw data is input to the computer where it is *processed* to give the required information. Third, the required information is *output* in a suitable format. These three stages are outlined below.

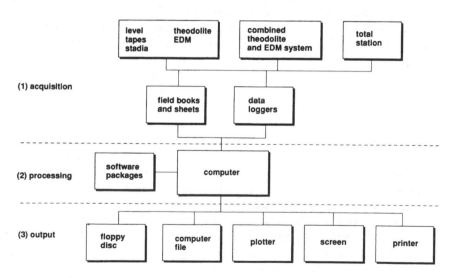

Figure 9.13   *Computer-aided plotting system*

(1) Acquisition of the raw data

The raw data is normally acquired using conventional surveying equipment as indicated in figure 9.13. Traditionally, all observations were recorded by hand on field sheets or in field books. Nowadays, however, observations are normally recorded automatically on data loggers plugged into total stations (see chapter 5). The computer can accept data from either source although it is really designed for observations which have been recorded automatically. These can be very quickly transferred from the data logger into the computer through a standard RS232 interface or similar. If field books have been used, the observations must be carefully keyed in, again by hand, and this can be very time consuming as well as a potential source of error.

If observations are being recorded automatically, the most important aspect of this occurs during the acquisition stage. Each point surveyed in the field is given a unique number, the angle and/or distance observations taken to it are recorded and a code is assigned to it which indicates the type of detail being observed, for example, tree, fence, building and so on. The code is vital to the success of the subsequent processing and output stages. Each code is recognised by the software which acts accordingly and uses it to build up the plan of the area. Codes can take several forms. Some simply assign a particular symbol and/or annotation to a point while others cause interactions between two or more points. An example of the former would be the code *tree* which would cause a tree symbol to be plotted at the point in question while an example of the latter would be the code *fence* which would cause a predetermined line type to be joined between the point in question and the previous point having the code *fence*. Although the observations are recorded automatically, the use of field sketches to illustrate any unusual areas or features is strongly recommended. Once all the observations and codes have been logged in the data recorder, they are downloaded into the computer ready for processing.

If traditional field sheets are being used, the codes must be recorded by hand together with the observations. Again, field sketches are strongly recommended. These can then be used to help with the interpretation of the field observations and codes when they are input by hand to the computer via its keyboard.

The acquisition stage is the most important part of the whole process. The quality of the field observations and coding will greatly influence the outcome of the survey. If the acquisition stage is carried out properly with all the correct codes being assigned to the points, the subsequent processing and output stages should proceed quickly without any problems. Consequently, the onus is on the surveyor and engineer to *get it right on site* and this requires a thorough knowledge of the coding system being used. With practice, the codes can soon be mastered and considerable satisfaction can be gained from completing the fieldwork and watching the computer produce the finished plan with a minimum of editing and correction being required.

(2) Processing the raw data

When they are downloaded or keyed into the computer, the software package stores the raw data (point numbers, field observations and codes) in a computer file. This can be viewed, printed and edited by the user as required before any processing of the data is initiated. At this stage, any changes can be made, although a file containing the original data is normally always preserved in the computer for legal purposes.

Once the data has been edited it is then automatically processed by the software. Usually this involves a number of steps.

(a) The three-dimensional coordinates (E, N and RL) for each point of detail surveyed in the field are computed.
(b) The coordinates, point numbers and codes are then stored in a *database* which can be accessed as required.
(c) In order to enable a plot to be created, the codes are checked against those contained in the software library. If they are present, the software activates them and assigns the correct symbol and/or line type together with any associated command to join (or not to join) to any other point.

Having processed the data in this way, it is now possible to view it on the computer screen either in *textual form*, for example, as a list of point numbers, E, N, RL and codes or in *graphical form* as a plot. Both can be displayed on the screen and edited as required with any selected section of the database being viewed. Either the keyboard or a *mouse* can be used to change, add or erase information as shown in figure 9.14. Some systems utilise *touch screens* activated by a special pen. As new points are added, coordinates are generated and appropriate codes are assigned. All changes are recorded in the database. Annotation can be added to any size, at any rotation and in any position on the drawing. Plots of any part of the database can be generated at any scale and in a wide range of colours. Title blocks, borders, keys, grids, north signs and so on can all be created within the software as a series of different *planforms* and added to the plots, as required.

The software packages have a wide range of features. They are supplied with libraries of symbols, line types, control codes and so on. These can themselves be amended and extended to allow users to create their own libraries and codes. The technique of *layering* is one very important feature. In this, either at the coding stage or during subsequent editing, particular features can be placed in their own unique layer, for example, all spot heights could be placed in a *spot height layer*, all trees in a *tree layer* and so on. The software allows layers to be turned on or off as required. This enables

Figure 9.14 *Graphical editing with a mouse*

users to set up layers of detail and allows them to select and plot only those of particular interest. For example, a plot showing only the detail contained in the road network layer and the underground services layer may be required. This is easily achieved by turning off all the unwanted layers and turning on all the required layers. Figure 9.15 shows a plot in which the spot heights layer has been turned on together with several of the detail layers.

Contours can also be added and placed in a contour layer. These are generated using the RL information stored in the database and involve the creation of a *surface* and a *digital terrain model* (DTM). DTMs are discussed further in section 9.12. Before the contours are formed, the user chooses those points from the database which are to be included in the surface and stipulates any areas over which contours should not be drawn. Normally, contours are only shown on natural surfaces and they should not cross embankments or cuttings (which have their own distinct symbol). Features which may influence the shape of the contours can also be specified, for example, details of any ridges in the area or other lines defining changes of slope. Once the points to be included in the surface have been defined, the computer creates a DTM which approximates to the shape of the actual ground surface. How this is done is described in section 9.12. The accuracy of the DTM depends greatly on the amount, accuracy and location of the points used in its formation. When the DTM has been created, the required contour interval is defined by the user and the computer interpolates the posi-

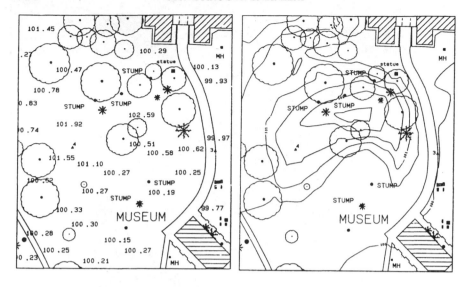

Figure 9.15 *Section of uncontoured plan showing*   Figure 9.16 *Section of a contoured plan*
spot heights and several detail layers   *without spot heights*

tions of the contours and plots them on the screen. Further editing is possible to allow different colours and different contour intervals to be used.
Once it is acceptable, it can be incorporated into a plot to provide a contour overlay. Figure 9.16 shows the contoured version of figure 9.15 in
which the spot heights layer has been turned off to avoid giving the drawing a cluttered appearance.

(3) Output of the required information

After all the editing has been carried out and all the contours have been
generated, the information required from the survey can be obtained. Usually, this is either in *graphical form* as drawings produced on a plotter or
in *textual form* as a listing of data on a printer or on the screen. However,
output can also be in the form of data files stored on floppy discs. Most
of the software systems can generate files in a number of different formats.
This enables data to be transferred to another computer which is either
loaded with similar software or software requiring a different data format,
for example, in the form of a drawing exchange file (DXF) as used by a
number of CAD software packages.

   The usual output, however, is a plot and, as with the software packages,
there is now a wide range of plotters from which to choose. All enable
multi-coloured drawings to be produced on either paper or film (see section
9.3). The actual mode of plotting usually involves pens, although some use

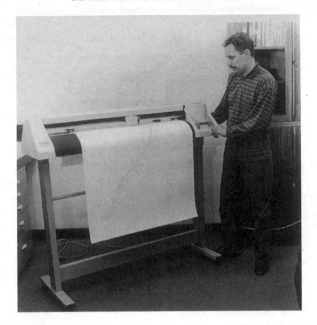

Figure 9.17 *Rolling drum plotter*

inkjets while others use thermal and electrostatic techniques. The most accurate and the most expensive are the *flat-bed* plotters on which, as their name implies, the drawing medium is laid flat while the plot is produced. In these, the plotting mechanism moves over the paper which remains stationary at all times. The most popular for the majority of surveying applications is the *rolling drum* type in which both the plotting mechanism and the paper move. Figure 9.17 shows such a plotter. They are manufactured in a range of sizes up to A0 (see table 9.2) and have the advantages of being cheaper and usually not occupying as much space as an equivalent sized flat-bed. Smaller *table-top* plotters are also available, usually at A2 or A3 size.

9.12 Digital Terrain Models (DTMs)

As discussed in section 9.11, the enormous development of computer technology in recent years coupled with their widespread acceptance and availability has led to their adoption in many surveying activities. However, not only have they improved techniques which existed long before they did, they have also caused techniques to evolve which would not have been possible but for the computers themselves. One example of this is in the subject of *terrain modelling* which is the technique of trying to represent the natural surface of the Earth as a mathematical expression. In surveying and engineering it has a large number of applications including contouring, highway design and earthwork calculations.

The name often given to a mathematical representation of part of the Earth's natural surface is a *digital terrain model* or *DTM* since the data is stored in digital form, that is *E*, *N* and RL. However, as discussed by Petrie and Kennie and referenced in Further Reading at the end of the chapter, there are a number of other names which have been applied to such a representation, for example, digital elevation model (DEM) and digital ground model (DGM). In fact, each applies to a slightly different type of surface representation. If the exact definitions are studied, the one most applicable to engineering surveying is *digital terrain model* because this is considered by most people to include both planimetric data (*E*, *N*) and relief data (RL, geographical elements and natural features such as rivers, ridge lines and so on). Since this is the exact type of information collected during a detail survey, the term DTM is used throughout this book, where applicable.

Earth surface data for the formation of DTMs can be acquired in a number of ways, usually involving either ground based methods or photogrammetric techniques. Photogrammetry using aerial photography is particularly well suited to obtaining three-dimensional and geographical/natural information over large areas where ground techniques would become laborious. However, ground methods are ideal for creating DTMs of smaller areas and the computer aided techniques discussed in section 9.11 involving total stations and automatic data loggers are widely employed for this purpose. Many of the software packages listed in sections 9.11 and 11.23 have the capability of producing DTMs and other PC-based DTM software has been specially developed, for example, *DGM* from L.M. Technical Services and *PC-Surface Modelling* from Survey Software.

Once the field observations have been processed and stored in a database by the software as described in section 9.11, a DTM can normally be formed from them using one of the following techniques.

(1) Grid-based terrain modelling

In this technique, a regular square grid is established over the site and the RL values at each of the grid nodes are interpolated from the field data points.

A grid size is chosen such that it is small enough to give an accurate representation of the irregular surface on which it is based. One disadvantage of this method is that the grid nodes do not coincide with the actual field data points. This tends to smooth out any surface irregularities and causes any contours generated from the grid to be less representative of the true surface. A further drawback is that any ridges and other changes of slope which have been carefully surveyed in the form of a *string* feature on site cannot be accurately reproduced on the grid. This will also affect the shape of any contours generated.

(2) Triangulation-based terrain modelling

This method uses the actual field survey points as node points in the DTM. The software joins together all the data points as a series of non-overlapping contiguous triangles with a data point at each node.

Such a technique has none of the disadvantages of the grid-based method outlined above. A much truer representation of the surface is obtained and features which have been carefully surveyed as *strings*, such as the tops and bottoms of embankments, are faithfully reproduced and can be taken into consideration if contours are generated. It is also possible to set up areas of the surface from which contours can be excluded; this is not possible with the grid-based system.

In addition to the generation of contours, DTMs have numerous applications in engineering surveying. They can be viewed from different angles and presented as wireframe and triangular mesh surface *perspective views* which can highlight areas of specific interest. Examples of these are shown in figures 9.18 and 9.19. Shading can be used to give them an added dimension. Once a DTM has been created, volumes of features such as lakes, spoil-heaps, stockpiles and quarries can easily be obtained and longitudinal and cross-sections quickly produced. This is discussed further in section 13.24. DTMs are also widely used in highway design as discussed in section 11.23. Further information on the creation and application of DTMs can be found in the publication by Petrie and Kennie referenced in Further Reading at the end of the chapter.

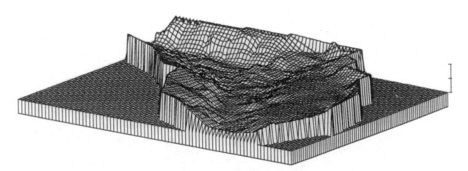

Figure 9.18   *Wire frame view of a DTM (courtesy LM Technical Services Ltd)*

Figure 9.19   *Triangular mesh surface of a DTM (courtesy Blue Moon Systems Ltd)*

Further Reading

M.J. Fort, 'Software for Surveyors', in *Civil Engineering Surveyor*, Vol. 18, No. 3, Electronic Surveying Supplement, pp. 19–27, April 1993.

M.J. Fort, 'Surveying by Computer', in *Engineering Surveying Showcase '93*, pp. 24, 27–31 (PV Publications, 101 Bancroft, Hitchin, Hertfordshire, January 1993).

G. Petrie and T.J.M. Kennie, *Terrain Modelling in Surveying and Civil Engineering* (Whittles Publishing in association with Thomas Telford, London, 1990).

The Royal Institution of Chartered Surveyors, *Specification for Mapping at Scales between 1:1000 and 1:10 000*, 2nd Edition (Surveyors Publications, London, 1988).

The Royal Institution of Chartered Surveyors, *Specification for Surveys of Land, Buildings and Utility Services at Scales of 1:500 and Larger* (Surveyors Publications, London, 1986).

10

Circular Curves

In the design of roads and railways, straight sections of road or track are connected by curves of constant or varying radius as shown in figure 10.1. The purpose of the curves is to deflect a vehicle travelling along one of the straights safely and comfortably through the angle θ to enable it to continue its journey along the other straight. For this reason, θ is known as the *deflection angle*.

The curves shown in figure 10.1 are *horizontal curves* since all measurements in their design and construction are considered in the horizontal plane. The two main types of horizontal curve are

(1) *circular curves*, which are curves of constant radius as shown in figure 10.1*a*
(2) *transition curves*, which are curves of varying radius as shown in figure 10.1*b*.

This chapter covers the design and setting out of circular curves and chapter 11 covers transition curves.

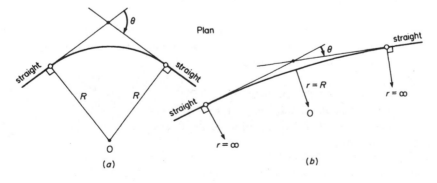

Figure 10.1   *Horizontal curves: (a) circular curve; (b) transition curve*

10.1 Types of Circular Curve

A *simple circular curve* consists of one arc of constant radius, as shown in figure 10.2.

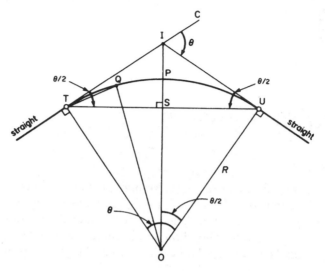

Figure 10.2 *Circular curve geometry*

A *compound circular curve* consists of two or more circular curves of different radii. The centres of the curves lie on the same side of the common tangent, as shown in figure 10.17 in section 10.17.

A *reverse circular curve* consists of two consecutive circular curves, which may or may not have the same radii, the centres of which lie on opposite sides of the common tangent, as shown in figure 10.18 in section 10.18.

10.2 Terminology of Circular Curves

Figure 10.2 illustrates some of the terminology of horizontal curves and it is important that these terms are fully understood before proceeding with the derivations of the formulae used.

In figure 10.2

    Q is any point on the circular curve TPU

    S is the mid-point of the long chord TSU

    P is the mid-point of the circular curve TPU

    intersection point = I

    tangent points = T and U

    deflection angle = θ = external angle at I = angle CIU

    radius of curvature = R

centre of curvature = O

intersection angle = (180° − θ) = internal angle at I = angle TIU

tangent lengths = IT and IU (IT = IU)

long chord = TU

tangential angle = for example, angle ITQ = angle from the tangent length at T (or U) to any point on the curve

mid-ordinate = PS

radius angle = angle TOU = deflection angle CIU

external distance = PI

10.3 Important Relationships in Circular Curves

In figure 10.2, triangle ITU is isosceles, therefore angle ITU = angle IUT = $\theta/2$. Hence, referring to figure 10.3:

The tangential angle, α, at T to any point, X, on the curve TU is equal to half the angle subtended at the centre of curvature, O, by the chord from T to that point.

Similarly, in figure 10.4:

The tangential angle, β, at any point, X, on the curve to any forward point, Y, on the curve is equal to half the angle subtended at the centre by the chord between the two points.

Another useful relationship is illustrated in figure 10.5, which is a combination of figures 10.3 and 10.4.

From figure 10.3, angle TOX = 2α, hence angle ITX = α.

From figure 10.4, angle XOY = 2β, hence angle AXY = β.

Therefore, in figure 10.5, angle TOY = $2(\alpha + \beta)$ and it follows that angle ITY = $(\alpha + \beta)$. In words, this can be stated as

The tangential angle to any point on the curve is equal to the sum of the tangential angles from each chord up to that point.

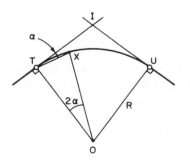

Figure 10.3

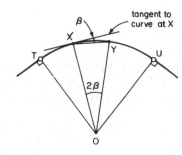

Figure 10.4

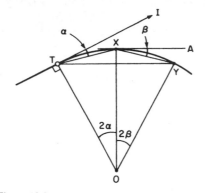

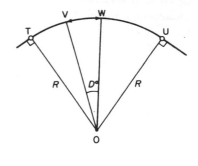

Figure 10.5   Figure 10.6 *Degree curve*

The relationships illustrated in figures 10.3, 10.4 and 10.5 are used when setting out the curves by the method of tangential angles (see section 10.13).

10.4 Useful Lengths

From the geometry of figure 10.2 the following can be derived

tangent length (IT and IU) $= R \tan \theta/2$
external distance (PI) $= R (\sec (\theta/2) - 1)$
mid-ordinate (PS) $= R (1 - \cos(\theta/2))$
long chord (TU) $= 2R \sin \theta/2$

10.5 Radius and Degree Curves

A circular curve can be referred to in one of two ways.

(1) In terms of its radius, for example, a 750 m curve. This is known as a *radius curve*.

(2) In terms of the angle subtended at its centre by a 100 m arc, for example, a 2° curve. This is known as a *degree curve*, and is shown in figure 10.6.

In figure 10.6 arc VW = 100 m and subtends an angle of $D°$ at the centre of curvature O. The curve TU is, therefore, a $D°$ degree curve.

The relationship between the two types of curve is given by the formula $DR = (18\ 000/\pi)$, in which D is in degrees and R in metres, for example, a curve of radius 1500 m is equivalent to

$$D° = \frac{18\ 000}{(1500\ \pi)} = \frac{12}{\pi} = 3.820°$$

that is, a 1500 m radius curve = a 3.820° degree curve.

10.6 Length of Circular Curves (L_c)

(1) For a radius curve, $L_c = (R\ \theta)$ m, where R is in metres and θ is in radians.
(2) For a degree curve, $L_c = (100\ \theta/D)$ m, where θ and D are in the same units, that is, degrees or radians.

10.7 Through Chainage

Through chainage or *chainage* is simply a distance and is usually in metres. It is a measure of the length from the starting point of the scheme to the particular point in question and is used in road, railway, pipeline and tunnel construction as a means of referencing any point on the centre line.

Figure 10.7 shows a circular curve, of length L_c and radius R running between two tangent points T and U, which occurs in the centre line of a new road. As shown in figure 10.7, chainage increases along the centre line and is measured from the point (Z) at which the new construction begins. Z is known as the *position of zero chainage*.

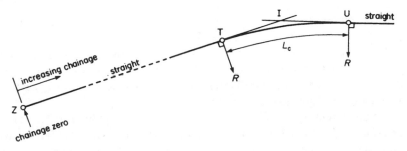

Figure 10.7   *Chainage along a circular curve*

Chainage continues to increase from Z along the centre line until a curve tangent point such as T is reached. At T, the chainage can continue to increase in two directions, either along the curve (that is, from T towards U) or along the straight (that is, from T along the line TI produced). Hence it is possible to calculate the chainage of the intersection point I.

At the beginning of the design stage, when only the positions of the straights will be known, chainage is considered along the straights. However, once the design has been completed and the lengths of all the curves are known, the centre line becomes the important feature and chainage values must be calculated from the position of zero chainage along the centre line only. This is done in order that pegs can be placed at regular intervals along the centre line to enable earthwork quantities to be calculated (see chapter 13).

Hence if the chainage of the intersection point, I, is known and the curve is then designed, the chainages of tangent points T and U, which both lie on the centre line, can be found as follows with reference to figure 10.7

through chainage of T = through chainage of I − IT
through chainage of U = through chainage of T + L_c

A common mistake in the calculation of through chainage is to assume that (TI + IU) = L_c. This is not correct. Similarly, the chainage of U does not equal the chainage of I + IU. To avoid such errors the following rule must be obeyed:

When calculating through chainage from a point which does not lie on the centre line (for example, point I in figure 10.7) it is necessary to first calculate the chainage of a point which lies further back on the centre line (that is, in the direction of zero chainage) before proceeding in a forward direction on the centre line.

10.8 A Design Method for Circular Curves

In circular curve design there are three main variables: the deflection angle θ, the radius of curvature R and the design speed v.

All new roads are designed for a particular speed and the chosen value depends on the type and location of the proposed road (see section 11.3). The Department of Transport (DTp) stipulates design speeds for particular classes of road. This leaves θ and R to be determined.

When designing new roads, there is usually a specific area (often referred to as a *band of interest*) within which the proposed road must fall to avoid certain areas of land and unnecessary demolition. When improving roads, this band of interest is usually very clearly defined and is often limited to the immediate area next to the road being improved. Hence, in both cases, there will be a limited range of values for both θ and R in order that the finished road will fall within this band of interest.

If at all possible, θ must be measured accurately in the field before the design begins. If this is not immediately possible, an approximate value of θ can be measured using a protractor from the two straights drawn on the plan of the area. This value is then used for an initial design which is later amended once θ has been accurately determined. The alteration will, however, be slight and the approximate value of θ is ideal for ensuring that the design will fit adequately into the area.

R is chosen with reference to design values again stipulated by the DTp. These values are discussed in much greater detail in chapter 11 but basically they limit the value of the minimum radius which can be used at a particular speed for a wholly circular curve. If a radius value below the minimum is used it is necessary to incorporate transition curves into the design.

An initial radius value, greater than the minimum without transitions is chosen, the tangent lengths are calculated using R and θ and they are fitted on the plan. If there are no problems of fit this initial design can be used, otherwise a new radius value would be chosen and a new fit obtained. Eventually, a suitable R value would be selected. The design is completed by calculating the *superelevation* required for the curve. This is fully discussed in section 11.2.

This trial and error method is suitable if any value of R above the minimum without transitions can be used and literally thousands of designs are possible and will all be perfectly acceptable. However, if a curve has to have particular tangent lengths, the following procedure can be adopted

(1) exact tangent length $= R \tan \theta/2$
(2) only R is unknown, hence it can be calculated
(3) R should be checked against the DTp values (see chapter 11) to ensure that it is greater than the minimum without transitions. If it is not, transitions must be incorporated.

The design procedure continues until θ and R have been finalised. Once this has been done, the setting out of the centre line of the curve can begin. This is discussed in section 10.10.

10.9 The Use of Computers in the Design Procedure

Before considering the methods by which the centre line can be set out, it is important to discuss how the design procedure just described for circular curves and those described in chapters 11 and 12 for transition curves and vertical curves, respectively, are nowadays often done using software produced for desktop, laptop and notebook computers. A wide range is now available with many different manufacturers offering complete suites of highway design and volume analysis packages. These cover all aspects of the design process from the initial choice and subsequent refining of the horizontal and vertical alignments to the calculation of the volumes and the planning of the movement of the earthworks necessary for the final designed route.

Although a detailed description of the ways in which such packages work is beyond the scope of this book, they are all based on the fundamental principles of curve design and earthwork calculations discussed in this chapter and in chapters 11, 12 and 13 which follow. Consequently, throughout these chapters, references are made to software packages where appropriate. Sections 11.23 and 12.15, in particular, discuss their applications in highway design in greater detail.

Highway design and volume analysis packages cannot function on their own. They need data and this is usually provided by a basic mapping package onto which the highway design and volume analysis software can be added as *modules* to extend the capabilities of the system. The basic package usually consists of a series of land surveying modules in which the control is established and the detail and the contours are located. These enable a contoured site plan to be produced and a three-dimensional digital terrain model (DTM) of the existing ground surface to be generated. Such DTMs provide the basic data required for the highway design and volume analysis modules. Further information on DTMs and other aspects of computer-aided surveying and mapping are given in sections 9.11 and 9.12 in the Detail Surveying and Plotting chapter.

10.10 Establishing the Centre Line on Site

The centre line provides an important reference line on site. Once it has been pegged out, other features such as channels, verges, tops and bottoms of embankments, edges of cuttings and so on, can be fixed from it. Consequently, it is vital that

(1) the centre line pegs are established to a high degree of accuracy;
(2) they are protected and marked in such a way that site traffic can clearly see them and avoid disturbing them accidentally;
(3) in the event of them being disturbed, they can be re-located easily and quickly to the same accuracy to which they were initially set out.

There are a number of methods by which the centre line can be set out, all of which fall into one of two categories.

(1) *Traditional* methods which involve working along the centre line itself using the straights, intersection points and tangent points for reference. These usually require some combination of tapes and/or theodolites.
(2) *Coordinate* methods which use control points situated some distance away from the centre line as reference. These normally require theodolite/ EDM systems or total stations.

Although both categories are still used, coordinate methods have virtually superseded traditional ones for all major curve setting-out operations for a number of reasons:

(a) There is now widespread use of theodolite/EDM systems and total stations on construction sites.
(b) The increasing adoption of highway design software packages which are invariably based on coordinate methods has eliminated the tedious nature of the calculations involved in such methods and enables setting-out data to be produced in a form ready for immediate use by total stations.

(c) Coordinate methods have the advantage that re-locating pegs on the centre line which have been disturbed is usually easier to carry out than by traditional methods.

However, coordinate methods are not always the most appropriate and traditional techniques are still widely used for less important curves, for example, housing estates, minor roads, kerb lines, boundary walls and so on, where they are often more convenient and quicker to use than coordinate methods. They also represent the only possibility in cases where no nearby control points are available. The relative merits of traditional and coordinate methods are discussed further in chapter 11, section 11.17.

Whichever method is used, the first setting-out operation is normally to fix the position of the intersection point on site in order that an accurate measurement of the deflection angle, θ, can be obtained for use in the design calculations. Once the design has been finalised, the tangent points can then be pegged out. These procedures are described in sections 10.11 and 10.12.

If traditional methods are to be used, setting out of the centre line can then be undertaken from the tangent points by a number of different methods. Section 10.13 deals with these traditional techniques and the first worked example in section 10.20 shows the use of one such method.

If coordinate methods are to be used, the coordinates of the tangent points are measured and used in the calculations to enable points on the centre line to be established directly on site from nearby control points. Section 10.14 deals with coordinate methods and the second worked example in section 10.20 shows how the coordinates of points on the centre line can be calculated prior to them being set out.

10.11 Location of the Intersection and Tangent Points in the Field

It is not sufficiently accurate to scale the position of the tangent points from a plan, they must be accurately set out on site. The procedure is as follows with reference to figure 10.8.

(1) Locate the two tangent lines AC and BD and define them by means of a suitable target (see section 3.7).
(2) Set a theodolite up on one of the lines (say AC) and sight towards the intersection of the two tangents at I.
(3) Drive in two pegs x and y on the line AC such that BD will intersect the line xy. The exact position of the tangent line should be marked by nails in the top of the pegs (see figure 7.3).
(4) Join pegs x and y by means of a string line.
(5) Set up the theodolite on BD pointing towards I and fix the position of I by driving in a peg where the line of sight from BD intersects the string line.

(6) Set up the theodolite over I and measure angle AIB, hence angle θ.
(7) Calculate tangent lengths IT and IU using $R \tan \theta/2$.
(8) Measure back from I to T and U, drive in pegs and mark the exact points by nails in the tops of the pegs.
(9) Check the setting out by measuring angle ITU, which should equal $\theta/2$.

The use of two theodolites simplifies the procedure by eliminating steps (3) and (4).

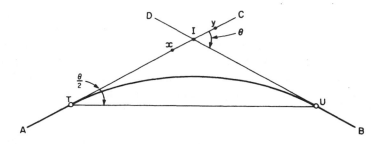

Figure 10.8 *Location of intersection and tangent points*

10.12 Location of the Tangent Points when the Intersection Point is Inaccessible

Occasionally, it is impossible to use the method described in section 10.7 owing to the intersection point falling on a very steep hillside, in marshy ground or in a lake or river and so on. In such cases, the following procedure should be adopted to determine θ and locate the tangent points T and U. Consider figure 10.9.

(1) Choose points A and B somewhere on the tangents such that it is possible to sight from A to B and B to A and also to measure AB.
(2) Measure AB.
(3) Measure angles α and β, deduce γ and hence θ.
(4) Use the Sine Rule to calculate IA and IB.
(5) Calculate IT and IU from $R \tan \theta/2$.

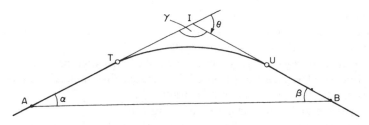

Figure 10.9 *Location of tangent points when intersection point is inaccessible*

(6) AT = IA − IT and BU = IB − IU, hence set out T and U. If A and B are chosen to be on the other side of T and U, AT and BU will have negative values.

(7) If possible, sight from T to U as a check. Measure angle ITU which should equal $\theta/2$.

10.13 Setting Out Circular Curves by Traditional Methods

This section describes the traditional methods of setting out circular curves from their tangent points. Modern methods, involving coordinates, are discussed in section 10.14.

The Tangential Angles Method

This is the most accurate of the traditional methods. It can be carried out using either one theodolite and a tape or two theodolites if no tape is available. The formula used for the tangential angles is derived as follows and uses the relationships developed in section 10.3. Consider figure 10.10, in which tangential angles α_1 and α_2 are required. The assumption is made that arc TK = chord TK if chord ⩽ $R/20$.

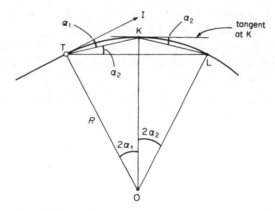

Figure 10.10 *Tangential angles method*

Therefore

$$\text{chord TK} = R2\alpha_1 \ (\alpha_1 \text{ in radians})$$

Hence

$$\alpha_1 = (\text{TK}/2R) \times (180/\pi) \text{ degrees}$$

Similarly

$$\alpha_2 = (KL/2R) \times (180/\pi) \text{ degrees}$$

Note that the chord for α_2 is KL not TL.
In general

$$\alpha = (180/2\pi) \times (\text{chord length/radius}) \text{ degrees}$$
$$= 1718.9 \times (\text{chord length/radius}) \text{ minutes}$$

(a) Using a theodolite and a tape

Calculation procedure

(1) Determine the total length of the curve.
(2) Select a suitable chord length $\leqslant (R/20)$, for example, 10 m, 20 m. This
will leave a sub-chord at the end and it is usually necessary to have an
initial sub-chord in order to maintain equal chord lengths.
This is very important since pegs are usually placed on the centre line
of the curve at exact multiples of through chainage to help in subsequent
earthwork calculations, for example, pegs would be required at chainages
0 m, 20 m, 40 m, 60 m and so on if a 20 m chord has been selected.
The chord must be $\leqslant (R/20)$ in order that the assumptions made in the
derivation of the formula still apply.
(3) A series of tangential angles is obtained from the formula previously
derived, for example, α_1, $(\alpha_1 + \alpha_2)$, $(\alpha_1 + \alpha_2 + \alpha_3)$ and so on corresponding to chords TK, KL, LM and so on.
In practice, $\alpha_2 = \alpha_3 = \alpha_4$ and so on, since all the chords except the
first and the last will be equal. Therefore, usually only three tangential
angles need to be calculated.
(4) All the *cumulative* angles are measured from the tangent point with
reference to the tangent line IT but the chord lengths swung are *individual*, not cumulative.
(5) The results are normally tabulated before setting out the curve on site.

Setting-out procedure

(1) The tangent points are fixed and the theodolite is set up at one of them,
preferably the one from which the curve swings to the right. This ensures that the tangential angles set on the horizontal circle will increase
from 0°.
(2) The intersection point is sighted such that the horizontal circle is reading zero.
(3) The tangential angle for the first chord is set on the horizontal circle.
(4) The first chord is then set out by lining in the tape with the theodolite

and marking off the length of the chord from the tangent point.

The chord lengths used in the calculations are considered in the horizontal plane, therefore the chord lengths set out must be either stepped or slope lengths must be calculated and used.

(5) The horizontal scale of the theodolite is set to the value of the first two tangential angles, that is ($\alpha_1 + \alpha_2$), and the tape again lined in.

With one end of the tape on the point fixed for the first chord, the length of the second chord is marked off. In practice, points are normally located by a peg in the top of which a nail is driven to within 5 to 10 mm of the top to mark the exact position. The end of the tape can be secured over the nail while the next point is located.

(6) This procedure is repeated until point U is set out. As a check, the tangential angle ITU should equal $\theta/2$.

An example based on this method is given in section 10.20.

(b) Using two theodolites

This variation is used when the ground between the tangent points is of such a character that taping proves difficult, for example, very steep slopes, undulating ground, ploughed fields or if the curve is partly over marshy ground. The method is as follows and is shown in figure 10.11.

Two theodolites are used, one being set at each tangent point. One disadvantage of the method is that two of everything are required, for example, two engineers, two instruments and, preferably, two assistants to locate the pegs.

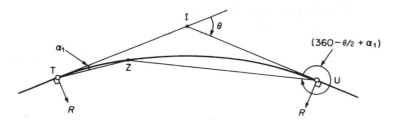

Figure 10.11   *Tangential angles method with two theodolites*

The method adopted is one of intersecting points on the curve with the theodolites.

In figure 10.11, to fix point Z

α_1 is set out from T relative to IT and

$(360° - (\theta/2) + \alpha_1)$ is set out from U relative to UI.

The two lines of sight intersect at Z where an assistant drives in a peg. Good liaison between the groups is essential and, for large curves, two-way radios are a very useful aid.

Offsets from the Tangent Lengths

This traditional method requires two tapes or a chain and a tape. It is suitable for short curves and it may be used to set out additional points between those previously established by the tangential angles method or by coordinate methods. This is often necessary to give a better definition of the centre line. Consider figure 10.12.

Required   The offset AB, from a point A on the tangent, to the curve.
In triangle OBC

$$OB^2 = OC^2 + BC^2$$

Therefore

$$R^2 = (R - X)^2 + Y^2$$

From here there are two routes
either

$$(1)\ R - X = \sqrt{(R^2 - Y^2)}\ \text{hence}\ X = R - \sqrt{(R^2 - Y^2)}$$

or

$$(2)\ R^2 = R^2 - 2RX + X^2 + Y^2$$

Dividing through by $2R$ gives

$$X = (Y^2/2R) + (X^2/2R)$$

but $(X^2/2R)$ will be very small since R is very large compared with X, therefore it can be neglected. Therefore

$$X = (Y^2/2R) \tag{10.1}$$

Equation (10.1) is accurate only for large radii curves and will give errors for small radii curves where the effect of neglecting the second term cannot be justified.

Once the tangent points are fixed, the lines of the tangents can be defined using a theodolite or ranging rods and the offsets (X) set off at right angles at distances (Y) from T and then from U. Half the curve is set out from each tangent point.

Offsets from the Long Chord

This traditional method also uses two tapes or a chain and tape. It is suitable for curves of small radius such as boundary walls and kerb lines at road intersections. Also, it is a very useful method when the tangent lengths are inaccessible and offsets from them cannot be used. Consider figure 10.13.

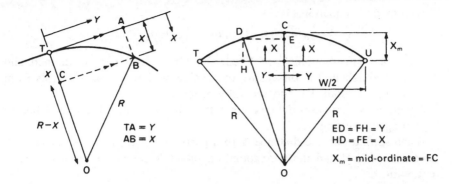

Figure 10.12 *Setting out a circular curve from the tangent length*

Figure 10.13 *Setting out a circular curve from the long chord*

Required The offset HD from the long chord TU at a distance Y from F. In this method all offsets are established from the mid-point F of the long chord TU. Let the length of chord TU $= W$. In triangle TFO

$$OT^2 = OF^2 + TF^2$$

Therefore

$$R^2 = (R + X_m)^2 + (W/2)^2$$

Hence

$$(R - X_m) = \sqrt{(R^2 - (W/2)^2)}$$

Therefore

$$X_m = R - \sqrt{(R^2 - (W/2)^2)} \qquad (10.2)$$

In triangle ODE

$$OD^2 = OE^2 + DE^2$$

Therefore

$$R^2 = (OF + X)^2 + Y^2$$

Hence

$$(OF + X) = \sqrt{(R^2 - Y^2)} \qquad (10.3)$$

But

$$OF = (R - X_m)$$

Therefore from equation (10.2)

$$OF = \sqrt{(R^2 - (W/2)^2)}$$

Therefore from equation (10.3)

$$X = \sqrt{(R^2 - Y^2)} - \sqrt{(R^2 - (W/2)^2)}$$

Once the tangent points are fixed, the long chord can be defined and point F established. The offsets are then calculated at regular intervals from point F, firstly along FT and secondly along FU.

Again, it is very useful to tabulate the offsets from FT and FU before beginning the setting out.

When setting out, the distance Y to a particular point is measured from F towards T and U and the corresponding offset X set out at right angles at that point.

10.14 Setting Out Circular Curves by Coordinate Methods

As discussed in section 10.10, these methods are nowadays used in preference to traditional techniques.

In such methods, which are suitable for all horizontal curves, the National Grid or local coordinates of points on the curve are calculated and these points are then fixed by *either*

(1) *intersection* using two theodolites from two of the control points in the main survey network surrounding the proposed scheme (see figure 10.14); *or*

(2) *bearing and distance (polar rays)* using EDM instruments or total stations from control points in the main survey network (see figure 10.15). To fix point A, α is turned off from direction PQ and distance PA measured and to fix point B, β is turned off and distance PB measured.

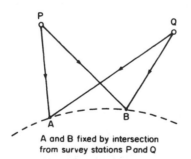

A and B fixed by intersection
from survey stations P and Q

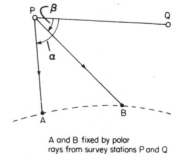

A and B fixed by polar
rays from survey stations P and Q

Figure 10.14   *Setting out a circular curve by intersection*

Figure 10.15   *Setting out a circular curve by bearing and distance*

For a complete curve, consider figure 10.16. Points A, B, C, D, E and F are points to be set out at regular intervals of through chainage on the curve from control points P and Q.

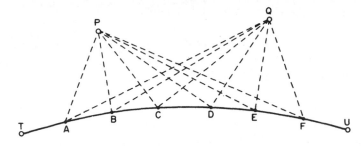

Figure 10.16 *Setting out a complete curve using coordinates*

Procedure
(1) Locate T and U as discussed in sections 10.11 and 10.12.
(2) Obtain coordinates of T and U either by taking intersection observations from P and Q or by including T and U in a traverse with P and Q.
(3) Calculate chord lengths TA, AB, BC and so on and respective tangential angles as normal.
(4) Calculate bearings TA, AB, BC and so on.
(5) Calculate the coordinates of points A to U from T, treating TABCDEFU as a closed route traverse.
(6) Derive bearings PA and QA, PB and QB, PC and QC, and so on from their respective coordinates.
(7) Calculate the lengths PA and QA, PB and QB, PC and QC, and so on from their respective coordinates.
(8) Set out the curve by *either*
    (i) *intersection* from P and Q using bearings PA and QA, PB and QB, and so on; *or*
    (ii) *polar rays* from P or Q, using bearings PA or QA, PB or QB, and so on, and lengths PA or QA, PB or QB, and so on.

The second worked example in section 10.20 illustrates the setting out of a circular curve by intersection from nearby traverse stations.

10.15 Obstructions to Setting Out

If care has been taken in route location and in the choice of a suitable radius there should be no obstruction to setting out other than the need to clear the ground surface. However, should obstructions arise, one of the coordinate methods of setting out described in section 10.14 can be used to set out the sections of the curve on either side of the obstruction to allow work to proceed. Once the obstruction has been removed, the same method can be employed to establish the missing section of the centre line.

10.16 Plotting the Centre Line on a Drawing

Although most highway alignment drawings are now produced in conjunction with highway design software packages on one of the wide range of
multi-pen plotters currently available, there are still occasions during the
initial design process when it is necessary to undertake a hand drawing of
the centre line – for example, to see if it is acceptable and falls correctly
within the band of interest as discussed in section 10.8. In order to do this,
the following procedure is recommended for each of the circular curves used
in the design. It assumes that there is an existing plan of the area available.

(1) Draw the intersecting straights in their correct relative positions on a
    sheet of tracing paper.
(2) Calculate the length of each tangent using the formula $R \tan(\theta/2)$.
(3) Plot the tangent points by measuring this distance along each straight
    on either side of the intersection point at the same scale as the existing
    plan.
(4) Using either *offsets from the tangent length* or *offsets from the long
    chord* as described in section 10.13, draw up a table of offset values
    (X) for suitable Y values using the appropriate formula. Ensure that the
    Y values chosen will provide a good definition of the centre line.
(5) At the scale of the existing plan, plot the X and Y values on the tracing
    paper to establish points on the centre line and carefully join these to
    define the curve. A set of French curves is useful for this purpose al-
    though, with care, a flexicurve can be used.
(6) Superimpose the tracing paper on the existing plan and decide whether
    or not the design is acceptable. If it is not, change the design and repeat
    the plotting procedure.

10.17 Compound Circular Curves

These consist of two or more consecutive circular curves of different radii
without any intervening straight section.

The object of such curves is to avoid certain points, the crossing of which
would involve great expense and which cannot be avoided by a simple circular curve.

Today they are uncommon since there is a change in the *radial force* (see
section 11.1) at the junction of the curves which go to make up the compound curve. The effect of this, if the change is marked, can be to give a
definite jerk to passengers, particularly in trains.

To overcome this problem, either very large radii should be used to minimise the forces involved or transition curves should be used instead of the
compound curve.

A typical two-curve compound curve is shown in figure 10.17. In figure 10.17, AB = common tangent through T_c and $(\alpha + \beta) = \theta$.

The design of such a curve is best done by treating the two sections separately and choosing suitable values for α, β, R_1 and R_2 and proceeding as for two simple circular curves, that is, T_1T_c and T_cT_2.

In compound circular curves, the tangent lengths IT_1 and IT_2 are not equal.

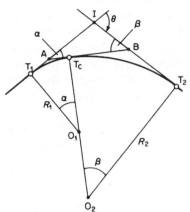

Figure 10.17 *Compound curve*   Figure 10.18 *Reverse curve*

10.18 Reverse Circular Curves

These curves consist of two consecutive curves of the same or different radii without any intervening straight section and with their centres of curvature falling on opposite sides of the common tangent. They are much more common than compound circular curves and, like such compound curves, they can be used to avoid obstacles. Often, however, they are used to connect two straights which are very nearly parallel and which would otherwise require a very long simple circular curve.

A typical reverse circular curve is shown in figure 10.18. In order to connect the two straights T_1I_1 and T_2I_2 it is necessary to introduce a third straight I_1I_2. A trial and error method using several different straights is employed until a suitable point, T_c, is chosen.

Once the point T_c has been decided, the reverse curve can be considered as two separate simple curves with no intermediate straight section, that is, T_1T_c and T_cT_2.

With reference to figure 10.18, $T_1I_1 = I_1T_c$ and $T_cI_2 = I_2T_2$ but I_1T_c does not necessarily equal T_cI_2.

10.19 Summary of Circular Curves

Although circular curves are straightforward in nature, much of their terminology also applies to transition curves and it is vital, therefore, that a good
understanding of circular curves is attained before proceeding to study transitions.

With the current widespread use of highway design software packages as
discussed in section 10.9, the design of circular curves tends to be done by
computer with the deflection angle and radius value being input and amended
as necessary until a suitable design is finalised. Of all the various types of
horizontal curves available, those with a constant radius are the easiest to
design and have the simplest setting-out calculations. As a result, they tend
to be tried first to see if they are suitable. However, they cannot always be
used owing to limitations on their minimum radii as specified by the Department of Transport (DTp).

If they cannot be used in isolation, circular curves can be combined with
transition curves to form *composite curves*. It is usually possible to design a
composite curve to fit any reasonable combination of deflection angle and
radius. Transition curves, composite curves and the restrictions on radii specified by the DTp are discussed much more fully in chapter 11.

Increasingly, however, because of the use of highway design software
packages, there is a tendency to eliminate circular curves and transition curves
from the design altogether and instead to use curves of constantly changing
radius. These are known as *polynomials* because their equations take the
form of cubic polynomials, an example of such a curve being a *cubic spline*.
These are true computer based curves which are generated by the highway
design software once any design constraints have been input, for example,
the positions of intersection points, the coordinates of points through which
the curve must pass and the locations of points which must be avoided.
Further information on these types of curves is also given in section 11.23
of chapter 11.

10.20 Worked Examples

(1) Setting Out by the Tangential Angles Method

Question
It is required to connect two straights whose deflection angle is 13°16'00"
by a circular curve of radius 600 m.

Make the necessary calculations for setting out the curve by the tangential
angles method if the through chainage of the intersection point is 2745.72 m.

Use a chord length of 25 m and sub-chords at the beginning and end of the curve to ensure that the pegs are placed at exact 25 m multiples of through chainage.

Solution
Consider figure 10.19

$$tangent\ length = R \tan \theta/2 = 600 \tan 6°38'00'' = 69.78\ m$$

Therefore

$$through\ chainage\ of\ \mathrm{T} = 2745.72 - 69.78 = 2675.94\ m$$

To round this figure to 2700 m (the next multiple of 25 m) an initial sub-chord is required. Hence

$$length\ of\ initial\ sub\text{-}chord = 2700 - 2675.94 = 24.06\ m$$
$$length\ of\ circular\ curve = R\ \theta = (600 \times 13.2667 \times \pi)/180$$
$$= 138.93\ m$$

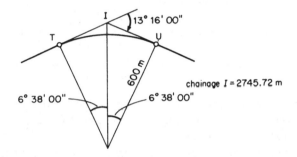

Figure 10.19

Therefore

$$through\ chainage\ of\ \ \mathrm{U} = 2675.94 + 138.93 = 2814.87\ m$$

Hence a final sub-chord is also required since 25 m chords can only be used up to chainage 2800 m. Therefore

$$length\ of\ final\ sub\text{-}chord = 2814.87 - 2800 = 14.87\ m$$

Hence three chords are necessary

initial sub-chord of 24.06 m
general chord of 25.00 m
final sub-chord of 14.87 m

The tangential angles for these chords are obtained from the formula $\alpha = 1718.9 \times$ (chord length/radius) min as follows

TABLE 10.1

Point	Chainage (m)	Chord length (m)	Individual tangential angle	Cumulative tangential angle
T	2675.94	0	00° 00′ 00″	00° 00′ 00″
C_1	2700.00	24.06	01° 08′ 56″ (α_1)	01° 08′ 56″
C_2	2725.00	25.00	01° 11′ 37″ (α_2)	02° 20′ 33″
C_3	2750.00	25.00	01° 11′ 37″ (α_3)	03° 32′ 10″
C_4	2775.00	25.00	01° 11′ 37″ (α_4)	04° 43′ 47″
C_5	2800.00	25.00	01° 11′ 37″ (α_5)	05° 55′ 24″
U	2814.87	14.87	00° 42′ 36″ (α_6)	06° 38′ 00″
		Σ 138.93 (checks)		

for initial sub-chord = 1718.9 × (24.06/600) = 68.93′ = 01°08′56″

for general chord = 1718.9 × (25.00/600) = 71.62′ = 01°11′37″

for final sub-chord = 1718.9 × (14.87/600) = 42.60′ = 00°42′36″

Applying these to the whole curve, the tabulated results are shown in table 10.1. The points on the centre line are designated C_1, C_2, C_3, C_4 and C_5 for use in the next worked example.

As a check, the final cumulative tangential angle shown in table 10.1 should equal $\theta/2$ within a few seconds. Also the sum of the chords should equal the total length of the circular arc.

Note that since α is proportional to the chord length any chords of equal length will have the same tangential angle and this is simply added to the cumulative total.

(2) Setting Out from Coordinates by Intersection

Question

The circular curve designed in the previous worked example is to be set out by intersection methods from two nearby traverse stations A and B. The position of the tangent point, T, is set out on the ground and its coordinates are obtained by taking observations to it from A and B. Observations taken from T to the intersection point, I, enable the whole-circle bearing of TI to be calculated as 63°27′14″.

The coordinates of A, B and T are as follows

A 829.17 m E, 724.43 m N
B 915.73 m E, 691.77 m N
C 798.32 m E, 666.29 m N

Using the relevant data from the previous worked example, calculate

(a) the coordinates of all the points on the centre line of the curve which lie at exact 25 m multiples of through chainage
(b) the bearing AB and the bearings from A required to establish the directions to all these points
(c) the bearing BA and the bearings from B required to establish the directions to all these points.

Solution
Figure 10.20 shows all the points to be set out together with traverse stations A and B.

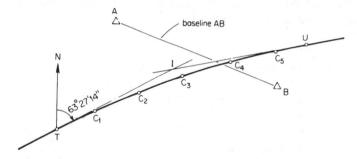

Figure 10.20

(a) Coordinates of all the points on the centre line

Coordinates of C_1
With reference to figure 10.21 and table 10.1

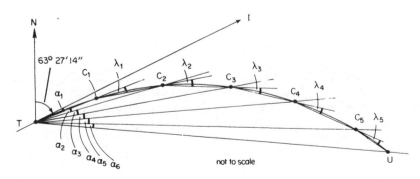

Figure 10.21

$$\text{bearing TC}_1 = \text{bearing TI} + \alpha_1$$
$$= 63°27'14'' + 01°08'56''$$
$$= 64°36'10''$$
$$\text{horizontal length TC}_1 = 24.06 \text{ m}$$

Therefore

$$\Delta E_{TC_1} = 24.06 \sin 64°36'10'' = +21.735 \text{ m}$$
$$\Delta N_{TC_1} = 24.06 \cos 64°36'10'' = +10.319 \text{ m}$$

Hence

$$\boldsymbol{E}_{C_1} = E_T + (\Delta E_{TC_1})$$
$$= 798.32 + 21.735 = \boldsymbol{820.055 \text{ m}}$$

$$\boldsymbol{N}_{C_1} = N_T + (\Delta N_{TC_1})$$
$$= 666.29 + 10.319 = \boldsymbol{676.609 \text{ m}}$$

These are retained with three decimal places for calculation purposes but are finally rounded to two decimal places.

Coordinates of C_2
With reference to figure 10.22 and table 10.1

$$\lambda_1 + (90° - \alpha_1) + (90° - \alpha_2) = 180°$$

Hence

$$\lambda_1 = \alpha_1 + \alpha_2$$
$$= 01°08'56'' + 00°11'37'' = 02°20'33''$$

Therefore

$$\text{bearing } C_1 C_2 = \text{bearing TC}_1 + \lambda_1$$
$$= 64°36'10'' + 02°20'33'' = 66°56'43''$$

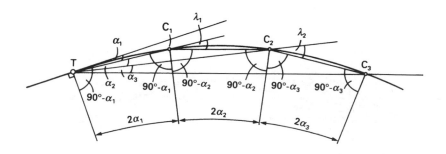

Figure 10.22

From table 10.1, horizontal length $C_1C_2 = 25.00$ m, therefore

$$\Delta E_{C_1C_2} = 25.00 \sin 66°56'43'' = +23.003 \text{ m}$$
$$\Delta N_{C_1C_2} = 25.00 \cos 66°56'43'' = + 9.790 \text{ m}$$

Hence

$$E_{C_2} = E_{C_1} + (\Delta E_{C_1C_2})$$
$$= 820.055 + 23.003 = \textbf{843.058 m}$$

$$N_{C_2} = N_{C_1} + (\Delta N_{C_1C_2})$$
$$= 676.609 + 9.790 = \textbf{686.399 m}$$

Coordinates of C_3
With reference to figures 10.21, 10.22 and table 10.1

$$\lambda_2 = \alpha_2 + \alpha_3$$
$$= 01°11'37'' + 01°11'37'' = 02°23'14''$$
bearing C_2C_3 = bearing $C_1C_2 + \lambda_2$
$$= 66°56'43'' + 02°23'14'' = 69°19'57''$$

From table 10.1, horizontal length $C_2C_3 = 25.00$ m, therefore

$$\Delta E_{C_2C_3} = 25.00 \sin 69°19'57'' = + 23.391 \text{ m}$$
$$\Delta N_{C_2C_3} = 25.00 \cos 69°19'57'' = + 8.824 \text{ m}$$

Therefore

$$E_{C_3} = E_{C_2} + (\Delta E_{C_2C_3})$$
$$= 843.058 + 23.391 = \textbf{866.449 m}$$

$$N_{C_3} = N_{C_2} + (\Delta N_{C_2C_3})$$
$$= 686.399 + 8.824 = \textbf{695.223 m}$$

Coordinates of C_4 and C_5
These are calculated by repeating the procedure used to calculate the coordinates of C_3 from those of C_2. The values obtained are

$$C_4 = \textbf{890.187 m E, 703.065 m N}$$
$$C_5 = \textbf{914.231 m E, 709.911 m N}$$

Coordinates of U
These are calculated twice to provide a check.

Firstly, they are calculated from point C_5 by repeating the procedure used to calculate the coordinates of C_3 from those of C_2. The values obtained are

$$U = \textbf{928.660 m E, 713.505 m N}$$

Secondly, they are calculated by working along the straights from T to I to U as follows

bearing TI = 63°27′14″
horizontal length TI = 69.78 m (see the previous worked example)

Hence

$$\Delta E_{TI} = 69.78 \sin 63°27′14″ = +62.423 \text{ m}$$
$$\Delta N_{TI} = 69.78 \cos 63°27′14″ = +31.186 \text{ m}$$

Therefore

$$E_I = E_T + (\Delta E_{TI})$$
$$= 798.32 + 62.423 = 860.743 \text{ m}$$

$$N_I = N_T + (\Delta N_{TI})$$
$$= 666.29 + 31.186 = 697.476 \text{ m}$$

From the previous worked example, $\theta = 13°16′00″$, hence

bearing IU = bearing TI + θ
$$= 63°27′14″ + 13°16′00″ = 76°43′14″$$
horizontal length IU = 69.78 m

Therefore

$$\Delta E_{IU} = 69.78 \sin 76°43′14″ = + 67.914 \text{ m}$$
$$\Delta N_{IU} = 69.78 \cos 76°43′14″ = + 16.029 \text{ m}$$

From which

$$E_U = E_I + (\Delta E_{IU})$$
$$= 860.743 + 67.914 = \mathbf{928.657 \text{ m}}$$

$$N_U = N_I + (\Delta N_{IU})$$
$$= 697.476 + 16.029 = \mathbf{713.505 \text{ m}}$$

These check, within a few millimetres, the values obtained for the coordinates of U calculated around the curve.

All the coordinates are listed in table 10.2 and have been rounded to two decimal places.

(b) Bearing AB and the bearings to the points from A

These are calculated from the coordinates of the points using either the quadrants method or by using rectangular/polar conversions as discussed in section 1.5. The bearings are listed in table 10.2.

(c) Bearing BA and the bearings to the points from B

Again, one of the methods discussed in section 1.5 is used. The bearings are listed in table 10.2.

TABLE 10.2

Point	Chainage (m)	Coordinates m E	m N	Bearing from A °	'	"	Bearing from B °	'	"
T	2675.94	798.32	666.29	207	56	59	257	45	23
C$_1$	2700.00	820.05(5)	676.61	190	47	34	260	59	44
C$_2$	2725.00	843.06	686.40	159	56	24	265	46	26
C$_3$	2750.00	866.45	695.22	128	04	39	274	00	33
C$_4$	2775.00	890.19	703.06(5)	109	17	51	293	51	17
C$_5$	2800.00	914.23	709.91	99	41	09	355	16	36
U	2814.87	928.66	713.50(5)	96	15	55	30	44	59

Bearing AB = 110° 40' 19"   Bearing BA = 290° 40' 19"

Further Reading

Department of Transport, Roads and Local Transport Directorate, *Departmental Standard TD 9/81, Road Layout and Geometry: Highway Link Design* (Department of Transport, 1981).

Department of Transport, Highways and Traffic Directorate, *Departmental Advice Note TA 43/84: Highway Link Design* (Department of Transport, 1984).

HMSO Publications, *Roads and Traffic in Urban Areas* (Institution of Highways and Transportation, with the Department of Transport, 1987).

11

Transition Curves

A transition curve differs from a circular curve in that its radius is constantly changing. As may be expected, such curves involve more complex formulae than curves of constant radius and their design can be complicated. Circular curves are unquestionably more easy to design than transition curves – they are easily set out on site – and so the questions naturally arise, why are transition curves necessary, and why is it not possible to use circular curves to join all intersecting straights?

11.1 Radial Force and Design Speed

The reason for the two types of curve is due to the *radial force* acting on the vehicle as it travels round the curve.

A vehicle travelling with a constant speed v along a curve of radius r is subjected to a radial force P such that $P = (mv^2/r)$, where m is the mass of the vehicle.

This force is, in effect, trying to push the vehicle back on to a straight course.

On a straight road, $r = $ infinity, therefore $P = 0$

On a circular curve of radius R, $r = R$,

therefore $P = (mv^2/R)$

Roads and railways are designed for particular speeds and hence v, the *design speed*, is constant for any given road; v is, in fact, the 85 percentile speed, that is, the speed not normally exceeded by 85 per cent of the vehicles using the road.

Similarly, the mass of the vehicle can be assumed constant, therefore $P \propto 1/r$, that is, the smaller the radius, the greater the force.

358

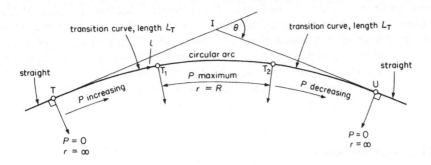

Figure 11.1 *Composite curve*

Therefore, any vehicle leaving a straight section of road and entering a circular curve section of radius R will experience the full force (mv^2/R) instantaneously.

If R is small, the practical effect of this is for the vehicle to skid sideways, away from the centre of curvature, as the full radial force is applied.

To counteract this, the Department of Transport (DTp) lay down minimum radii for wholly circular curves. These are discussed in section 11.3. If it is necessary to go below the minimum stipulated radius at a particular speed, transition curves must be incorporated into the design.

Transition curves are curves in which the radius changes from infinity to a particular value. The effect of this is to gradually increase the radial force P from zero to its highest value and thereby reduce its effect.

Consider a road curve consisting of two transitions and a circular curve as shown in figure 11.1.

For a vehicle travelling from T to U, the force P gradually increases from zero to its maximum on the circular curve and then decreases to zero again. This greatly reduces the tendency to skid and reduces the discomfort experienced by passengers in the vehicles. This is one of the purposes of transition curves; by introducing the radial force gradually and uniformly they minimise passenger discomfort. However, to achieve this they must have a certain property. Consider figure 11.1.

For a constant speed v, the force P acting on the vehicle is (mv^2/r). Since any given curve is designed for a particular speed and the mass of a vehicle can be assumed constant, it follows that $P \propto 1/r$.

However, if the force is to be introduced uniformly along the curve, it also follows that P must be proportional to l, where l is the length along the curve from the entry tangent point to the point in question.

Combination of these two requirements gives $l \propto 1/r$ or $rl = K$, where K is a constant. If L_T is the total length of each transition and R the radius of the circular curve, then $RL_T = K$.

Hence, if the transition curve is to introduce the radial force in a gradual and uniform manner it must have the property that the product of the radius of curvature at any point on the curve and the length of the curve up to that point is a constant value. This is the definition of a spiral and because of this, transition curves are also known as *transition spirals*. The types of curves used are discussed further in section 11.7.

A further purpose of transition curves is to gradually introduce *superelevation* and this is discussed in section 11.2.

11.2 Superelevation

Although transition curves can be used to introduce the radial force gradually in an attempt to minimise its effect, this effect can also be greatly reduced and even eliminated by raising one side of the roadway or one side of the track relative to the other. This procedure is shown in figure 11.2 and the difference in height between the road channels is known as the *superelevation*. By applying such superelevation, the resultant force (see figure 11.2) can be made to act perpendicularly to the road surface, thereby forcing the vehicle down on to the road surface rather than throwing it off.

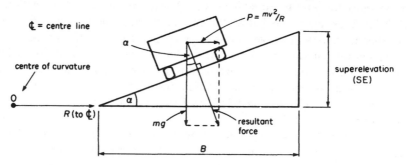

Figure 11.2 *Superelevation*

The maximum superelevation (SE) occurs when r is a minimum. With reference to figure 11.2

$$\tan \alpha = \frac{mv^2/R}{mg}$$

Therefore

$$\tan \alpha = \frac{v^2}{gR}$$

But

$$SE = B \tan \alpha$$

Therefore

$$maximum \ SE = \frac{Bv^2}{gR}$$

This value is constant on the circular curve and is gradually introduced on the entry transition curve and gradually reduced on the exit transition curve. If a wholly circular curve has been designed, between one-half and two-thirds of the superelevation should be introduced on the approach straight and the remainder at the beginning of the curve. The superelevation should be run out into the straight at the end of the curve in a similar manner.

For high design speeds, wide carriageways and small radii, the maximum superelevation will be very large and if actually constructed will be alarming to drivers approaching the curve. Also, any vehicle travelling below the design speed will tend to slip down the road surface and the driver will have to understeer to compensate. Should the maximum SE be constructed then any vehicle travelling at the design speed will travel round the curve without the driver needing to adjust the steering wheel.

Therefore, because of these aesthetic effects, the DTp stipulate maximum and minimum values for superelevation.

The DTp lay down the following rules for *maximum* superelevation.

(1) It should normally balance out only 45 per cent of the radial force P.
(2) It should not normally be steeper than 7 per cent (approximately 1 in 14.5) and, wherever possible, should be kept within the desirable value of 5 per cent (1 in 20).
(3) On sharp curves in urban areas superelevation shall be limited to 5 per cent.

Therefore, although the *maximum theoretical* SE $= (Bv^2/gR)$, in practice the *maximum allowable* SE $= 0.45 \ (Bv^2/gR)$. Also, once calculated, if this maximum allowable value gives a cross slope greater than 7 per cent then only 7 per cent should be used, for example, even if the design requires a superelevation of 10 per cent, only 7 per cent should be used.

The other 55 per cent of the radial force and any extra superelevation not accounted for in the final design are assumed to be taken by the friction between the road surface and the tyres of the vehicle. Hence the reason for vehicles skidding in wet or greasy conditions.

Expressing v in kph, R in metres and substituting for g gives the *maximum allowable superelevation* as

$$SE = \frac{Bv^2}{282.8R} \ metres$$

or, expressing the maximum allowable superelevation as a percentage, s, such that $s = 100(SE)/B$ gives

$$s \text{ per cent} = \frac{v^2}{2.828R} = \frac{v^2}{(2\sqrt{2})\,R}$$

These expressions for maximum allowable superelevation hold for values of R down to the absolute minimum values only (see section 11.3).

The *minimum allowable* SE, to allow for drainage, is 1 in 40, that is, 2.5 per cent.

Further details on superelevation can be found in the references given in Further Reading at the end of the chapter.

11.3 Current Department of Transport Design Standards

The DTp stipulates allowable radii for particular design speeds. These are discussed in a number of DTp publications; in particular *Departmental Standard TD 9/81, Road Layout and Geometry: Highway Link Design* which replaces previous DTp publications including *Layout of Roads in Rural Areas* and *Roads in Urban Areas*.

An advice note, TA 43/84, provides a useful guide to TD 9/81 and the application of both of these to highway link design is summarised in chapter 37 of the DTp publication *Roads and Traffic in Urban Areas*, from which table 11.1 has been reproduced. The full references for these publications are given in Further Reading at the end of the chapter.

11.4 Use of the Design Standards

It is strongly recommended that the DTp publication from which table 11.1 has been taken, together with its advice note, be studied in great detail before the commencement of any highway link design (see Further Reading at the end of the chapter).

The following example is included merely to illustrate the use of the standards in horizontal curve design work. Many factors, which are discussed in great detail in the standards themselves, may influence the final choice of design.

Examples of the use of table 11.1 in vertical curve design are discussed in chapter 12.

Question
Two intersecting straights on a section of a highway designed for a speed of 85 kph are to be joined using a horizontal curve. With reference to the current DTp design standards, summarise the various choices of radii that are available.

TABLE 11.1

*Current Department of Transport Highway Design Standards
(published here by permission of the Controller of Her Majesty's Stationery Office)*

DESIGN SPEED kph.		120	100	85	70	60	50	v²/R
A. STOPPING SIGHT DISTANCE m.								
A1 Desirable Minimum		295	215	160	120	90	70	
A2 Absolute Minimum		215	160	120	90	70	50	
B. HORIZONTAL CURVATURE m.								
B1 Minimum R * without elimination of Adverse Camber and Transitions		2880	2040	1440	1020	720	510	5
B2 Minimum R * with Superelevation of	2.5%	2040	1440	1020	720	510	360	7.07
B3 " " " "	3.5%	1440	1020	720	510	360	255	10
B4 Desirable Minimum R " "	5%	1020	720	510	360	255	180	14.14
B5 Absolute Minimum R " "	7%	720	510	360	255	180	127	20
B6 Limiting Radius " "	7%	510	360	255	180	127	90	28.28
at sites of special difficulty (Category B Design Speeds only)								
C. VERTICAL CURVATURE								
C1 FOSD Overtaking Crest K Value		*	400	285	200	142	100	
C2 Desirable Minimum * Crest K Value		182	100	55	30	17	10	
C3 Absolute Minimum " " "		100	55	30	17	10	6.5	
C4 Absolute Minimum Sag K Value		37	26	20	20	13	9	
D. OVERTAKING SIGHT DISTANCE								
D1 Full Overtaking Sight Distance FOSD m.		*	580	490	410	345	290	

* Not recommended for use in the design of single carriageways

Solution

From table 11.1, if a wholly circular curve is to be used, the minimum value of R must be 1440 m (row B1).

If transition curves are to be included in the design then the following radii are permissible for various superelevation values (rows B2 to B5):

for superelevation = 2.5 per cent, R must be ≥ 1020 m

for superelevation ≤ 3.5 per cent, R must be $\geq$ 720 m

for superelevation $\leq$ 5 per cent, R must be $\geq$ 510 m (Desirable Minimum)

for superelevation $\leq$ 7 per cent, R must be ≥ 360 m (Absolute Minimum)

In certain special cases, R can be lowered to 255 m (row B6) provided that 7 per cent superelevation is used. However, wherever possible, radii values greater than the desirable minimum ones should be used.

11.5 Use of Transition Curves

Transition curves can be used to join intersecting straights in one of two ways, either (1) in conjunction with circular curves to form *composite curves*, or (2) in pairs to form *wholly transitional curves*.

(1) Composite Curves

Figure 11.1 shows an example of a composite curve. In these, transition curves of equal length are used on either side of a circular curve of radius R.

Although this type of design has widespread use, it has the disadvantage that the radius and hence the radial force is constant on the circular section and, if this force is large, the length of the circular section represents a danger length over which the maximum force applies. The values given for limiting radii in table 11.1 do greatly reduce this occurrence but the use of transitions on their own with no intervening circular curve is sometimes preferred. A design method for composite curves is discussed in section 11.19.

(2) Wholly Transitional Curves

Figure 11.3 shows a wholly transitional curve consisting of two transitions of equal length.

Each of the transitions in this curve has a constantly changing radius and hence a constantly changing force, therefore there is only a short length over which the force is high and hence safety is increased. It is not always possible, however, to fit this type of curve between the two intersecting straights owing to DTp limitations on the minimum radii values that can be used. A design method for wholly transitional curves is discussed in section 11.20.

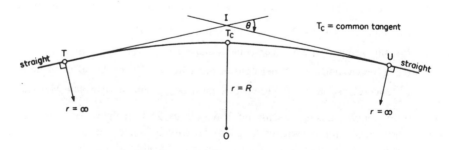

Figure 11.3 *Wholly transitional curve*

11.6 Length of Transition Curve to be Used (L_T)

Whatever length of transition curve is used, it must be checked to ensure that passenger discomfort is minimised. This depends on a parameter known as the *rate of change of radial acceleration* (*c*).

In practice, the value of *c* is kept below a certain maximum value and the length of curve is calculated from it. Consider figure 11.4.

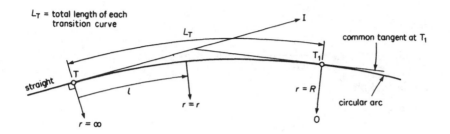

Figure 11.4 *Rate of change of radial acceleration*

The radial force at any point on the curve is given by $P = (mv^2/r)$, but force = mass × acceleration, hence the radial acceleration at any point on the curve is given by (v^2/r).

Since *v* is constant for any given curve, the radial acceleration is inversely proportional to the radius. Therefore, the rate at which the radial acceleration changes is inversely proportional to the rate at which the radius changes. The faster the change in radius, the greater the rate of change of radial acceleration and hence the faster the introduction of the radial force, resulting in a greater passenger discomfort. In effect, the shorter the transition, the greater the potential danger.

The transition curve must, therefore, be long enough to ensure that the radius can be changed at a slow enough rate in order that the radial force can change at a rate which is acceptable to passengers.

The rate of change of radial acceleration, therefore, should be treated as a safety or comfort factor the value of which has an upper limit beyond which discomfort is too great. The DTp recommended maximum value of *c* is 0.3 m/s³ although this can be increased to 0.6 m/s³ in difficult cases. In practice, whenever possible, long transitions with *c* values below 0.3 m/s³ should be used.

A summary of the design method together with the final choice of a *c* value is given in sections 11.19 and 11.20.

The length of transition to be used (L_T) can be obtained by consideration of the rate of change of radial acceleration (*c*) as follows. In figure 11.4

the radial acceleration at $T_1 = (v^2/R)$ and

the radial acceleration at T = zero

Therefore, the change in radial acceleration from T to $T_1 = (v^2/R)$, but the time taken to travel along the transition curve $= L_T/v$. Hence, the rate of change of radial acceleration $= c = (v^2/R)/(L_T/v)$. Therefore

$$c = \frac{v^3}{L_T R}$$

Hence

$$L_T = \frac{v^3}{cR} \quad \text{where } v \text{ is in m s}^{-1}$$

If v is in kph

$$L_T = \frac{v^3}{3.6^3 cR} \quad \text{metres}$$

and this is the formula used in the design of transition curves as shown in sections 11.19 and 11.20.

11.7 Type of Transition Curve to be Used

Although an expression for the length of the transition curve is now known, it is still not possible to set out the curve on site. This requires the equation of the transition.

In the following sections, two different transitions are considered, the *clothoid* and the *cubic parabola*.

In section 11.1 it was shown that for a transition curve the expression $rl = K$ must apply, that is, the radius of curvature must decrease in proportion to the length. The clothoid is such a curve and, because of this it is usually referred to as the *ideal transition curve* or the *ideal transition spiral*. It is discussed in section 11.8.

Another transition curve in common use is the cubic parabola and, although it does not have the property that rl is always constant, it can be used over a certain range and the design calculations involved are much simpler than those for the clothoid. The cubic parabola is discussed in section 11.9.

It is recommended that the clothoid section be studied first since much of the cubic parabola theory is based on it.

11.8 The Clothoid

Figure 11.5 shows two points M and N close together on the transition curve. ϕ is the *deviation angle* between the tangent at M and the straight

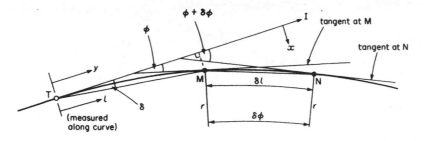

Figure 11.5 *Clothoid geometry*

IT; δ is the *tangential angle* to M from T with reference to IT; x is the offset to M from the straight IT at a distance y from T; l is the length from T to any point, M, on the curve.

The distance MN on the curve is considered short enough to assume that the radius of curvature at both M and N is the same. Therefore

$$\delta l = r\delta\phi$$

but it has been shown that $rl = K$ is required, hence substituting $r = l/K$ gives

$$\delta\phi = \frac{l}{K}\,\delta l$$

Integration gives $\phi = l^2/2K + $ constant, but when $l = 0$, $\phi = 0$, hence the constant $= 0$. Therefore

$$\phi = \frac{l^2}{2K}$$

but $K = rl = RL_\mathrm{T}$ hence

$$\phi = \frac{l^2}{2RL_\mathrm{T}} \quad (\phi \text{ being in radians})$$

This is the basic equation of the clothoid. If its conditions are satisfied and speed is constant, radial force will be introduced uniformly.

The maximum value of ϕ will occur at the common tangent between the transition and the circular curve, that is, when $l = L_\mathrm{T}$, hence

$$\phi_\mathrm{max} = \frac{L_\mathrm{T}}{2R} \quad (\text{in radians})$$

Setting Out the Clothoid by Offsets from the Tangent Length

Figure 11.6 shows an enlarged section of figure 11.5. Since M and N are close, it can be assumed that curve length MN is equal to chord length

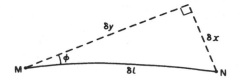

Figure 11.6

MN and expressions for δx and δy can be derived as follows.

$$\delta x = \delta l \sin \phi = \left(\phi - \frac{\phi^3}{3!} + \frac{\phi^5}{5!} - \ldots\right) \delta l$$

$$= [(l^2/2K) - (l^2/2K)^3/3! + (l^2/2K)^5/5! - \ldots] \delta l$$

Integration gives $x = \left(\frac{l^3}{6K}\right) - \left(\frac{l^7}{336\ K^3}\right) + \left(\frac{l^{11}}{42\ 240\ K^5}\right) - \ldots$

$$\delta y = \delta l \cos \phi = \left(1 - \frac{\phi^2}{2!} + \frac{\phi^4}{4!} - \ldots\right)\delta l$$

$$= [1 - (l^2/2K)^2/2! + (l^2/2K)^4/4! - \ldots]\delta l$$

Integration gives $y = \left[l - \left(\frac{l^5}{40\ K^2}\right) + \left(\frac{l^9}{3546\ K^4}\right) - \ldots\right]$

There are no constants of integration since $x = y = 0$ when $l = 0$. These formulae can be used to set out the clothoid as follows. In figure 11.7

(i) choose l and calculate x and y;
(ii) set out x at right angles to the tangent length a distance y from T towards I.

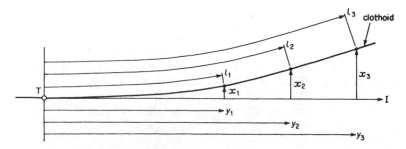

Figure 11.7 *Setting out a clothoid by offsets from the tangent length*

Since the formulae for x and y are both in the form of infinite series, their exact calculation is difficult. However, the third and subsequent terms in each expression tend to be very small and can be neglected. In the past special tables were produced listing values for these series up to and including the second terms. Nowadays, however, they can easily be evaluated using hand calculators and even these have been superseded by the wide range of highway design software packages currently available. These usually incorporate clothoids in their design procedures together with the option to use another type of curve, a *cubic spline*. The use of such packages is discussed in section 11.23.

Setting Out the Clothoid by Tangential Angles

With reference to figure 11.5, tan $\delta = x/y$, hence, by calculating x and y for a particular length l along the curve, δ can be calculated. Infinite series are again involved but, as discussed in (1) above, the third and subsequent terms can be neglected and calculators and computer packages have greatly simplified the calculations involved.

The calculation procedure and the method of setting out are identical to those for the cubic parabola and are dealt with in section 11.9.

11.9 The Cubic Parabola

As discussed in section 11.7, the cubic parabola is not a true spiral but approximates very closely to one over a certain range. Because of this, it cannot always be used (see (3) below) but its advantage of having simpler formulae and hence easier calculations has led to it being widely adopted. In practice, for the ranges over which it tends to be used, it can be considered to be identical to the clothoid from which its formulae are derived by making certain assumptions. The validity of these is discussed in (3) below.

Setting Out the Cubic Parabola by Offsets from the Tangent Length

In section 11.8, formulae involving infinite series were developed for setting out the clothoid by means of offsets from the tangent. The assumption is now made that the second and subsequent terms in these formulae can be neglected. Hence $x = (l^3/6K)$ and $y = l$. Substituting for l gives $x = (y^3/6K)$. But $K = rl = RL_T$ hence

$$x = \frac{y^3}{6RL_T}$$

This is the basic equation of the cubic parabola and it can be used to

set out the curve by offsets from the tangent lengths in a similar manner
to that shown for the clothoid in figure 11.7. In this case, however, since
it is assumed that the length is the same measured along the
curve or along the tangent, the offset, x, is calculated for different values
of y and set out as shown in figure 11.7.

Setting Out the Cubic Parabola by Tangential Angles

With reference to figure 11.5, $\tan \delta = x/y$. But $x = y^3/6RL_T$ for the cubic
parabola, hence

$$\tan \delta = (y^3/6RL_T)/y = \frac{y^2}{6RL_T}$$

Here another assumption is made in that only small angles are considered.
Therefore $\tan \delta = \delta$ radians, hence

$$\delta = \frac{y^2}{6RL_T} \text{ radians}$$

A useful relationship can be developed between δ, the tangential angle,
and ϕ, the deviation angle, as follows. $\phi = l^2/2RL_T$ is the basic equation
of the clothoid, but for the cubic parabola $y = l$. Therefore

$$\phi = \frac{y^2}{2RL_T} \text{ for the cubic parabola}$$

However

$$\delta = \frac{y^2}{6RL_T} \text{ for the cubic parabola}$$

Hence it follows that for the cubic parabola

$$\delta = \frac{\phi}{3} \text{ and } \delta_{max} = \frac{\phi_{max}}{3}$$

This relationship is shown in figure 11.8.

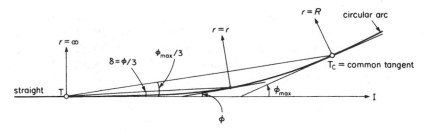

Figure 11.8 *Relationship between ϕ and δ*

In order that the tangential angles can be set out by theodolite an expression in terms of degrees or minutes is necessary, therefore

$$\delta = (y^2/6RL_T)(180/\pi)\,60 \text{ minutes}$$

Hence

$$\delta = \frac{1800}{\pi\,RL_T} \cdot l^2 \text{ minutes}$$

since $y = l$.
The actual setting-out procedure is as follows.

Setting out the first peg

(i) l_1 is chosen as a chord length such that it is $\leq R/20$, where R is the minimum radius of curvature.
(ii) δ_1 is calculated from l_1.
(iii) A theodolite is set at T, aligned to I with a reading of zero and δ_1 is turned off.
(iv) A chord of length l_1 is swung from T and lined in at point A as shown in figure 11.9.

Setting out the second and subsequent pegs

(i) δ_2 is calculated from l_2.
(ii) δ_2 is set on the horizontal circle of the theodolite.
(iii) A chord of length $(l_2 - l_1)$ is swung from A and lined in at point B using the theodolite as shown in figure 11.10.

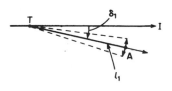

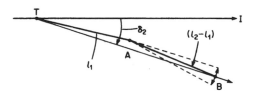

Figure 11.9   Figure 11.10

(iv) The system is repeated for all subsequent setting out points. Often, as with circular curves, a sub-chord is necessary at the beginning of the curve to maintain pegs at exact multiples of through chainage and hence a final sub-chord is often required to set out the common tangent between the transition and circular curves.

Validity of the Assumptions made in the Derivation of the Cubic Parabola Setting-out Formulae

Three assumptions are made during the derivation of the formulae.

(i) The second and subsequent terms in the expansion of sin ϕ and cos ϕ are neglected as being too small. This will depend on the value of ϕ.

(ii) Tan δ is assumed to equal δ radians. Since $\delta = \phi/3$ this will also depend on the value of ϕ.

(iii) y is assumed to equal l, that is, the length along the tangent is assumed to equal the length along the curve. Again, the value of ϕ will be critical since the greater the deviation, the less likely is this assumption to be true.

Hence, all the assumptions are valid and the cubic parabola can be used as a transition curve only if ϕ is below some acceptable value.

If the deviation angle remains below approximately 12°, there is no difference between the clothoid and the cubic parabola. However, beyond 12° the assumptions made in the derivation of the cubic parabola formulae begin to break down and, to maintain accuracy, further terms must be included, thereby losing the advantage offered by the simple equations. In fact, as shown in figure 11.11, once the deviation angle reaches 24° 06′ γ no longer equals ϕ, even if the formulae are expressed as infinite series, and the cubic parabola becomes useless as a transition curve because its radius of curvature begins to increase with its length, that is, rl is no longer constant. Hence, in theory, the cubic parabola can be used as a transition curve only if ϕ_{max} is less than approximately 24° but, in practice, it tends to be restricted to curves where ϕ_{max} is less approximately 12° and δ_{max} is less than approximately 4° (since $\phi_{max} = \delta_{max}/3$) in order that the simple formulae can be used.

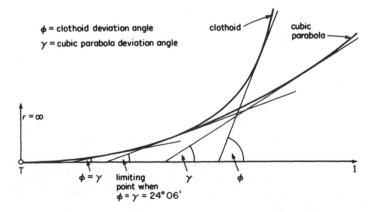

ϕ = clothoid deviation angle

γ = cubic parabola deviation angle

$r = \infty$

$\phi = \gamma$ limiting point when $\phi = \gamma = 24° 06'$ γ ϕ

Figure 11.11

11.10 Choice of Transition Curve

In practice both the clothoid and the cubic parabola are used. The angles involved are usually well below the limiting values for the cubic parabola and hence the final choice is usually one of convenience or habit. The clothoid can be used in any situation but has more complex formulae, the cubic parabola is easier to calculate by hand but cannot always be used. Nowadays, if highway design software packages are used, the question of choosing between these two becomes irrelevant since the clothoid, being the ideal transition curve, would always take precedence over the cubic parabola. However, another type of curve, a *cubic spline* is normally offered as an alternative to the clothoid in such packages. This is discussed further in section 11.23.

The remainder of the chapter is devoted to the cubic parabola simply because its formulae and calculations are easier to show in written form.

It would appear that there is enough known about the cubic parabola to enable it to be set out on the ground. This is not true as one parameter remains to be calculated and this is known as the *shift*. It is necessary to calculate the shift in order that a value can be obtained for the **tangent lengths**.

11.11 The Shift of a Cubic Parabola

Figure 11.12 shows a typical composite curve arrangement. The dotted arc between V and W represents a circular curve of radius $(R + S)$ which has been replaced by a circular curve T_1T_2 of radius R plus two transition curves, entry TT_1 and exit T_2U. By doing this the original curve VW has been shifted inwards a distance S, where $S = VG = WK$. This distance S is known as the shift.

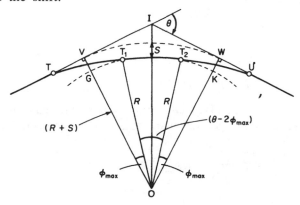

Figure 11.12 *Shift of a cubic parabola*

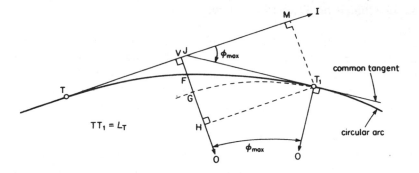

Figure 11.13

The tangent points and the lengths of the original curve and the new curve are not the same and the lengths of the circular arcs are not the same.

Figure 11.13 shows an enlargement of the left-hand side of figure 11.12. In quadrilateral VJT_1O

$$\text{angle OVJ} = \text{angle } JT_1O = 90°$$

Hence

$$\text{angle } IJT_1 = \text{angle } T_1OV = \phi_{max}$$

$$shift = S = VG = (VH - GH) = (MT_1 - (GO - HO))$$

But, from the cubic parabola equation

$$x = (y^3/6RL_T)$$

When $y = L_T$, $x = MT_1$, therefore

$$MT_1 = L_T^3/6RL_T$$

Hence

$$S = L_T^3/6RL_T - (R - R \cos \phi_{max})$$
$$= L_T^2/6R - R [1 - (1 - \phi^2_{max}/2! + \phi^4_{max}/4! - \dots)]$$

This expression for S involves an infinite series but, again assuming small deviation angles, terms greater than ϕ^2_{max} can be neglected as being too small. Hence

$$S = L_T^2/6R - R\phi^2_{max}/2!$$

Therefore

$$S = L_T^2/6R - (R/2)(L_T/2R)^2$$

Hence, the formula for the shift of a cubic parabola transition curve is

$$S = \frac{L_T^2}{24R}$$

The shift is an important parameter in the design and setting out of composite and wholly transitional curves. Once its value is known, the tangent lengths can be calculated. Consider figure 11.13 in which F is the point where the shift and the transition curve cross each other.

Since the angles involved are small, it is assumed that $FT_1 = GT_1$ and since GT_1 forms part of the circular curve and is equal to $R\phi_{max}$ it follows that $FT_1 = R\phi_{max}$. But $\phi_{max} = L_T/2R$, hence $FT_1 = R(L_T/2R) = L_T/2$. Hence FT must also equal $L_T/2$. Using the formula $x = y^3/6RL_T$ and the assumption that $y = l$, when $y = L_T/2$ and $x = VF$ then

$$VF = (L_T/2)^3/6RL_T = L_T^2/48R$$

But

$$VG = S = L_T^2/24R$$

Therefore

$$VF = \tfrac{1}{2} \times \text{shift} = FG$$

This gives the property that the shift at VG is bisected by the transition curve and the transition curve is bisected by the shift.

11.12 Tangent Lengths and Curve Lengths

Figure 11.14 shows the geometry of a composite curve with each shift bisecting each transition. With reference to this

$$IT = IV + VT = IW + WU = IU$$

In triangle IVO

$$IV = (R + S) \tan\left(\frac{\theta}{2}\right)$$

From section 11.11

$$VT = \frac{L_T}{2} = WU$$

Hence

$$IT = (R + S) \tan\left(\frac{\theta}{2}\right) + \frac{L_T}{2}$$

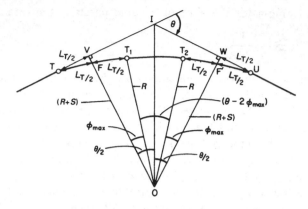

Figure 11.14 *Tangent and curve lengths*

This formula applies not only to a composite curve but also to a wholly transitional curve as inspection of figure 11.18 will show.

If the total length of the composite curve (L_{total}) in figure 11.14 is required, it can be obtained from either

$$L_{total} = TF + FF' + F'U = L_T/2 + R\theta + L_T/2$$

or

$$L_{total} = TT_1 + T_1T_2 + T_2U = L_T + R(\theta - 2\phi_{max}) + L_T$$

The total length of a wholly transitional curve is simply given by $2L_T$ since it does not contain a central circular section.

11.13 Establishing the Centre Line on Site

As discussed in section 10.10 of the circular curves chapter, the centre line provides an important reference on site from which other features can be established and it can be set out either by *traditional* or *coordinate* methods. Although these were defined in the circular curves chapter, they apply equally to composite and wholly transitional curves.

Again, the initial step is to obtain an accurate value of the deflection angle, θ, for use in the design calculations. In order to do this, it may be necessary first to set out the intersection point, as described in section 10.11, so that θ can be measured. Once the design has been completed, the tangent points which lie on each straight can be pegged out on site.

If traditional methods are used, the centre line is then set out from these tangent points as discussed in section 11.15 and as shown in the first worked example in section 11.24.

If coordinate methods are used, the coordinates of the tangent points are measured and then used in the calculations required to enable pegs to be located on the centre line from nearby control points. The procedures involved are described in section 11.16 and the second worked example in section 11.24 shows a typical set of calculations.

Both traditional and coordinate methods are used nowadays although coordinate techniques are normally preferred for all major curves for the reasons outlined in section 10.10. Traditional methods, however, still have their place and they are often more convenient and quicker to use when defining the centre lines of less important curves, for example, minor roads, boundaries, kerbs, housing estates and so on. If there are no control points nearby then only traditional methods can be used. The relative merits of coordinate and traditional methods are discussed in section 11.17.

11.14 Locating the Tangent Points on the Straights (T and U)

This method assumes that the intersection point, I, has been located (see section 10.11). With reference to figure 11.14

(1) Calculate the shift from $S = L_T^2/24R$.
(2) Calculate the tangent lengths from $IT = (R + S) \tan \theta/2 + L_T/2 = IU$.
(3) Measure back from I to locate T and forward from I along the other straight to locate U.

11.15 Setting Out the Curves by Traditional Methods

This section describes the traditional methods of setting out composite and wholly transitional curves from their tangent points. Modern methods, involving coordinates, are discussed in section 11.16.

The Entry and Exit Transition Curves

With reference to figure 11.14, the *entry* transition curve runs from tangent point T to the common tangent point T_1 and the *exit* transition curve runs from the common tangent T_2 to tangent point U. These can be set out using either offsets from the tangent lengths or by tangential angles as described in section 11.9.

In the case of a composite curve, the tangential angles method is preferred since a theodolite is required to set out the circular arc between the two common tangent points T_1 and T_2.

In the case of a wholly transitional curve, either can be used since there is only one common tangent point T_c and no central circular arc (see figure 11.18. However, the tangential angles method is the more accurate.

(a) The entry transition is set out from point T with point T_1 being the last to be pegged out on a composite curve and point T_c being the last to be pegged out on a wholly transitional curve.

(b) The exit transition is set out from point U to point T_2. If offsets are used, they are set out on the opposite side of the tangent length to those used for the entry transition curve. If tangential angles are used, the theodolite is set at U, aligned on I with the horizontal circle reading 360°00′00″ and the method carried out with the tangential angles being subtracted from 360°00′00″. For composite curves, point T_2 is established. For wholly transitional curves, point T_c is set out again, having already been fixed at the end of the entry transition; the difference between its two positions gives a measure of the accuracy of the setting out.

The Central Circular Arc

This is normally set out from T_1 to T_2 and it is first necessary to establish the line of the common tangent at T_1.

The final tangential angle from T to T_1 will be $\delta_{max} = \phi_{max}/3$. This is shown in figure 11.15. The procedure is as follows.

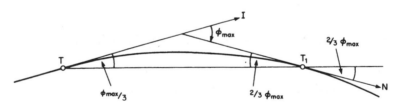

Figure 11.15

(i) Move the theodolite to T_1, align back to T with the horizontal circle reading $180° - (2\phi_{max}/3)$.

(ii) Rotate the telescope in azimuth until a reading of 00°00′ 00″ is obtained. This is the common tangent along T_1N.

(iii) Set out the circular arc from T_1 to T_2 using tangential angles calculated from the circular curve formula

$$\alpha = 1718.9 \times (\text{chord length/radius}) \text{ minutes}$$

(iv) Finally, point T_2, the second common tangent point is established. Since T_2 is also fixed when setting out the exit transition curve from point U, the difference between its two positions gives a measure of the accuracy of the setting out.

11.16 Setting Out the Curves by Coordinate Methods

Section 11.15 described the traditional methods of establishing the various curves involved in the centre line from their tangent points. Although these traditional techniques are still used, they have been virtually super-seded for all major curves by methods which use control points located some distance from the proposed centre line. In these, the coordinates of points at regular intervals along the centre line are calculated and the points are then pegged out usually by bearing and distance methods from nearby control points using theodolite/EDM instruments or total stations. They are generally referred to as *coordinate* methods and they can be used for any type of curve, being equally applicable to transitions and circulars (as discussed in section 10.10). Nowadays, the calculations involved are often done within computer software highway design packages and section 11.23 discusses the wide range of these which is currently available. The results of such computations are normally presented in the form of com-puter printouts ready for immediate setting out use on site.

Table 11.2 shows a typical format for a printout and gives all the infor-mation required to set out the curve shown in figure 11.16. The curve is to be set out by polar coordinates from nearby traverse stations, each centre line point being established from one station and checked from another. The calculations undertaken to produce table 11.2 are summarised as follows.

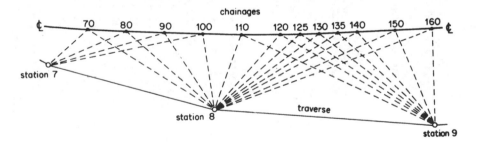

Figure 11.16

(1) The coordinates of the traverse stations are found from the original site traverse.
(2) The horizontal alignment is designed and the intersection and tangent points located on the ground. They are incorporated into the original site traverse and their coordinates calculated.
(3) Suitable chord lengths are chosen to ensure that the centre line is pegged at exact multiples of through chainage and the tangential angles are calculated for the entry and exit transition curves and the central circular arc.

TABLE 11.2
Example Computer Printout Format

JOB REFERENCE JO1777
PORTSMOUTH RAIL BRIDGE MAIN ALIGNMENT CH 70 TO CH 160

CENTRELINE CHAINAGE TO BE SET OUT	STATION REFERENCE NUMBER	WHOLE CIRCLE BEARING DEG. MIN. SEC.			HORIZONTAL DISTANCE FROM STATION TO CENTRELINE (M)	STATION REFERENCE NUMBER	WHOLE CIRCLE BEARING DEG. MIN. SEC.			HORIZONTAL DISTANCE FROM STATION TO CENTRELINE (M)
70	7	24	10	57	13.695	8	276	02	31	38.734
80	7	41	38	08	22.183	8	285	31	19	30.504
90	7	49	00	38	31.585	8	301	13	45	23.726
100	7	52	53	24	41.281	8	325	47	54	19.939

WHOLE CIRCLE BEARING FROM STATION 7 TO STATION 8 = 79 DEG. 12 MIN. 09 SEC.
WHOLE CIRCLE BEARING FROM STATION 8 TO STATION 7 = 259 DEG. 12 MIN. 09 SEC.

CENTRELINE CHAINAGE TO BE SET OUT	STATION REFERENCE NUMBER	WHOLE CIRCLE BEARING DEG. MIN. SEC.			HORIZONTAL DISTANCE FROM STATION TO CENTRELINE (M)	STATION REFERENCE NUMBER	WHOLE CIRCLE BEARING DEG. MIN. SEC.			HORIZONTAL DISTANCE FROM STATION TO CENTRELINE (M)
110	8	354	04	27	20.844	9	270	04	27	54.792
120	8	15	22	10	25.954	9	275	26	51	45.969
125	8	22	43	13	29.480	9	278	59	41	41.768
130	8	28	24	28	33.392	9	283	19	50	37.772
135	8	32	51	00	37.571	9	288	40	14	34.052
140	8	36	22	09	41.937	9	295	15	48	30.709
150	8	41	30	43	51.034	9	313	00	40	25.732
160	8	45	01	39	60.436	9	335	51	19	24.166

WHOLE CIRCLE BEARING FROM STATION 8 TO STATION 9 = 68 DEG. 34 MIN. 09 SEC.
WHOLE CIRCLE BEARING FROM STATION 9 TO STATION 8 = 248 DEG. 34 MIN. 09 SEC.

(4) The coordinates of the points to be established on the centre line are calculated using the chord lengths, tangential angles and the coordinates of the intersection and tangent points.

(5) The nearest two traverse stations which are visible from and which will give a good intersection to each proposed centre line point are found and the polar coordinates calculated from each traverse station to the centre line point.

(6) The computer repeats this procedure for all the points on the centre line and a printout is obtained with a format similar to that shown in table 11.2.

An example showing the calculations involved when a composite curve is to be set out from coordinates is given in section 11.24.

11.17 Coordinate Methods Compared with Traditional Methods

When compared with the traditional methods of setting out from the tangent points, coordinate methods have a number of important advantages. However, they are not always the most appropriate. Some of the relative merits of the two categories of technique are listed below.

(1) Coordinate methods can be carried out by anyone who is capable of using a theodolite/EDM system or a total station. Since the data is in the form of bearings and distances, no knowledge of curve design is necessary. This is not the case with traditional methods.

(2) The increased use of highway design computer software packages in which the setting-out data is presented ready for use in coordinate form has produced a corresponding increase in the adoption of such methods.

(3) The widespread use of calculators and computers has also greatly speeded up the calculation procedures associated with coordinate methods which were always perceived to be much more difficult to perform by hand when compared with those associated with the traditional methods.

(4) Traditional methods require tangent points to be occupied. This can delay the construction process. Coordinate methods enable the work to proceed unhindered. This is very important since on any site involving centre lines, the pegs will inevitably be disturbed by the construction process and they will need to be re-established as each stage of the work is completed.

(5) Any disturbed centre line pegs can quickly be relocated from the control stations in coordinate methods since the control points will be well protected and located away from site traffic. In the tangential angles method, however, the tangent points will have to be re-

occupied and these can often themselves be lost during construction, requiring extra work in their relocation.

(6) In coordinate methods, each peg on the centre line is fixed independently of all the other pegs on the centre line. This ensures that any error made when locating one peg is not carried forward to the next peg as can occur in the tangential angles method.

(7) Coordinate methods enable key sections of the centre line to be set out in isolation, for example, a bridge centre line, in order that work can progress in more than one area of the site.

(8) Obstacles on the proposed centre line which may be the subject of disputes can easily be by-passed using coordinate methods to allow work to proceed while arbitration takes place. Once the obstacle is removed, it is an easy process to establish the missing section of the centre line. This is not usually possible with traditional methods.

(9) Coordinate methods have the disadvantage that there is very little check on the final setting out. Large errors will be noticed when the centre line does not take the designed shape but small errors could pass unnoticed. In traditional methods, checks are provided by locating common tangent points from two different positions as described in section 11.15.

(10) Although the widespread use of theodolite/EDM systems and total stations on sites encourages the use of coordinate techniques, such equipment may not always be available and it may be simpler to use traditional methods which work along the centre line. This will particularly be the case if the control points from which the coordinate methods are to be used are located long distances from the centre line where accurate taping becomes very difficult. Although intersection from two control points is possible in such cases, this can be a slow process and three people are required, one on each theodolite and one on the centre line locating the pegs.

11.18 Plotting the Centre Line on a Drawing

As discussed in section 10.16 for circular curves, despite the widespread use of computer plotting facilities, there are still occasions during the initial horizontal alignment design process when it is necessary to undertake a hand drawing of the centre line. For composite and wholly transitional curves, the following procedure is recommended. It assumes that there is an existing plan of the area available.

(1) Draw the intersecting straights in their correct relative positions on a sheet of tracing paper.

(2) Calculate the length of each tangent using the formula $(R + S)\tan(\theta/2) + L_T/2$.

(3) Plot the tangent points by measuring this distance along each straight on either side of the intersection point at the same scale as the existing plan.

(4) To plot the *entry and exit transition curves*, use offsets from the tangent length as described in section 11.8. Draw up a table of offset values (x) for suitable y values using the formula $(x = y^3/6RL_T)$. Ensure that the y values chosen will provide a good definition of the centre line.

(5) At the scale of the existing plan, plot the x and y values on the tracing paper from the tangent lengths to establish points on the entry and exit transition curves.

(6) To plot the *central circular arc* (where appropriate), carefully join the ends of the entry and exit transition curves plotted in (5). This is the long chord of the central circular arc.

(7) Using the formula given in section 10.13 for offsets from the long chord, draw up a table of offset values (X) for appropriate Y values. Again ensure that the Y values chosen will provide a good definition of the centre line.

(8) At the scale of the existing plan, plot the X and Y values from the long chord to establish points on the central circular arc.

(9) Carefully join the points plotted in (5) and (8) to define the complete centre line. A set of French curves is useful for this purpose although, with care, a flexicurve can be used.

(10) Superimpose the tracing paper on the existing plan and decide whether or not the design is acceptable. If it is not, change the design and repeat the plotting procedure.

11.19 A Design Method for Composite Curves

Composite curves were defined in section 11.5. Consider figure 11.17 which shows a composite curve that is to be designed. The design is based on the fact that the composite curve must deflect the road through angle θ. Of this, the circular curve takes $(\theta - 2\phi_{max})$ and each transition takes ϕ_{max}.

Given: design speed, v, and the road type.
Problem: to calculate a suitable curve to fit between the straights TI and IU.
Solution: Before detailing the design method, it must be noted that there are many solutions to this problem, all of them perfectly acceptable. Hence, the following method can only be a guide to design from which a suitable rather than a unique solution can be found. The procedure is based on the DTp design standards given in table 11.1 and is as follows.

(1) The deflection angle, θ, must, if possible, be accurately measured on site. This is discussed in section 11.13.

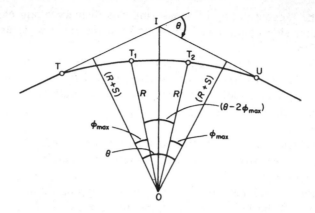

Figure 11.17

(2) Use a value of R greater than the desirable minimum radius for the design speed and road type in question and let $c = 0.3$ m/s³, that is, start off with the recommended limiting values for both R and c so that they can be amended later if necessary.

(3) Calculate the length of each transition from $L_T = v^3/3.6^3 \, cR$.

(4) Calculate the shift, S, from $S = L_T^2/24R$.

(5) Calculate the tangent lengths IT and IU from $(R + S) \tan \theta/2 + L_T/2$.

(6) The working drawings should show the two straights superimposed on the existing area. The calculated lengths IT and IU should now be fitted on the plan to see if they are acceptable. Owing to the *band of interest* discussed in section 10.8, it may be necessary to alter the lengths of IT and IU in order to obtain a suitable fit. This can be done by altering R and/or c.

Ideally, R should be greater than the appropriate desirable minimum value (given in row B4 of table 11.1) and c should normally not exceed 0.3 m/s³. Wherever possible, the value of c should be kept well below 0.3 m/s³ in order to ensure long transition curves and thereby increase the safety aspects of the design. However, it may not always be possible to fit in long transitions and, in difficult cases, R can be reduced to the limiting value with 7 per cent superelevation (given in row B6 of table 11.1) to reduce the effect of the large radial force that may result and c can be increased to a value of 0.6 m/s³.

The process is an iterative one and ends when the tangent lengths are of an acceptable length to fit the given situation.

(7) Once a suitable radius has been found, calculate ϕ_{max} from $L_T/2R$ radians.

(8) Calculate $(\theta - 2\phi_{max})$, hence the length of the circular arc from $R(\theta - 2\phi_{max})$.

(9) Calculate the superelevation (see section 11.2).

(10) Set out the curve on site using one of the methods discussed earlier.

The first two examples given in section 11.24 show the calculations involved when setting out a composite curve.

11.20 A Design Method for Wholly Transitional Curves

As defined in section 11.5, these are curves which consist only of transitions. They can be considered as a composite curve which has a central circular arc of zero length. Figure 11.18 shows such a curve.

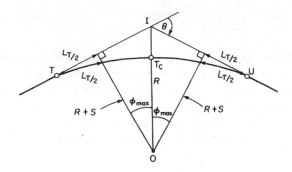

Figure 11.18 *Wholly transitional curve*

Wholly transitional curves have the advantage that there is only one point at which the radial force is a maximum and, therefore, the safety is increased. Unfortunately, it is not always possible to fit a wholly transitional curve into a given situation.

This section deals with the design of wholly transitional curves with equal tangent lengths only. Although it is possible to design and construct wholly transitional curves with unequal tangent lengths by using a different rate of change of radial acceleration for each half of the curve, they are rarely used and space does not permit a discussion on their method of design.

Wholly transitional curves with equal tangents have a very interesting property. With reference to figure 11.18, since the circular arc is missing, it follows that

$$\theta = 2\phi_{max} \text{ but } \phi_{max} = L_T/2R$$

Hence

$$\theta = 2L_T/2R = L_T/R$$

Therefore, for a wholly transitional curve

$$\theta = \frac{L_T}{R} \tag{11.1}$$

In addition, all the other transition curve equations still apply and consequently the equation for length must still apply, that is

$$L_T = v^3/3.6^3 \ cR \tag{11.2}$$

From equations (11.1) and (11.2)

$$R\theta = v^3/3.6^3 \ cR$$

Therefore

$$R = \sqrt{\frac{v^3}{3.6^3 \ c\theta}} \ \text{metres}$$

This leads to the property of wholly transitional curves that for any given two straights there is only one symmetrical wholly transitional curve that will fit between them for a given design speed if the rate of change of radial acceleration is maintained at a particular value, that is, since v and θ are usually fixed, R has a unique value if c is maintained at, say, 0.3 m/s^3.

This is, in fact, the method of designing such curves and it is summarised as follows.

(1) Choose a value for c, ideally less than 0.3 m/s^3 as discussed in section 11.19.
(2) Substitute this into the equation $R = (v^3/3.6^3 \ c\theta)^{\frac{1}{2}}$ and hence calculate the minimum radius of curvature.
(3) The radius value must be checked against the DTp values given in table 11.1. R must, if possible, be greater than the desirable minimum value and must always be greater than the limiting value.
   If R checks, L_T can be calculated using either equation (11.1) or equation (11.2).
   If R does not check then the value of c must be reduced and the calculation repeated.
   The third example given in section 11.24 shows the way in which the radius value is checked.
(4) Having calculated L_T it is necessary to ensure that the curve will fit within the band of interest (see section 10.8). For a quick check, the assumption can be made with wholly transitional curves that the length along the tangent is equal to the length of the transition curve. This is shown in figure 11.19. However, when the design is finalised, the formula previously derived for IT and IU should be used, namely

$$IT = IU = (R + S) \tan \theta/2 + L_T/2$$

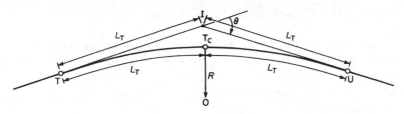

Figure 11.19

Hence, the tangent length is checked for fit on the working drawings. If it does not fit, it is necessary to return to the start of the calculations and change some of the variables, either v, θ or c. It is not always possible to fit a wholly transitional curve between straights within the limits stipulated by the DTp.

(5) The superelevation is calculated and the curve is set out by either tangential angles, offsets or coordinates.

11.21 Phasing of Horizontal and Vertical Alignments

It is very unusual for a horizontal alignment to be designed in isolation since allowance must also be made for the change in height of the ground surface along the proposed centre line. This requires consideration of the vertical shape of the centre line and leads to the design of the *vertical alignment*. Just as horizontal curves are used to join intersecting straights in the design of the horizontal alignment, *vertical curves* are used to join intersecting *gradients* in the design of the vertical alignment.

A full description of the design and setting out of vertical curves is given in chapter 12 but, because of the interdependence of the horizontal and vertical alignments, each must be considered during the design of the other. In practice, they should be correctly *phased*, that is, their tangent points should coincide to ensure that they are the same length. If this is the case, it will avoid the creation of optical illusions in the road surface which could distract drivers.

Hence, before finalising the design of the horizontal alignment, the total length of each composite or wholly transitional curve involved should be calculated using one of the formulae given in section 11.12 and compared with the length of its equivalent vertical curve as appropriate. The length of either the horizontal or vertical curve should then be changed to ensure that the two curves are equal. Normally, it is the vertical curve length that is changed since this is usually easier to do. This need to equate the horizontal and vertical alignments is discussed in section 12.10.

11.22 Summary of Horizontal Curve Design

In sections 10.8, 11.19 and 11.20, methods for designing wholly circular, composite and wholly transitional curves were discussed. Usually, these three techniques are combined into one general design and considered as possible solutions to the same problem, the aim being to design the best curve to fit a particular set of conditions.

Often, only the design speed and class of road are known and the problem becomes one of choosing the ideal combination of θ, R and c to fit into the band of interest concerned while maintaining current design standards.

If a vertical curve is designed in conjunction with the horizontal curve, the problem is further complicated by the need to phase the two curves correctly.

Hence, when undertaken manually, the design can be tedious and time consuming. Fortunately, the iterative processes involved are ideal for solution by computer and such methods are now in widespread use. The basic steps of the design are written into the computer program and the curve parameters, v, θ, c and R together with chainage values, reduced levels and any external constraints are fed into the computer which runs the program and calculates suitable values for the radius of curvature, deflection angle, rate of change of radial acceleration, superelevation values, tangential angles, chord lengths and so on. These results are presented in list form on a printout from the computer.

In addition, if the program is suitably modified and coordinates of nearby traverse stations are fed in, as discussed in section 11.16, polar coordinates for setting out the curve can be obtained similar to those shown in table 11.2. The first two examples given in section 11.24 show the steps involved.

Although many engineers and surveyors have written personal horizontal and vertical alignment design programs for use with their own computers and programmable calculators, there has been such a huge development in commercially available highway design software packages in recent years that these are now in widespread use. They cover all stages of the design procedure, starting with trial alignments and continuing to the production of long-section and cross-sectional drawings, the listing of setting-out data, the calculation of volumes, the planning of the movement of materials and, in some cases, even the preparation of a computer graphics 'driver's-eye view' of how the proposed design will look on completion.

These packages have revolutionised the design process by freeing engineers from the tedious calculations and allowing them to concentrate on the important design concepts. Different parameters can be tried and different designs compared in a very short period of time.

The following section details the evolution of such packages and discusses the general concepts on which they are based.

11.23 Computer-Aided Road Design

In highway alignment design, many factors such as design standards, topography, environment and the visual impact of the road have to be considered. This creates a demand for a number of alternative routes to be studied for any given road scheme and, for each route, the ability to produce a visual representation or model of the proposed road is highly desirable as a means of checking design work and for presentation at public hearings.

The preparation of different alignments by hand methods involves much work and the production of the various drawings for each design manually is an almost impossible task. However, by using a computer system in road design, these problems can be overcome to such an extent that many trial designs can be studied and presented with relative ease.

As a consequence, considerable emphasis has been placed on the development of highway design software packages and there is now an enormous range of these available from commercial surveying and software companies. This has not always been the case and it was not until the 1970s that a number of packages for highway design appeared. At that time, the two most widely used in Great Britain were BIPS (The British Integrated Program Suite for Highway Design) which ran on mainframe computers and MOSS (MOdelling SyStems) which took advantage of the minicomputers (as opposed to the later microcomputers) which evolved towards the end of that decade. However, the arrival of the relatively inexpensive PC microcomputers in the 1980s and the parallel development of integrated Total Stations for surveying fieldwork, as discussed in section 9.11, have caused a fundamental change in surveying practices.

Computers and software are now involved in all stages of survey work from the initial data collection on site, through all the analysis and design procedures to the final output of graphical, numerical and setting-out information. This has led to the development of the present wide range of commercially produced software packages for many surveying activities (see section 9.11). Those currently available for highway design include names such as *ProSURVEYOR* from Applications in CADD, *MicroBIPS* and *VALOR* from Brockwood Systems, *STRINGS* from Geodetic Software Systems, *panTERRA* from Ground Modelling Systems, *LSS* from Hall & Watts Systems, *Cadsite* from JTC Computer Systems, *MOSS* from MOSS Systems, *NRG* from NRG Surveys, *PC Road Engineer* from National Survey Software, *STARDUST for Windows* from Softcover International, *SDRDesign* from Sokkia and *CIVILCAD* from Survey Supplies. Given this wide choice, it is not possible to give a detailed review of such individual packages in a general textbook such as this. For further information on their capabilities and costs, however, the reader is recommended to study the two excellent articles by Mike Fort referenced in Further Reading at the end of this chapter.

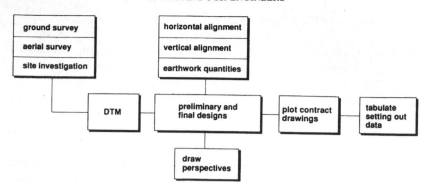

Figure 11.20 *Stages involved in computer–aided road design*

Although individual packages are not reviewed here, the general concepts on which they are based tend to be similar and the block diagram of figure 11.20 shows the various stages involved. These are briefly described as follows.

Initially, a digital terrain model (DTM) is produced of the area covered by the corridor or band of interest. The DTM is formed using air or ground survey methods as described in section 9.12 and is essentially a map of the area stored digitally in a computer. In addition to surface information, the results of any site investigations can also be stored in the DTM. Such data may include ground-water conditions, geotechnical characteristics of the area and any other properties which may affect the design.

After the DTM has been completed, many trial alignments can be studied by the computer. For horizontal alignments, two methods are used by the computer: the conventional method in which straight sections of road are joined by circular and/or transition curves (see chapter 10 and previous sections of this chapter) or a new technique based on curves known as *cubic splines*. A cubic spline is a curve of continually changing radius, the equation of which takes the form of a cubic polynomial. Cubic splines are specified to fit between given location points, for example, straights, end points of a scheme, points the curve must pass through to avoid obstacles and so on.

For vertical alignment design (see chapter 12), three methods are used by the computer: the traditional method based on intersecting gradients and parabolic curves, an extension of this traditional method in which parabolic curves are fitted to various fixed elements along the horizontal alignment such as sections of gradient, bridges, tie-ins to existing road junctions and so on, and the cubic spline method mentioned above.

Each combined horizontal and vertical alignment, as designed by the computer, is passed through the DTM and the computer produces a longitudinal section, as many cross-sections as desired and an estimate of the earthwork quantities involved. This considerably shortens the time required to carry out these procedures by manual methods, details of which can be found in chapter 13. In addition, for any alignment, the computer system can also produce perspective drawings showing views along the proposed road. Such drawings can be used for visually checking the design and for preparing material for reports, exhibitions and public enquiries. The flexibility of highway design software packages enables any amount of design data to be combined with DTMs and it is possible to carry out a much more thorough preliminary design than that which could ever be undertaken by conventional methods.

As soon as the optimum alignment has been chosen, further data is entered into the DTM to enable a set of contract drawings to be produced by the computer interfaced with a suitable plotter. If all the relevant information for the optimum road alignment is computerised, these drawings will consist of a series of plans showing all aspects of the road construction including longitudinal and cross sections along the main alignment and also at interchanges, junctions, sliproads and so on. Based on these, schedules of earthwork quantities can be produced by the computer along any section of road and setting-out tables can be computed giving angles and distances relative to existing survey stations.

The greatest benefits of using a computer system in road design are the ability to investigate different alignments and a reduction in the overall time taken for the design and production of contract drawings. If the design should change at any time, these changes can be entered reasonably quickly into the system and modified drawings produced.

The main drawback to the use of computer systems, that of the perceived high cost of purchasing the necessary hardware and software, has been eliminated. Most of the current packages are available for very reasonable costs and virtually all can be run on microcomputers, the prices of which continue to fall despite their constantly improving specifications. With capabilities far greater than those which until a few years ago would have required a mainframe or minicomputer for their operation, modern highway design software packages are now well within the budget of any engineering and surveying practice, no matter how small.

11.24 Worked Examples

(1) Setting Out a Composite Curve by the Tangential Angles Method

Question
The deflection angle between two straights is measured as 14°28'26". The straights are to be joined by a composite horizontal curve consisting of a central circular arc and two transition curves of equal length.

The design speed of the road is 85 kph and the radius of the circular curve is 600 m.

If the through chainage of the intersection point is 461.34 m, draw up the setting-out table for the three curves at exact 20 m multiples of through chainage using the tangential angles method. The rate of change of radial acceleration should be taken as 0.3 m/s³.

Solution
Consider figure 11.21.

Design of entry transition, from T to T$_1$

$$L_T = v^3/3.6^3 \; cR = (85^3/3.6^3 \times 0.3 \times 600) = 73.13 \text{ m}$$

$$S = L_T^2/24R = (73.13^2/24 \times 600) = 0.37 \text{ m}$$

$$IT = (R + S) \tan\theta/2 + L_T/2 = 76.24 + 36.56 = 112.80 \text{ m}$$

Therefore

$$through \; chainage \; of \; T = 461.34 - 112.80 = 348.54 \text{ m}$$

$$through \; chainage \; of \; T_1 = 348.54 + 73.13 = 421.67 \text{ m}$$

Therefore, to keep to exact 20 m through chainage values, the chord lengths for the entry transition curve are as follows

$$initial \; sub\text{-}chord \; length = 11.46 \text{ m}$$

$$general \; chord \; length \quad = 20.00 \text{ m}$$

$$final \; sub\text{-}chord \; length \quad = 1.67 \text{ m}$$

Using the formula for tangential angles, $\delta = (1800 \; l^2)/(\pi R L_T)$, table 11.3 is obtained. As a further check on table 11.3, $\phi_{max}/3$ should be calculated and compared with δ_{max}

Data

$R = 600 \text{ m}$
$\theta = 14° 28' 26''$
Ch.$I = 461.34 \text{ m}$
$V = 85 \text{ kph}$
$C = 0.3 \text{ m/s}^3$

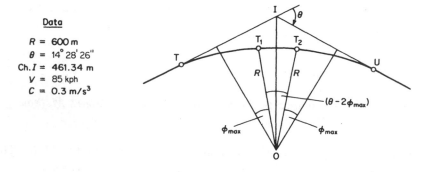

Figure 11.21

TABLE 11.3

Through chainage (m)	Chord length (m)	l (m)	δ (minutes)	Cumulative clockwise tangential angle from T relative to TI		
				°	′	″
348.54 (T)	0	0	0	00	00	00
360.00 (C$_1$)	11.46	11.46	1.715	00	01	43 (δ_1)
380.00 (C$_2$)	20.00	31.46	12.924	00	12	55 (δ_2)
400.00 (C$_3$)	20.00	51.46	34.579	00	34	35 (δ_3)
420.00 (C$_4$)	20.00	71.46	66.681	01	06	41 (δ_4)
421.67 (T$_1$)	1.67	73.13	69.834	01	09	50 (δ_{max})
	Σ73.13	(checks)				

$$\phi_{max} = (L_T/2R) \text{ rad} = 209.50 \text{ min} = 03°29'30''$$

Hence

$$\phi_{max}/3 = 01°09'50'' \text{ (checks)}$$

Design of the central circular arc, from T_1 to T_2

The circular arc takes $(\theta - 2\phi_{max})$; $\phi_{max} = 03°29'30''$, hence $2\phi_{max} = 06°59'00''$
Therefore

$$(\theta - 2\phi_{max}) = 07°29'26'' = 0.13073 \text{ rad}$$

Therefore

$$\text{length of circular arc} = L_c = R (\theta - 2\phi_{max})$$
$$= 600 (0.13073) = 78.44$$

Hence

$$\text{through chainage of } T_2 = 421.67 + 78.44 = 500.11 \text{ m}$$

Therefore, using 20 m chords and keeping to exact 20 m multiples of through chainage, the chord lengths for the circular arc are as follows

$$\text{initial sub-chord length} = 18.33 \text{ m}$$
$$\text{general chord length} = 20.00 \text{ m}$$
$$\text{final sub-chord length} = 0.11 \text{ m}$$

Using the formula for circular curve tangential angles, $\alpha = 1718.9$ (chord length/radius) min, table 11.4 is obtained. As a check on table 11.4, the

TABLE 11.4

Through chainage (m)	Chord length (m)	Tangential angle for each chord			Cumulative clockwise tangential angle from T relative to the common tangent		
		°	′	″	°	′	″
421.67 (T$_1$)	0	00	00	00	00	00	00
440.00 (C$_5$)	18.33	00	52	31 (α_1)	00	52	31
460.00 (C$_6$)	20.00	00	57	18 (α_2)	01	49	49
480.00 (C$_7$)	20.00	00	57	18 (α_3)	02	47	07
500.00 (C$_8$)	20.00	00	57	18 (α_4)	03	44	25
500.11 (T$_2$)	0.11	00	00	19 (α_5)	03	44	44
	Σ78.44 (checks)						

final cumulative tangential angle should equal $(\theta - 2\phi_{max})/2$ within a few seconds

$$(\theta - 2\phi_{max})/2 = 03°44'43'' \text{ (checks)}$$

Design of the exit transition curve, from U to T$_2$ (that is, in the opposite direction)

Since the curve is symmetrical, the length of the exit transition again equals 73.13 m. Therefore

$$\text{through chainage of U} = 500.11 + 73.13 = 573.24 \text{ m}$$

To keep to exact 20 m multiples of through chainage, the chord lengths for the exit transition curve, working from U to T$_2$, are as follows

$$\text{initial sub-chord length from U} = 13.24 \text{ m}$$

$$\text{general chord length} = 20.00 \text{ m}$$

$$\text{final sub-chord length to T}_2 = 19.89 \text{ m}$$

Again, using $\delta = (1800l^2)/(\pi RL_T)$ min, the setting-out table shown in table 11.5 is obtained.
The check that $\delta_{max} = \phi_{max}/3$ must again be applied

$$\delta_{max} = (360° - 358°50'10'') = 01°09'50''$$

$$\phi_{max}/3 = 01°09'50'' \text{ (checks)}$$

The tangential angles for the exit transition curve are subtracted from 360° since it is set out from U to T$_2$. The two positions of T$_2$ provide a check on the setting out.

TABLE 11.5

Through chainage (m)	Chord length (m)	l (m)	δ °	δ ′	δ ″	Cumulative clockwise tangential angle from U relative to UI °	′	″
573.24 (U)	0	0	00	00	00	360	00	00
560.00 (C_{11})	13.24	13.24	00	02	17 (δ_{11})	359	57	43
540.00 (C_{10})	20.00	33.24	00	14	26 (δ_{10})	359	45	34
520.00 (C_9)	20.00	53.24	00	37	01 (δ_9)	359	22	59
500.11 (T_2)	19.89	73.13	01	09	50 (δ_{max})	358	50	10
	$\Sigma73.13$ (checks)							

TABLE 11.6

Point	m E	m N
G	727.61	893.83
H	940.57	886.28
I	789.14	863.72
T	704.95	788.64

(2) Setting Out a Composite Curve by Coordinate Methods

Question

The composite curve calculated in the previous worked example is to be set out by bearing and distance methods from two horizontal control points G and H. The intersection point, I, and the entry tangent point, T, have been set out on site and the coordinates of these, together with those of points G and H are listed in table 11.6. Using the data calculated in the previous worked example, calculate

(a) the coordinates of all the pegs that are to be placed along the centre line,
(b) the bearing GH that must be set on the theodolite at G and the bearings and horizontal lengths from G that are necessary to set out all the pegs on the centre line using a combined theodolite and EDM system.

Solution

Figure 11.22 shows all the points to be set out. Their chainage values and the required tangential angles and chords are listed in tables 11.3, 11.4 and 11.5.

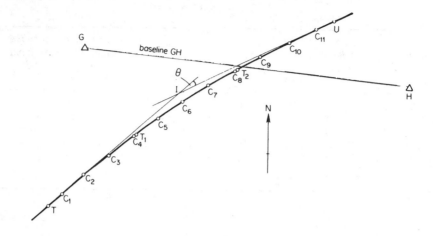

Figure 11.22

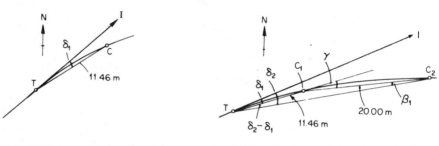

Figure 11.23   Figure 11.24

(a) Coordinates of all the points on the centre line

Coordinates of C_1
From figure 11.23 and table 11.3

$$\text{bearing } TC_1 = \text{bearing } TI + \delta_1$$

But

$$\Delta E_{TI} = E_I - E_T = 789.14 - 704.95 = 84.19 \text{ m}$$

$$\Delta N_{TI} = N_I - N_T = 863.72 - 788.64 = 75.08 \text{ m}$$

and, from a rectangular/polar conversion

$$\text{bearing } TI = 48°16'25''$$

Hence

$$\text{bearing } TC_1 = 48°16'25'' + 00°01'43'' = 48°18'08''$$

Therefore, since the horizontal length of TC_1 = 11.46 m

$$\Delta E_{TC_1} = 11.46 \sin 48°18'08'' = + 8.557 \text{ m}$$

$$\Delta N_{TC_1} = 11.46 \cos 48°18'08'' = + 7.623 \text{ m}$$

Therefore, the *coordinates of C_1* are

$$E_{C_1} = E_T + (\Delta E_{TC_1}) = 704.95 + 8.557 = \mathbf{713.507 \text{ m}}$$

$$N_{C_1} = N_T + (\Delta N_{TC_1}) = 788.64 + 7.623 = \mathbf{796.263 \text{ m}}$$

These are retained with three decimal places for calculation purposes but are finally rounded to two decimal places.

Coordinates of C_2

With reference to figure 11.24, application of the Sine Rule in triangle TC_1C_2 gives

$$\frac{TC_1}{\sin \beta_1} = \frac{C_1 C_2}{\sin (\delta_2 - \delta_1)}$$

Substituting values from table 11.3 gives

$$\sin\beta_1 = \frac{11.46}{20.00} (\sin 00°11'12'')$$

Hence

$$\beta_1 = 00°06'25''$$

Therefore

$$\gamma = \beta_1 + (\delta_2 - \delta_1) = 00°17'37''$$

and

$$\text{bearing } C_1C_2 = \text{bearing } TC_1 + \gamma$$

$$= 48°18'08'' + 00°17'37'' = 48°35'45''$$

Therefore, the *coordinates of C_2* are obtained as follows

$$\Delta E_{C_1C_2} = 20.00 \sin 48°35'45'' = + 15.001 \text{ m}$$

$$\Delta N_{C_1C_2} = 20.00 \cos 48°35'45'' = + 13.227 \text{ m}$$

$$E_{C_2} = E_{C_1} + 15.001 = \mathbf{728.508 \text{ m}}$$

$$N_{C_2} = N_{C_1} + 13.227 = \mathbf{809.490 \text{ m}}$$

Coordinates of C_3

With reference to figure 11.25, the chord length TC_2 can be taken to equal the curve length TC_2, that is

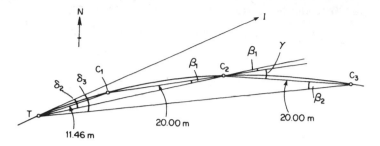

Figure 11.25

$$TC_2 = TC_1 + C_1C_2 = 31.46 \text{ m}$$

In triangle TC_2C_3

$$\frac{TC_2}{\sin \beta_2} = \frac{C_2 C_3}{\sin (\delta_3 - \delta_2)}$$

$$\sin\beta_2 = \frac{31.46}{20.00} (\sin 00^\circ 21'40'')$$

$$\beta_2 = 00^\circ 34'05''$$

And, since $(\gamma + \beta_1) = (\delta_3 - \delta_2) + \beta_2$

$$\gamma = 00^\circ 21'40'' + 00^\circ 34'05'' - 00^\circ 06'25''$$

$$= 00^\circ 49'20''$$

and

$$\text{bearing } C_2C_3 = \text{bearing } C_1C_2 + \gamma$$

$$= 48^\circ 35'45'' + 00^\circ 49'20''$$

$$= 49^\circ 25'05''$$

Therefore, the *coordinates of* C_3 are obtained as follows

$$\Delta E_{C_2C_3} = 20.00 \sin 49^\circ 25'05'' = +15.190 \text{ m}$$

$$\Delta N_{C_2C_3} = 20.00 \cos 49^\circ 25'05'' = +13.011 \text{ m}$$

$$E_{C_3} = E_{C_2} + 15.190 = \textbf{743.698 m}$$

$$N_{C_3} = N_{C_2} + 13.011 = \textbf{822.501 m}$$

Coordinates of C_4 *and* T_1
The coordinates of C_4 and T_1 are calculated from those of C_3 and C_4 respectively by repeating the procedure used to calculate the coordinates of C_3 from those of C_2. The values obtained are as follows

$$C_4 = 759.188 \text{ m E}, \ 835.152 \text{ m N}$$

$$T_1 = 760.498 \text{ m E}, \ 836.187 \text{ m N}$$

Coordinates of C_5

Point C_5 lies on the central circular arc as shown in figure 11.26. From this figure and the data in table 11.4.

$$\text{bearing } T_1 Z = \text{bearing } TI + \phi_{max}$$

$$= 48°16'25'' + 03°29'30'' = 51°45'55''$$

and

$$\text{bearing } T_1 C_5 = \text{bearing } T_1 Z + \alpha_1$$

$$= 51°45'55'' + 00°52'31'' = 52°38'26''$$

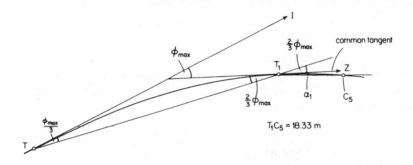

Figure 11.26

Hence the *coordinates of C_5* are obtained as follows

$$\Delta E_{T_1 C_5} = 18.33 \sin 52°38'26'' = +14.569 \text{ m}$$

$$\Delta N_{T_1 C_5} = 18.33 \cos 52°38'26'' = +11.123 \text{ m}$$

$$E_{C_5} = E_{T_1} + 14.569 = \textbf{775.067 m}$$

$$N_{C_5} = N_{T_1} + 11.123 = \textbf{847.310 m}$$

Coordinates of C_6

With reference to figure 11.27 and table 11.4

$$\lambda = (\alpha_1 + \alpha_2) = 01°49'49''$$

$$\text{bearing } C_5 C_6 = \text{bearing } T_1 C_5 + \lambda$$

$$= 52°38'26'' + 01°49'49'' = 54°28'15''$$

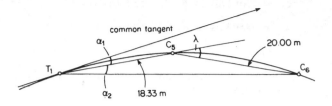

Figure 11.27

And the *coordinates of* C_6 are obtained as follows

$$\Delta E_{C_5 C_6} = 20.00 \sin 54°28'15'' = +16.276 \text{ m}$$

$$\Delta N_{C_5 C_6} = 20.00 \cos 54°28'15'' = +11.622 \text{ m}$$

$$E_{C_6} = E_{C_5} + 16.276 = \mathbf{791.343 \text{ m}}$$

$$N_{C_6} = N_{C_5} + 11.622 = \mathbf{858.932 \text{ m}}$$

Coordinates of C_7, C_8 *and* T_2
These coordinates are calculated using procedures similar to those used in the worked example in section 10.20 in the Circular Curves chapter. The values obtained are as follows

$$C_7 = \mathbf{807.998 \text{ m E, } 870.005 \text{ m N}}$$

$$C_8 = \mathbf{825.013 \text{ m E, } 880.517 \text{ m N}}$$

$$T_2 = \mathbf{825.108 \text{ m E, } 880.573 \text{ m N}}$$

Coordinates of U, C_{11}, C_{10}, C_9 *and* T_2
Using the deflection angle, bearing IU is calculated from

$$\text{bearing IU} = \text{bearing TI} + \theta$$

$$= 48°16'25'' + 14°28'26'' = 62°44'51''$$

Using the tangent length IU, the *coordinates of* U are calculated from those of I as follows

$$\Delta E_{IU} = 112.80 \sin 62°44'51'' = +100.279 \text{ m}$$

$$\Delta N_{IU} = 112.80 \cos 62°44'51'' = +51.653 \text{ m}$$

$$E_U = E_I + 100.279 = \mathbf{889.419 \text{ m}}$$

$$N_U = N_I + 51.653 = \mathbf{915.373 \text{ m}}$$

Starting from U and working back to T_2, the *coordinates of points* C_{11}, C_{10}, C_9 *and* T_2 are calculated by repeating the procedures used to calculate the coordinates of points C_1, C_2, C_3, C_4 and T_1. The data required for this is given in table 11.5. The coordinate values obtained are as follows

$$C_{11} = 877.653 \text{ m E, } 909.302 \text{ m N}$$

$$C_{10} = 859.933 \text{ m E, } 900.028 \text{ m N}$$

$$C_9 = 842.356 \text{ m E, } 890.486 \text{ m N}$$

$$T_2 = 825.109 \text{ m E, } 880.579 \text{ m N}$$

The coordinates of T_2 are calculated twice and this provides a check on the calculations. In this example, the two sets of coordinates for T_2 differ by 0.001 m in the eastings and by 0.006 m in the northings which is perfectly acceptable.

The coordinates of all the points on the curve are listed in table 11.7 and have been rounded to two decimal places.

(b) Setting-out data from point G

The bearings and lengths are calculated from the coordinates listed in table 11.7 using one of the methods discussed in section 1.5. The required bearings and lengths are also listed in table 11.7.

TABLE 11.7

Point	Through chainage (m)	Coordinates mE	mN	Bearing from G °	′	″	Horizontal length from G (m)
T	348.54	704.95	788.64	192	09	25	107.60
C_1	360.00	713.51	796.26	188	13	23	98.58
C_2	380.00	728.51	809.49	179	23	19	84.34
C_3	400.00	743.70	822.50	167	17	18	73.12
C_4	420.00	759.19	835.15	151	42	43	66.64
T_1	421.67	760.50	836.19	150	17	26	66.36
C_5	440.00	775.07	847.31	134	25	37	66.46
C_6	460.00	791.34	858.93	118	42	22	72.66
C_7	480.00	808.00	870.01	106	30	17	83.84
C_8	500.00	825.01	880.52	97	46	53	98.31
T_2	500.11	825.11	880.58	97	44	20	98.40
C_9	520.00	842.36	890.49	91	40	02	114.80
C_{10}	540.00	859.93	900.03	87	19	02	132.47
C_{11}	560.00	877.65	909.30	84	06	48	150.84
U	573.24	889.42	915.37	82	25	03	163.24

Bearing GH = 92° 01′ 50″

(3) A Wholly Transitional Curve

Question

Part of a proposed rural road consists of two straights which intersect at an angle of 168°16′. These are to be joined using a wholly transitional horizontal curve having equal tangent lengths. The design speed of the road is to be 100 kph and the rate of change of radial acceleration 0.20 m/s³.

Calculate the minimum radius of the curve and comment on its suitability.

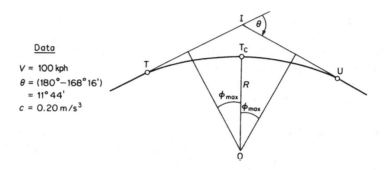

Data

V = 100 kph
θ = (180° − 168° 16′)
  = 11° 44′
c = 0.20 m/s³

Figure 11.28

Solution

Consider figure 11.28. From section 11.20

$$R = (v^3/3.6^3\ c\theta)^{\frac{1}{2}}$$

But

$$\theta = 11°44' = 0.20479\ \text{rad}$$

Therefore

$$R = (100^3/3.6^3 \times 0.20 \times 0.20479)^{\frac{1}{2}} = \textbf{723.40 m}$$

From table 11.1, the desirable minimum radius for this road is 720 m, the absolute minimum radius is 510 m and the limiting radius at sites of special difficulty is 360 m, hence this design is acceptable. If the radius had been less than the desirable minimum it would have been advisable to alter one of the variables, either v, θ or c, to increase the radius above the desirable minimum value.

Further Reading

Department of Transport, Roads and Local Transport Directorate, *Departmental Standard TD 9/81, Road Layout and Geometry: Highway Link Design* (Department of Transport, 1981).

Department of Transport, Highways and Traffic Directorate, *Departmental Advice Note TA 43/84: Highway Link Design* (Department of Transport, 1984).

M.J. Fort, 'Surveying by Computer', in *Engineering Surveying Showcase '93*, pp. 24, 27–31 (PV Publications, 101 Bancroft, Hitchin, Hertfordshire, January 1993).

M.J. Fort, 'Software for Surveyors', in *Civil Engineering Surveyor*, Vol. 18 No. 3, Electronic Surveying Supplement, pp. 19–27, April 1993.

HMSO Publications, *Roads and Traffic in Urban Areas* (Institution of Highways and Transportation, with the Department of Transport, 1987).

12

Vertical Curves

In the same way as horizontal curves are used to connect intersecting straights in the horizontal plane, vertical curves are used to connect intersecting straights in the vertical plane. These straights are usually referred to as *gradients*.

As with horizontal curves, vertical curves are designed for particular speed values and the design speed is constant for each particular vertical curve.

12.1 Gradients

These are usually expressed as percentages, for example, 1 in 50 = 2 per cent, 1 in 25 = 4 per cent. The Department of Transport (DTp) recommends *desirable* and *absolute maximum gradient values* for all new highways and these are shown in table 12.1. Wherever possible, the desirable maximum values should not be exceeded. Any gradient steeper than 4 per cent on motorways and 8 per cent on all other highways is considered to be substandard.

For drainage purposes the channels should have a minimum gradient of 0.5 per cent. This is achieved on level sections of road by steepening the channels between gullies while the road itself remains level.

TABLE 12.1
(published here by permission of the Controller of Her Majesty's Stationery Office)

Type of road	Desirable maximum gradient	Absolute maximum gradient
Motorways	3%	4%
Dual carriageways	4%	8%
Single carriageways	6%	8%

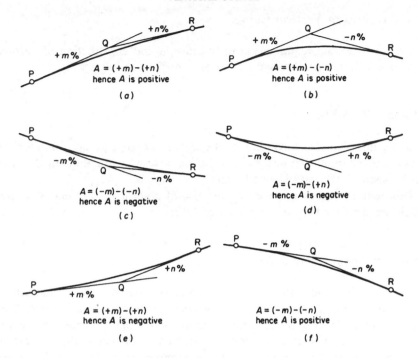

Figure 12.1 *Types of vertical curve*

Further details concerning gradients can be found in the DTp publications referenced in Further Reading at the end of the chapter.

In the design calculations, which are discussed later in the chapter, the *algebraic difference* between the gradients is used. This necessitates the introduction of the sign convention that gradients rising in the direction of increasing chainage are considered to be positive and those falling are considered to be negative.

This leads to the *six* different types of vertical curve. These are shown in figure 12.1 together with the value of the algebraic difference (*A*). Note that *A* can be either positive or negative and is calculated in the direction of increasing chainage.

Throughout the remainder of this chapter, reference will be made to the terms *crest curve* and *sag curve* and, in order to avoid confusion, these terms are defined as follows. A *crest curve*, which can also be referred to as a summit or hogging curve, is one for which the algebraic difference of the gradients is positive, and a *sag curve*, which can also be referred to as a valley or sagging curve, is one for which the algebraic difference of the gradients is negative.

Hence, in figure 12.1, (*a*), (*b*) and (*f*) are crest curves and (*c*), (*d*) and (*e*) are sag curves.

12.2 Purposes of Vertical Curves

There are two main requirements in the design and construction of vertical curves: *adequate visibility* and *passenger comfort and safety*.

Adequate Visibility

In order that vehicles travelling at the design speed can stop or overtake safely it is essential that oncoming vehicles or any obstructions in the road can be seen clearly and in good time.

This requirement is achieved by the use of *sight distances* and *K-values* which are discussed in sections 12.6 and 12.7.

Passenger Comfort and Safety

As the vehicle travels along the curve a radial force, similar to that which occurs in horizontal curves, acts on the vehicle in the vertical plane. This has the effect of trying to force the vehicle away from the centre of curvature of the vertical curve. In crest design this could cause the vehicle to leave the road surface, as in the case of hump-back bridges, while in sag design the underside of the vehicle could come into contact with the surface, particularly where the gradients are steep and opposed. This results in both discomfort and danger to passengers travelling and must, therefore, be minimised. This is achieved firstly by restricting the gradients (see table 12.1), which has the effect of reducing the force, and secondly by choosing a suitable type and length of curve such that this reduced force is introduced as gradually and uniformly as possible. The *K-values* discussed in section 12.7 also ensure that sufficient comfort is provided.

12.3 Type of Curve Used

In practice, owing to the restrictions placed on the gradients, vertical curves can be categorised as *flat*; the definition of a flat curve is that if its length is L_v and its radius R then $L_v/R<1/10$. This definition does assume that the vertical curve forms part of a circle of radius R but, again owing to the restricted gradients, there is no appreciable difference between a circular arc, an ellipse or a parabola and the definition can be applied to all three types of curve by approximating the value of R.

The final choice of curve is governed by the requirement for passenger safety and comfort discussed in section 12.2. In practice, a *parabolic curve* is used to achieve a uniform rate of change of gradient and therefore a uniform introduction of the vertical radial force. This uniformity of rate of change of gradient is shown as follows

$$x = cy^2, \quad dx/dy = 2cy, \quad d^2x/dy^2 = 2c = \text{constant}$$

12.4 Assumptions Made in Vertical Curve Calculations

The choice of a parabola simplifies the calculations and further simplifications are possible if certain assumptions are made. Consider figure 12.2, which is greatly exaggerated for clarity and shows a parabolic vertical curve having equal tangent lengths joining two intersecting gradients PQ and QR.

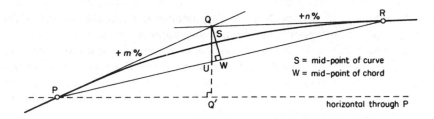

Figure 12.2

The assumptions are as follows

(1) Chord PWR = arc PSR = PQ + QR.
(2) Length along tangents = horizontal length, that is PQ = PQ'.

Assumptions (1) and (2) are very important since they are saying that the length is the same whether measured along the tangents, the chord, the horizontal or the curve itself.

(3) QU = QW, that is, there is no difference in lengths measured either in the vertical plane or perpendicular to the entry tangent length.
    In general, vertical curves are designed such that the two tangent lengths are equal, that is, PQ = QR, but it is possible to design vertical curves with unequal tangents and these are discussed in section 12.14.

These assumptions are valid if the DTp recommendations for gradients as listed in table 12.1 are adhered to.

12.5 Equation of the Vertical Curve

Since the curve is to be parabolic, the equation of the curve will be of the form $x = cy^2$, y being measured along the tangent length and x being set off

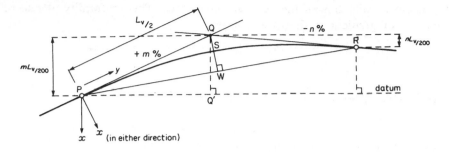

Figure 12.3

at right angles to it. In fact, from the assumptions, x can also be set off in a vertical direction without introducing any appreciable error.

The basic equation is usually modified to a general equation containing some of the parameters involved in the vertical curve design. This general equation will be developed for the equal tangent length crest curve shown in figure 12.3, but the same equation can be derived for sags and applies to all six possible combinations of gradient.

Consider figure 12.3. Let QS $= e$ and let the total length of the curve $= L_v$. Using the assumptions

$$\text{level of Q above P} = (m/100)(L_v/2) = (mL_v/200)$$

$$\text{level of R below Q} = (n/100)(L_v/2) = (nL_v/200)$$

Hence

$$\text{level of R above P} = (mL_v/200) - (nL_v/200) = (m - n)L_v/200$$

But, from the assumptions, PW $=$ WR, therefore,

$$\text{level of W above P} = (m - n)\,L_v/400$$

But, from the properties of the parabola

$$\text{QS} = \text{QW}/2 = \text{SW}$$

Therefore

$$\text{QS} = \tfrac{1}{2}(mL_v/200 - (m - n)L_v/400) = (m + n)L_v/800$$

But $(m + n) =$ algebraic difference of the gradients $= A$, therefore

$$\text{QS} = e = L_v A/800$$

The equation of the parabola is $x = cy^2$, therefore at point Q, when $y = L_v/2$, $x = e$, hence

$$e = c(L_v/2)^2$$

Therefore

$$c = e/(L_v/2)^2$$

Therefore, substituting in the equation of the parabola gives

$$x = ey^2/(L_v/2)^2$$

But, from above

$$e = L_v A/800$$

Therefore

$$x = Ay^2/200L_v$$

12.6 Sight Distances

The length of curve to be used in any given situation depends on the *sight distance*. This is simply the distance of visibility from one side of the curve to the other.

There are two categories of sight distance

(1) *Stopping Sight Distance (SSD)* which is the theoretical forward sight distance required by a driver in order to stop safely and comfortably when faced with an unexpected hazard on the carriageway, and
(2) *Full Overtaking Sight Distance (FOSD)* which is the length of visibility required by drivers of vehicles to enable them to overtake vehicles ahead of them in safety and comfort.

Since it requires a greater distance to overtake than to stop, the *FOSD* values are greater than the *SSD* values.

When designing vertical curves, it is essential to know whether safe overtaking is to be included in the design. If it is then the *FOSD* must be incorporated, if it is not then the *SSD* must be incorporated.

It is usually necessary to consider whether to design for overtaking only at crest curves on single carriageways since overtaking should not be a problem on dual carriageways and visibility is usually more than adequate for overtaking at sag curves on single carriageways.

The DTp specify sight distances for both stopping and overtaking at various design speeds and these are shown in sections A and D respectively of table 11.1. They were obtained as follows.

The SSD ensures that there is an envelope of clear visibility such that, at one extreme, drivers of low vehicles are provided with sufficient visibility to see low objects, while, at the other extreme, drivers of high vehicles are provided with visibility to see a significant portion of other vehicles. This envelope of visibility is shown in figure 12.4 in which 1.05 m represents

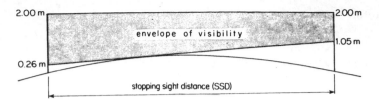

Figure 12.4   *Measurement of SSD*

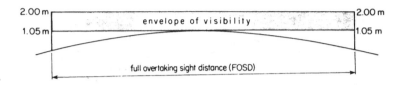

Figure 12.5   *Measurement of FOSD*

the drivers' eye height for low vehicles and 2.00 m that for high vehicles;
a lower object height of 0.26 m is used to include the rear tail lights of
other vehicles and an upper object height of 2.00 m ensures that a sufficient portion of a vehicle ahead can be seen to identify it as such.

The FOSD ensures that there is an envelope of clear visibility between
the 1.05 m and 2.00 m drivers' eye heights above the centre of the carriageway as shown in figure 12.5.

12.7 K-values

In the past it was necessary to use the appropriate sight distance for the
road type and design speed in question to calculate the minimum length of
the vertical curve required. Nowadays, however, constants, known as *K-
values*, have been introduced by the DTp and greatly simplify the calculations.

The minimum length of vertical curve (*min* L_v) for any given road is
obtained from the formula

$$min \ L_v = KA \text{ metres} \tag{12.1}$$

where K is the constant obtained from the DTp standards for the particular
road type and design speed in question and A is the algebraic difference of
the gradients, the absolute value (always positive) being used.

Section C of table 11.1 shows the current DTp K-values for various
design speeds. The K-values ensure that the minimum length of vertical
curve obtained from equation (12.1) contains adequate visibility and provides sufficient comfort. It must be noted that the length obtained from

equation (12.1) is the minimum required and it is perfectly acceptable to increase the value obtained. This may be necessary when trying to *phase* the vertical alignment with the horizontal alignment as discussed in section 12.10.

The units of K are metres and their values have been derived from the sight distances discussed in section 12.6. There are three categories of K-values for crests and one category of K-values for sags. These are discussed as follows.

Crest K-values

If a full overtaking facility is to be included in the design of single carriageways then the *FOSD crest K-values* given in row C1 of table 11.1 should be used in equation (12.1).

If overtaking is not considered in the design then, if possible, to ensure more than adequate visibility, K-values in excess of the *desirable minimum crest K-values* given in row C2 of table 11.1 should be used. If, owing to site constraints, this cannot be done, then it is permissible to use K-values as low as the *absolute minimum crest K-values* given in row C3 of table 11.1. These still ensure adequate visibility.

Sag K-values

Only one set of K-values is given for sags since overtaking visibility is usually unrestricted on this type of vertical curve. Row C4 of table 11.1 lists the *absolute minimum sag K-values* which will ensure adequate visibility and comfort.

Examples of the use of K-values are given in the following section.

12.8 Use of K-values

Example (1)

Dual carriageway, design speed 85 kph, Crest.
    From table 11.1

FOSD crest K-value	$= 285$ m
desirable minimum crest K-value	$= 55$ m
absolute minimum crest K-value	$= 30$ m

Since a dual carriageway is being designed, overtaking is not critical. Therefore, from equation (12.1)

$$\text{if possible, use } L_v \geqslant 55A \text{ metres}$$

$$\text{otherwise use } L_v \geqslant 30A \text{ metres}$$

Example (2)

Single carriageway, design speed 60 kph, Crest.
From table 11.1

FOSD crest K-value	$= 142$ m
desirable minimum crest K-value	$= 17$ m
absolute minimum crest K-value	$= 10$ m

Since a single carriageway is being designed, a decision has to be made as to whether or not full overtaking is to be allowed for in the design.
If full overtaking is to be included, equation (12.1) gives

$$min \; L_v = 142A \text{ metres}$$

If full overtaking is not to be included, it would appear that, from equation (12.1)

$$min \; L_v = 17A \text{ metres}$$

However, the current DTp standards state that for crests on single carriageways, unless FOSD crest K-values can be used, it is sufficient to use only the absolute minimum crest K-values since the use of the desirable minimum crest K-values may result in sections of road having dubious visibility for overtaking.

In summary, this means that on single carriageway crests, overtaking should be either easily achieved or not possible at all. Hence, in this example

$$\text{if possible, use } L_v \geqslant 142A \text{ metres}$$

$$\text{otherwise use } L_v = 10A \text{ metres}$$

Further details on restrictions involved in the design of single carriageways can be found in the current DTp standards referenced in Further Reading at the end of the chapter.

Example (3)

Single carriageway, design speed 100 kph, Sag.
From table 11.1

$$\text{absolute minimum sag } K\text{-value} = 26 \text{ m}$$

Therefore, from equation (12.1), use $L_v \geq 26A$ metres.

12.9 Length of Vertical Curve to be Used (L_v)

Often the value for the minimum length of curve obtained from the K-values is not used, a greater length being chosen. This may be done for several reasons, for example, it may be necessary to fit the curve into particular site conditions. However, there is another factor which must be considered before deciding on the final length of a vertical curve. This is the necessity to try to fit the vertical alignment of the road to the horizontal alignment, a procedure known as *phasing*, which is discussed in the following section.

12.10 Phasing of Vertical and Horizontal Alignments

Usually, when designing new roads or improving existing alignments, the procedure is as follows.

(1) Design or redesign the horizontal alignment.
(2) Take reduced levels at regular intervals along the proposed centre line and plot a longitudinal section (see section 2.20).
(3) Superimpose chosen gradients on the longitudinal section, altering their percentage gradient and position as necessary to try to balance out any *cut* and *fill* (see section 13.5).
(4) Design the vertical alignment to join the gradients such that, if possible, the vertical curve tangent points coincide with those of the horizontal curve.

It is stage (4) which often gives the final length of the vertical curve. The tangent points of the vertical curve must, wherever possible, coincide exactly with the tangent points of the horizontal curve, where applicable. This is to avoid the creation of optical illusions. If a vertical curve is started during a horizontal curve then to a driver travelling along the curve the road appears disjointed owing to the vertical directional change of the vertical alignment being inflicted on the horizontal curve at a point where the horizontal radial force and superelevation may be severe. This can lead to driver error and must be avoided wherever possible.

In most cases, the horizontal curve will be greater in length than the minimum required for the vertical curve and it will be necessary to increase the vertical curve length to that of the horizontal curve. Should the minimum vertical curve length be greater than the length of the horizontal curve then the opposite will apply.

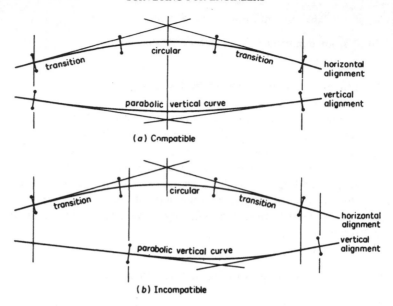

Figure 12.6   *Phasing of horizontal and vertical alignments*

When phasing vertical and horizontal alignments, the curves should run between the start and finish tangent points and not between any two tangent points. This is shown in figure 12.6. To introduce the two alignments at different tangent points would again create optical illusions.

12.11 Plotting and Setting Out the Vertical Curve

Once the length and gradients have been decided, it is necessary to plot the curve on the longitudinal section as a check on the design and then set it out on the ground.

In order that these can be done, it is necessary to calculate the reduced levels (RL) of points along the proposed centre line.

With reference to figure 12.7, if P is datum level, the level of any point Z on the curve with respect to P is given by ΔH, where

$$\Delta H = [(m)y/100 - (A)y^2/200L_v]$$
(12.2)

This is a general expression and ΔH can be either positive or negative, depending on the signs of m and A.

All ΔH values are related to the RL of P and should be added to or subtracted from this to obtain the reduced levels of points on the curve which lie a general distance y along the curve from P.

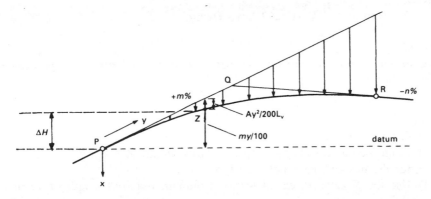

Figure 12.7

Hence, reduced levels of points on the curve can be calculated and plotted on the longitudinal section (see section 13.5). If the design is acceptable, the reduced levels of the points are set out on site using sight rails as described in chapter 14.

12.12 Highest Point of a Crest, Lowest Point of a Sag

In order that drainage gullies can be positioned effectively, it is necessary to know the through chainage and reduced level of the highest or lowest point of the vertical curve. The highest point of a crest occurs when ΔH is a maximum and the lowest point of a sag occurs when ΔH is a minimum.

For a maximum or minimum value of ΔH, $d(\Delta H)/dy = 0$, therefore

$$\frac{d}{dy}(\Delta H) = \frac{m}{100} - \frac{Ay}{100L_v} = 0$$

Hence, $m/100 = Ay/100L_v$, therefore $y = L_v m/A$ for a maximum or minimum value ΔH. This gives the point along the curve at which the maximum or minimum level occurs. To find the reduced level at this point it is necessary to substitute this expression for y back into the equation for ΔH. Therefore

$$\Delta H_{max/min} = (m/100)(L_v m/A) - (A/200L_v)(L_v^2 m^2/A^2)$$

Hence

$$\Delta H_{max/min} = L_v m^2/200A \tag{12.3}$$

above or below point P.

12.13 Summary of Vertical Curve Design

Problem
To design a vertical curve to fit between two gradients for a particular design speed.

Solution
(1) Calculate *A*.
(2) From the current DTp design standards obtain the appropriate *K*-value for the design speed and road type.
(3) Use the *K*-value to calculate the minimum required length of vertical curve.
   At this stage it may be necessary to phase the vertical and horizontal alignment as described in section 12.10 and an alteration in gradients may be necessary. An attempt should also be made on the longitudinal section to balance out cut and fill.
(4) The reduced levels of the entry and exit tangent points on the vertical curve should be calculated as shown in section 12.5.
(5) The formula for Δ*H* (equation 12.2) is used together with the reduced level of the entry tangent point to calculate the reduced levels of points on the curve itself. As a check on the calculations, the reduced level of the exit tangent point should be calculated using the formula and it should equal that found in (4).
(6) The curve is plotted on the longitudinal section by plotting the reduced levels calculated in (5) and, if acceptable, is set out on site using sight rails set some convenient height above the formation level.

12.14 Vertical Curves with Unequal Tangent Lengths

The foregoing discussion has been limited to vertical curves having equal tangent lengths (symmetrical). These are easy to design and can be fitted to the majority of cases but, occasionally, either to meet particular site conditions or to avoid large amounts of cut and/or fill, it becomes necessary to design a curve having unequal tangent lengths (asymmetrical).

With reference to figure 12.8, the easiest method of designing such a curve is to introduce a third gradient BCD which splits the total curve PR into two consecutive equal tangent length curves PC and CR. The common tangent line BCD is parallel to the chord PR and C is the common tangent point between the two curves.

The first curve PC is equal in length to the entry tangent length PQ and the second curve CR is equal in length to the exit tangent length QR. B is the mid-point of PQ and D is the mid-point of QR.

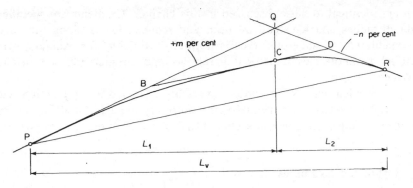

Figure 12.8

From figure 12.8

$$PC = L_1, \; CR = L_2$$

$$L_v = L_1 + L_2$$

$$PB = BC = \frac{PQ}{2} = \frac{L_1}{2}$$

$$CD = DR = \frac{QR}{2} = \frac{L_2}{2}$$

When calculating RLs at regular chainage intervals along the curves, each curve is treated as a separate equal tangent length vertical curve. The second worked example in section 12.16 shows how this is done.

12.15 Computer-Aided Road Design

Although calculations and drawings for highway design are still undertaken by hand, it is much more usual nowadays for computer software packages to be used. Many of these are available and their relatively low cost coupled with the falling price of desktop and laptop computers makes them ideal for even the smallest engineering or surveying practice.

They have a number of very significant advantages over hand methods for vertical alignment design. Their speed enables the calculations to be performed very quickly and their graphics capability gives an instantaneous on-screen view of the gradients and the vertical curves. The need for phasing, as discussed in section 12.10, is much easier to perform using the computer software. Editing facilities enable site constraints to be incorporated and different alignments to be tried until a suitable solution is achieved. Graphical, numerical and setting out data are easily provided, as required.

The principles on which these packages base their vertical alignment design are identical to those discussed in this chapter. Gradients are specified, sight distances and K-values are used and reduced levels along the curve are calculated. A range of curves is normally available, for example, symmetrical parabolas, asymmetrical parabolas and circular arcs are usually provided.

Since vertical curves are almost invariably designed in conjunction with horizontal curves, further details of the currently available highway design software packages are given in section 11.23 of the transition curves chapter.

12.16 Worked Examples

(1) Vertical Curve having Equal Tangent Lengths

Question
The reduced level at the intersection of a rising gradient of 1.5 per cent and a falling gradient of 1.0 per cent on a proposed road is 93.60 m AOD. Given that the K-value for this particular road is 55, the through chainage of the intersection point is 671.34 m and the vertical curve is to have equal tangent lengths, calculate

  (i) the through chainages of the tangent points of the vertical curve if the minimum required length is to be used
 (ii) the reduced levels of the tangent points and the reduced levels at exact 20 m multiples of through chainage along the curve
(iii) the position and level of the highest point on the curve.

Solution
Figure 12.9 shows the curve in question. From equation (12.1), minimum $L_v = KA = 55 \times 2.5 = 137.5$ m. Therefore

$$through \; chainage \; of \; P = 671.34 - (137.5/2) = \textbf{602.59 m}$$

$$through \; chainage \; of \; R = 671.34 + (137.5/2) = \textbf{740.09 m}$$

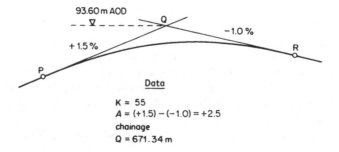

K = 55
A = (+1.5) − (−1.0) = +2.5
chainage
Q = 671.34 m

Figure 12.9

From the diagram it is obvious that P and R are both lower than Q, therefore, ignoring the signs of m and n

$$\text{reduced level of P} = 93.60 - (mL_v/200) = 93.60 - 1.03$$

$$= \textbf{92.57 m}$$

$$\text{reduced level of R} = 93.60 - (nL_v/200) = 93.60 - 0.69$$

$$= \textbf{92.91 m}$$

To keep to exact multiples of 20 m of through chainage there will need to be an initial short value of y of $620.00 - 602.59 = 17.41$ m. y will increase in steps of 20 m and the final y will be equal to the length of the curve, that is 137.5 m.

The reduced levels of points on the curve are given by equation (12.2) as

$$RL = 92.57 + [(m)y/100 - (A)y^2/200L_v]$$

working from P towards R.

The results are tabulated and shown in table 12.2.

As a check on the calculations, the reduced level of R should equal that calculated earlier as is the case in this example.

From section (12.12), the *highest point* on the curve occurs when

$$y = L_v m/A = (137.5 \times 1.5)/2.5 = \textbf{82.50 m}$$

From equation (12.3), the *highest level* on the curve $= 92.57 + L_v m^2/200A$

$$= 92.57 + (137.5 \times 1.5^2)/(200 \times 2.5) = \textbf{93.19 m}$$

These can be confirmed by inspection of table 12.2.
In this example both m and A are positive and the positive sign has been retained in the calculations. When either m or A is negative, the negative

TABLE 12.2
(all quantities are in metres)

Chainage	y	$(m)y/100$	$(A)y^2/200L_v$	ΔH	RL
602.59 (P)	0	0	0	0	92.57
620.00	17.41	+0.261	+0.028	+0.233	92.80
640.00	37.41	+0.561	+0.127	+0.434	93.00
660.00	57.41	+0.861	+0.300	+0.561	93.13
680.00	77.41	+1.161	+0.545	+0.616	93.19
700.00	97.41	+1.461	+0.863	+0.598	93.17
720.00	117.41	+1.761	+1.253	+0.508	93.08
740.00	137.41	+2.061	+1.717	+0.344	92.91
740.09 (R)	137.50	+2.063	+1.719	+0.344	92.91

sign should also be retained and taken into account in the equation for ΔH in order that ΔH can have the correct sign.

The reduced levels shown in table 12.2 are rounded to the nearest 10 mm since the initial data was only quoted to this precision.

(2) Vertical Curve having Unequal Tangent Lengths

Question

A parabolic vertical curve is to connect a -2.50 per cent gradient to a $+3.50$ per cent gradient on a highway designed for a speed of 100 kph. The K-value for the highway is 26 and the minimum required length is to be used.

The reduced level and through chainage of the intersection point of the gradients are 59.34 m AOD and 617.49 m respectively and, in order to meet particular site conditions, the through chainage of the entry tangent point is to be 553.17 m. Calculate

 (i) the reduced levels of the tangent points,
(ii) the reduced levels at exact 20 m multiples of through chainage along
       the curve.

Solution

$$A = (-2.50) - (+3.50) = -6.00$$

Hence

$$min \ L_v = 26 \times 6.00 = 156.00 \ m$$

Figure 12.10 shows the required curve and, from this, the tangent lengths are

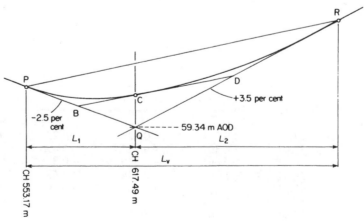

Figure 12.10

$$L_1 = PQ = 617.49 - 553.17 = 64.32 \text{ m}$$

$$L_2 = QR = 156.00 - 64.32 = 91.68 \text{ m}$$

Since these are unequal, a third gradient BCD is introduced as discussed in section 12.14.

(i) Reduced levels of P and R

From figure 12.10, it can be seen that

$$RL_P = RL_Q + \frac{(2.50)PQ}{100} = 59.34 + \frac{(2.50)64.32}{100} = \textbf{60.95 m AOD}$$

$$RL_R = RL_Q + \frac{(3.50)QR}{100} = 59.34 + \frac{(3.50)91.68}{100} = \textbf{62.55 m AOD}$$

(ii) Reduced levels at exact 20 m multiples of through chainage along the curve

through chainage of P = 553.17 m
through chainage of R = through chainage of P + L_v
= 553.17 + 156.00 = 709.17 m

Also, from figure 12.10, it can be seen that

gradient of BCD = gradient of PR

$$= \frac{(-2.50)L_1 + (+3.50)L_2}{L_v} \text{ per cent}$$

$$= \frac{(-2.50)64.32 + (+3.50)91.68}{156.00} \text{ per cent}$$

$$= +1.03 \text{ per cent}$$

For the vertical curve PC, the reduced levels are calculated from P to C with reference to RL_p using

$$RL = RL_p + [(m)y/100 - (A)y^2/200L_1]$$

where $m = -2.50$ per cent, $L_1 = 64.32$ m, $A = (-2.50) - (+1.03) = -3.53$ per cent, through chainage of C = through chainage of Q = 617.49 m. The RLs calculated along curve PC are shown in table 12.3.

For the vertical curve CR, the reduced levels are calculated from C to R with reference to RL_c using

$$RL = RL_c + [(m)y/100 - (A)y^2/200L_2]$$

where $m = +1.03$ per cent, $L_2 = 91.68$ m, $A = (+1.03) - (+3.50) = -2.47$.

TABLE 12.3
(all quantities are in metres)

Chainage	y	(m)y/100	(A)y²/200L₁	ΔH	RL
553.17 (P)	0	0	0	0	60.95
560.00	6.83	−0.17	−0.01	−0.16	60.79
580.00	26.83	−0.67	−0.20	−0.47	60.48
600.00	46.83	−1.17	−0.60	−0.57	60.38
617.49 (C)	64.32	−1.61	−1.14	−0.47	60.48

TABLE 12.4
(all quantities are in metres)

Chainage	y	(m)y/100	(A)y²/200L₂	ΔH	RL
617.49 (C)	0	0	0	0	60.48
620.00	2.51	+0.03	0.00	+0.03	60.51
640.00	22.51	+0.23	−0.07	+0.30	60.78
660.00	42.51	+0.44	−0.24	+0.68	61.16
680.00	62.51	+0.64	−0.53	+1.17	61.65
700.00	82.51	+0.85	−0.92	+1.77	62.25
709.17 (R)	91.68	+0.94	−1.13	+2.07	62.55

The RLs calculated along curve CR are shown in table 12.4. Note that the value obtained for the RL of point R in table 12.4 agrees with the value obtained in the solution to the first part of the question.

Further Reading

Department of Transport, Roads and Local Transport Directorate, *Departmental Standard TD/81, Road Layout and Geometry: Highway Link Design* (Department of Transport, 1981).

Department of Transport, Highways and Traffic Directorate, *Departmental Advice Note TA 43/84: Highway Link Design* (Department of Transport, 1984).

HMSO Publications, *Roads and Traffic in Urban Areas* (Institution of Highways and Transportation, with the Department of Transport, 1987).

13

Earthwork Quantities

In many engineering projects, large parcels of land are required for the site and huge amounts of material have to be moved in order to form the necessary embankments, cuttings, foundations, basements, lakes and so on, that have been specified in the design. Suitable land and materials can be very ex–pensive and, if a project is to be profitable to the construction company involved, it is essential that its engineers make as accurate a measurement as possible of any *areas* and *volumes* involved in order that appropriate cost estimates for such *earthwork quantities* can be included in the tender documents (see chapter 14).

In addition, for certain projects, such as the construction of a new high–way, where large amounts of material have to be excavated and moved around the site, careful planning of this movement is essential since charges may be levied not only on the volumes involved but also on the distances over which they are moved.

The purposes of this chapter, therefore, are to discuss some of the more important and most often used techniques for calculating the sizes of par–cels of land, the areas of cross-sections and the volumes of materials and to show how earth-moving can be planned. These are summarised as follows.

(1) Parcels of land are generally either straight-sided, irregular-sided or some combination of both. Methods by which their *plan areas* can be calcu–lated are covered in sections 13.2 to 13.4, inclusive.

(2) *Cross-sections* are often drawn to help with the volume calculations required for highway construction projects. They can take a number of different forms and they are normally based on *longitudinal sections*. These, and the methods by which *cross-sectional areas* can be calcu–lated are discussed in sections 13.5 to 13.11, inclusive.

(3) *Volumes* of materials can be calculated in a number of ways, depending on the project concerned. The three major methods involve the use of

cross-sections, spot heights and contours. These are covered in sections 13.12 to 13.15, inclusive.
(4) The movement of earth on a project is best planned with the aid of a
    mass haul diagram and this is discussed in sections 13.16 to 13.23,
    inclusive.

The chapter concludes with a brief discussion of the role of computer
software packages in the calculation of earthwork quantities. For ease of
studying, worked examples are introduced as necessary throughout the chapter
rather than in a separate section at the end.

13.1 Units

Although Système International (SI) units are used throughout this book,
there are times when other widely accepted units are equally appropriate.
Area calculation represents one such occasion since, although the standard
SI unit of area is the square metre, the figures involved can become very
big for large areas. To overcome this, the following unit system is often
adopted:

$$\begin{array}{llll}
100 \text{ m}^2 & = & 1 \text{ } are & \\
100 \text{ } ares & = & 1 \text{ } hectare & = & 10\,000 \text{ m}^2 \\
100 \text{ } hectares & = & 1 \text{ } square\ kilometre & = & 10^6 \text{ m}^2
\end{array}$$

For volumes, the SI unit is the cubic metre (m^3) and this is used throughout the book for all volumes of materials, no matter how large or small.

13.2 Areas Enclosed by Straight Lines

Into this category fall areas enclosed by traverse, triangulation, trilateration
or detail survey lines. The results obtained for such areas will be exact
since correct geometric equations and theorems can be applied.

Areas from Triangles

The straight-sided figures can be divided into well-conditioned triangles,
the areas of which can be calculated using one of the following formulae.
(1) Area = $\sqrt{[S(S - a)(S - b)(S - c)]}$ where a, b and c are the lengths of
    the sides of the triangle and $S = \frac{1}{2}(a + b + c)$.
(2) Area = $\frac{1}{2}$ (base of triangle x height of triangle).
(3) Area = $\frac{1}{2} a b \sin C$ where C is the angle contained between side lengths
    a and b.

The area of any straight-sided figure can be calculated by splitting it into triangles and summing the individual areas.

Areas from Coordinates

In traverse, triangulation and trilateration calculations, the coordinates of the junctions of the sides of a straight-sided figure are calculated and it is possible to use them to calculate the area enclosed by the control network lines. This is achieved using the *cross coordinate method*.

Consider figure 13.1, which shows a three-sided clockwise control network ABC. The required area = ABC.

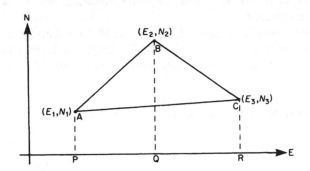

Figure 13.1   *Cross coordinate method*

$$\text{area of ABC} = \text{area of ABQP} + \text{area of BCRQ} - \text{area of ACRP} \qquad (13.1)$$

These figures are trapezia for which the area is obtained from

$$\text{area of trapezium} = (\text{mean height} \times \text{width})$$

Therefore

$$\text{area of ABQP} = \tfrac{1}{2}(N_1 + N_2)(E_2 - E_1)$$

Hence equation (13.1) becomes

$$\text{area ABC} = \tfrac{1}{2}(N_1 + N_2)(E_2 - E_1) + \tfrac{1}{2}(N_2 + N_3)(E_3 - E_2)$$
$$- \tfrac{1}{2}(N_1 + N_3)(E_3 - E_1)$$

Therefore

$$2 \times \text{area ABC} = N_1E_2 - N_1E_1 + N_2E_2 - N_2E_1 + N_2E_3 - N_2E_2$$
$$+ N_3E_3 - N_3E_2 - N_1E_3 + N_1E_1 - N_3E_3$$
$$+ N_3E_1$$

Rearranging, this gives

$$2 \times \text{area ABC} = (N_1E_2 + N_2E_3 + N_3E_1) -$$

$$(E_1N_2 + E_2N_3 + E_3N_1)$$

The similarity between the two brackets should be noted.

Although the example given is only for a three-sided figure, the formula can be applied to a figure containing N sides and the general formula for such a case is given by

$$2 \times \text{area} = (N_1E_2 + N_2E_3 + N_3E_4 + \ldots + N_{N-1}E_N + N_NE_1)$$

$$-(E_1N_2 + E_2N_3 + E_3N_4 + \ldots + E_{N-1}N_N + E_NN_1)$$

If the figure is numbered in the opposite direction, the signs of the two brackets are reversed.

The cross coordinate method can be used to subdivide straight-sided areas as shown in the following worked example and can also be used to calculate the area of irregular cross-sections as discussed in section 13.10.

Worked Example 13.1: Division of an Area Using the Cross Coordinate Method

Question
The polygon traverse PQRSTP shown in figure 13.2 is to be divided into two equal areas by a straight line that must pass through point R and which meets line TP at Z. The coordinates of the points are given in table 13.1. Calculate the coordinates of point Z.

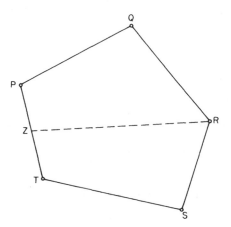

Figure 13.2 *Division of an area*

TABLE 13.1

Point	$m\ E$	$m\ N$
P	613.26	418.11
Q	806.71	523.16
R	942.17	366.84
S	901.89	203.18
T	652.08	259.26

Solution

The traverse is lettered and specified in a clockwise direction. Hence, using the clockwise version of the cross coordinate method gives

$$\text{area PQRSTP} = \tfrac{1}{2}\,(1\ 452\ 532 - 1\ 314\ 662) = 68\ 935\ \text{m}^2$$

Therefore

$$\text{area PQRZP} = \text{area ZRSTZ} = (68\ 935/2) = 34\ 467.5\ \text{m}^2$$

Let point Z have coordinates (E_z, N_z).

Applying the clockwise version of the cross coordinate method to area PQRZP gives

$$68\ 935 = 213\ 432.58 - 51.27E_z - 328.91N_z$$

from which

$$E_z = 2818.365 - 6.415\ 25N_z \tag{13.2}$$

A similar application to area ZRSTZ gives

$$E_z = 2.696\ 51N_z - 286.765 \tag{13.3}$$

Solving equations (13.2) and (13.3) gives

$$E_z = 632.16\ \text{m},\ N_z = 340.78\ \text{m}$$

As a check, since Z lies on the line PT

$$\frac{E_T - E_P}{N_T - N_P} \quad \text{should equal} \quad \frac{E_Z - E_P}{N_Z - N_P}$$

Substituting the coordinates of P, T and Z gives

$$-0.2444 = -0.2444$$

which checks the coordinates of Z as calculated above.

13.3 Areas Enclosed by Irregular Lines

For such cases only approximate results can be achieved. However, methods are adopted which will give the best approximations.

Give and Take Lines

In this method an irregular-sided figure is divided into triangles or trapezia, the irregular boundaries being replaced by straight lines such that any small areas excluded from the survey by the lines are balanced by other small areas outside the survey but included as shown in figure 13.3.

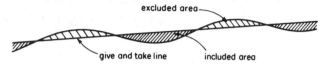

Figure 13.3   *Give and take line*

The positions of these lines can be estimated by eye on a survey plan. The area is then calculated using one of the straight-sided methods.

Graphical Method

This method involves the use of a transparent overlay of squared paper which is laid over the drawing or plan. The number of squares and parts of squares which are enclosed by the area is counted and, knowing the plan scale, the area represented by each square is known and hence the total area can be computed. This can be a very accurate method if a small grid is used.

Mathematical Methods

The following two methods make a mathematical attempt to calculate the area of an irregular-sided figure.

Trapezoidal rule   Figure 13.4 shows a control network contained inside an area having irregular sides. The shaded area is that remaining to be calculated after using one of the straight-sided methods to calculate the area enclosed by the control network lines.

Figure 13.5 shows an enlargement of a section of figure 13.4. The offsets $O_1, O_2, O_3 \ldots O_8$ are either measured directly in the field or scaled from a plan.

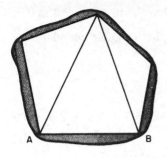

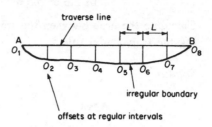

Figure 13.5

Figure 13.4

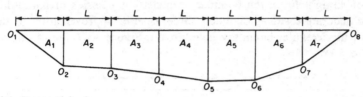

Figure 13.6 *Trapezoidal rule*

The trapezoidal rule assumes that if the interval between the offsets is small, the boundary can be approximated to a straight line between the offsets. Hence, figure 13.5 is assumed to be made up of a series of trapezia as shown in figure 13.6. Therefore, in figure 13.6

$$A_1 = \frac{(O_1 + O_2)}{2} L; A_2 = \frac{(O_2 + O_3)}{2} L, \text{ etc.}$$

Hence, for N offsets, the total area (A) is given by

$$A = \frac{(O_1 + O_2)}{2} L + \frac{(O_2 + O_3)}{2} L + \dots + \frac{(O_{N-1} + O_N)}{2} L$$

Which leads to the general trapezoidal rule shown below

$$A = \frac{L}{2} (O_1 + O_N + 2(O_2 + O_3 + O_4 + \dots + O_{N-1}))$$

The trapezoidal rule applies to any number of offsets.

Consider the following worked example.

Worked Example 13.2: Trapezoidal Rule

Question

The following offsets, 8 m apart, were measured at right angles from a traverse line to an irregular boundary.

0 m 2.3 m 5.5 m 7.9 m 8.6 m 6.9 m 7.3 m 6.2 m 3.1 m 0 m

Calculate the area between the traverse line and the irregular boundary using the trapezoidal rule.

Solution

$$\textbf{Area} = \frac{8.0}{2} \ [0 + 0 + 2 \ (2.3 + 5.5 + 7.9 + 8.6 + 6.9 \\ + 7.3 + 6.2 + 3.1)]$$

$$= \ 4 \times 2 \ (47.8) = \textbf{382.4 m}^2$$

Simpson's rule   This method assumes that instead of being made up of a series of straight lines, the boundary consists of a series of parabolic arcs. A more accurate result is obtained since a better approximation of the true shape of the irregular boundary is achieved. Figure 13.7 shows this applied to figure 13.6.

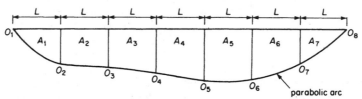

Figure 13.7   *Simpson's rule*

Simpson's rule considers offsets in sets of three and it can be shown that the area between offset 1 and 3 is given by

$$A_1 + A_2 = \frac{L}{3} \ (O_1 + 4O_2 + O_3)$$

Similarly

$$A_3 + A_4 = \frac{L}{3} \ (O_3 + 4O_4 + O_5)$$

Hence, in general

$$\text{Total area} = \frac{L}{3} \ (O_1 + O_N + 4 \ \Sigma \ \text{even offsets} + \\ 2 \ \Sigma \ \text{remaining odd offsets})$$

However, N MUST be an ODD number for Simpson's rule to apply.

When faced with an even number of offsets, as in figure 13.7, when using Simpson's rule, the final offset must be omitted (for example, O_8), the rest of the area calculated and the last small area calculated as a trapezium (that is, using the trapezoidal rule). Consider the following worked example.

Worked Example 13.3: Simpson's Rule

Question
Using the data given in worked example 13.2, calculate the area between the traverse line and the irregular boundary using Simpson's rule.

Solution
There are an even number of offsets, 10, hence calculate the area between 1 and 9 by Simpson's rule and the area between 9 and 10 by the trapezoidal rule.

$$\text{Area}_{1-9} = \frac{8.0}{3} [0 + 3.1 + 4(2.3 + 7.9 + 6.9 + 6.2) + 2(5.5 + 8.6 + 7.3)]$$

$$= \frac{8.0}{3} [3.1 + 4(23.3) + 2(21.4)]$$

$$= \frac{8.0 \times 139.1}{3} = 370.9 \text{ m}^2$$

$$\text{Area}_{9-10} = \frac{8.0}{2} (3.1 + 0) = 12.4 \text{ m}^2$$

Therefore

total area = 370.9 + 12.4 = 383.3 m²

Note the difference between this result and that obtained using the trapezoidal rule in worked example 13.2. Simpson's rule will give the more accurate result when the boundary is genuinely irregular and the trapezoidal rule will give the more accurate result when the boundary is almost a series of straight lines. In general for irregular-sided figures, Simpson's rule should be used.

13.4 The Planimeter

A planimeter is an instrument which automatically measures the area of any irregular-sided plane figure. Traditionally, *mechanical* devices were used but, although these are still manufactured, they have been largely superseded by *digital* instruments.

When using planimeters, a high degree of accuracy can be achieved no matter how complex the shape of the area in question.

Mechanical Planimeters

A mechanical planimeter consists of two arms, the *pole arm* and the *tracing arm* which are joined at a *pivot* as shown in figure 13.8. At the other end of the pole arm is a heavy weight known as the *pole block* and at the other end of the tracing arm is the *tracing point* which normally consists of magnifying eyepiece containing an *index mark*. The tracing arm also incorporates a *measuring unit* which contains an *integrating disc*.

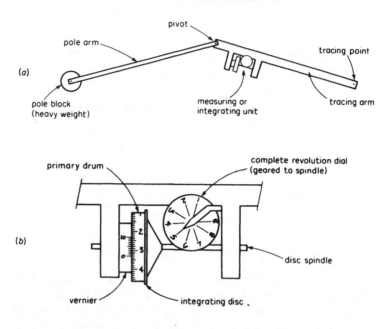

Figure 13.8   *Mechanical planimeter: (a) main features; (b) integrating unit*

The area is obtained from the integrating disc which revolves and alters the reading on the measuring unit as the tracing point is moved round the perimeter of the figure. It is possible to read to 1/1000th of a revolution of the disc. The reading obtained on the measuring unit is directly related to the length of the tracing arm. There are two types of mechanical planimeter, those with fixed tracing arms and those with movable tracing arms.

On a *fixed tracing arm instrument* the readings are obtained directly in mm^2 and then have to be converted according to the plan scale to obtain the ground area.

On a *movable arm instrument* the tracing arm length can be set to particular values depending on the plan scale such that the readings obtained give the ground area directly.

A mechanical planimeter can be used as follows.

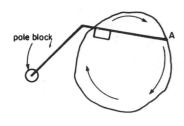

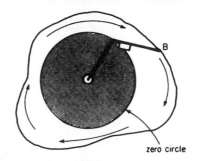

Figure 13.9 *Measuring area with pole block outside area*

Figure 13.10 *Measuring area with pole block inside area*

With the pole block outside the figure

This is the most common method of use and is shown in figure 13.9.

(1) Point A is marked and the scale read.
(2) The tracing point is moved clockwise round the perimeter of the figure back to point A and the scale is again read.
(3) The difference between the two readings multiplied by any necessary factor gives the ground area.

With the pole block inside the figure

This is essentially the same as for the pole block outside, but a constant must be added, as shown in figure 13.10. The shaded area is known as the *zero circle* and is that area which is not registered on the scale owing to the disc being dragged at right angles to its direction of rotation during the measurement. This constant is given with the planimeter and will also have to be converted as necessary.

Whichever method is used, the planimeter should always be checked over a known area and if a discrepancy is found a further correction factor should be computed and applied to all the planimeter readings. *Testing bars* are usually provided with the planimeter for this purpose.

Digital Planimeters

Digital planimeters, although based on the same principles as those described above for mechanical planimeters, incorporate many electronic features and can be extremely sophisticated. As with the mechanical devices they will measure areas of any shape no matter how complex. However, there are some models available which will also measure the circumference of the area, the length of any segment or arc within the circumference, the radius of the arcs, the total length along a boundary, and the rectangular coordi-

Figure 13.11   *Digital planimeter (courtesy Ushikata Mfg. Co. Ltd)*

nates of points based either on an arbitrary or an existing grid system. The readings are displayed digitally on a built-in screen and some instruments incorporate a printer, a calculator pad and an internal memory in which the data is stored prior to downloading into a computer via an RS232 interface. Figure 13.11 shows such a sophisticated digital planimeter.

Their method of operation is similar to that of the mechanical models in that it is still necessary to follow the circumference of the area using the index mark in the magnifying eyepiece. However, the problem of the zero circle is automatically accounted for in the values displayed on the screen.

Mechanical and digital planimeters can be used to measure any area that has been drawn on a map or a sheet of paper. As well as being ideal for *plan areas*, they are extremely useful for measuring *cross-sectional areas*. They are often used for this purpose in preference to one of special cross-sectional area formulae discussed in sections 13.6 to 13.9. Worked example 13.4 shows how a planimeter can be used to measure the area of a cross-section. They can also be used to help with *mass haul calculations* as shown in worked example 13.7.

13.5 Longitudinal Sections and Cross-sections

In the construction of a road, railway, large diameter underground pipeline or similar, having set out the proposed centre line on the ground, levels are taken at regular intervals both along it and at right angles to it to obtain the longitudinal and cross-sections. This is shown in figure 13.12, the fieldwork being described in detail in sections 2.20 and 2.21.

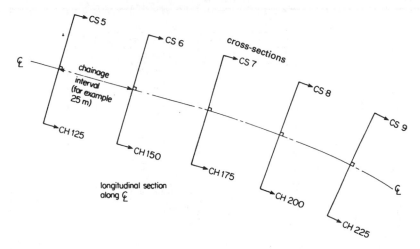

Figure 13.12 *Cross-section layout*

When plotting the *longitudinal section*, the vertical alignment is designed and the formation levels along the centre line are calculated. A typical longitudinal section showing the formation level is shown in figure 13.13.

Each *cross-section* (CS) is drawn and the area between the existing and proposed levels is calculated. Figure 13.14 shows typical cross-sections.

Both the longitudinal section and the cross-sections are usually drawn with their horizontal and vertical scales at different values, that is
Scales for longitudinal section

> horizontal – as road layout drawings, for example, 1 in 500
> vertical – exaggerated, for example, 1 in 100

Scales for cross-sections

> horizontal – exaggerated, for example, in 1 in 200
> vertical – exaggerated, for example, 1 in 50

The reason for exaggerating the vertical scales of both sections and the horizontal scale of the cross-sections is to give a clear picture of the exact shape of the sections.

If the cross-sections have different horizontal and vertical scales it is still possible to calculate their areas either by the graphical method or by using a planimeter as normal and applying a conversion factor. Consider the following worked example.

436

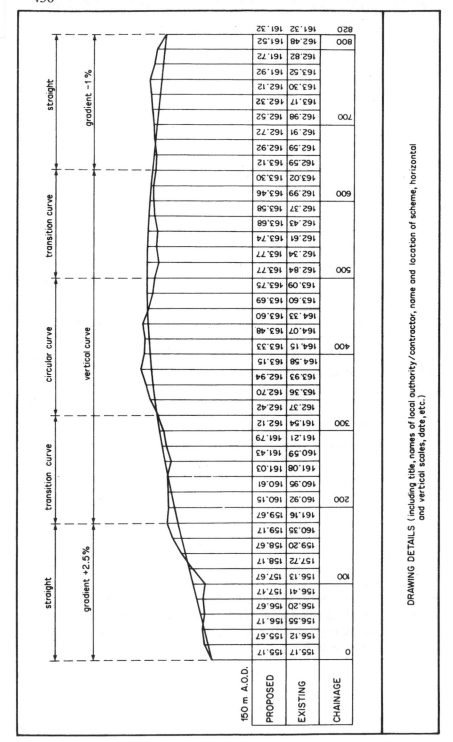

Figure 13.13 *Example of a longitudinal section*

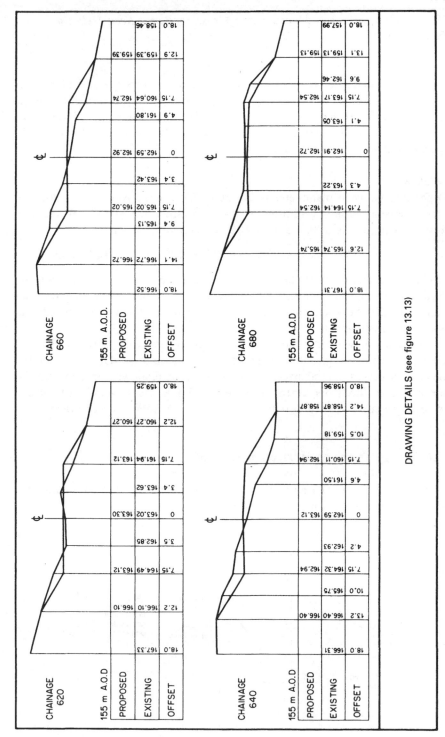

Figure 13.14 *Example cross-sections*

Worked Example 13.4: Measuring a Cross sectional Area Using a Planimeter

Question
A cross-sectional area was measured using a fixed arm mechanical planimeter which gave readings directly in mm^2. The initial planimeter reading was set to zero and the final reading was 7362. If the horizontal scale of the cross-section was 1 in 200 and the vertical scale 1 in 100, calculate the true area represented by the cross-section.

Solution
Difference between planimeter readings = 7362 mm^2. But 1 mm^2 does, in fact, represent an area (200 mm × 100 mm) since the horizontal and vertical scales are 1 in 200 and 1 in 100 respectively. Hence

$$7362 \text{ mm}^2 = (7362 \times 200 \times 100) \text{ mm}^2 = \textbf{147.24 m}^2$$

Once the areas of all the cross-sections have been obtained they are used to calculate the volumes of material to be either excavated (*cut*) or imported (*fill*) between consecutive cross-sections. Such volume calculations are considered in section 13.12.

Although, in worked example 13.4, a planimeter was used to measure the cross-sectional area, this is not the only method available; any of the methods discussed in sections 13.2 and 13.3 can be employed or, alternatively, one of the methods discussed in sections 13.6 to 13.10 which follow can be used. In these, five different types of cross-section are considered.

13.6 Cross-sections on Horizontal Ground

Figure 13.15 shows a sectional drawing of a cutting formed in an area where existing ground level is constant.
From figure 13.15

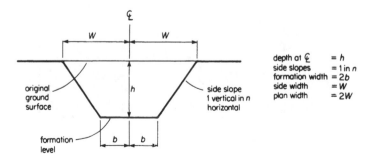

Figure 13.15 *Level section*

$$area \ of \ cross\text{-}section = h(2b + nh)$$
$$plan \ width = 2W = 2(b + nh)$$

For an embankment, the diagram is inverted and the same formulae apply.

The *side widths* (*W* values) are used to show the extent of the embankments and cuttings on the road drawings as discussed in section 13.11.

13.7 Two-level Cross-sections

A cutting with a constant transverse slope is shown in figure 13.16, where W_G = greater side width, W_L = lesser side width, h = depth of cut on centre line from existing to proposed level, 1 in n = side slope, 1 in s = ground or transverse slope.

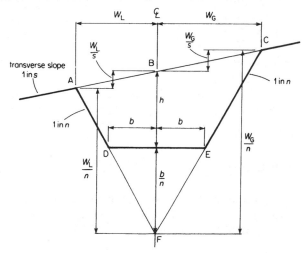

Figure 13.16 *Two-level section*

Considering vertical distances at the centre line gives

$$\frac{W_L}{n} = \frac{b}{n} + h - \frac{W_L}{s}$$

and

$$\frac{W_G}{n} = \frac{b}{n} + h + \frac{W_G}{s}$$

Multiplying by sn gives

$$W_L s = bs + hsn - W_L n$$

and

$$W_G s = bs + hsn + W_G n$$

Hence

$$W_L = \frac{s(b + nh)}{s + n}$$

and

$$W_G = \frac{s(b + nh)}{s - n}$$

The *plan width* is given by $(W_L + W_G)$. The *cross-sectional area* (A) of the cutting is given by

$$A = \text{area ABF} + \text{area BCF} - \text{area DEF, hence}$$

$$A = \frac{1}{2} \left[h + \frac{b}{n} \right] (W_L + W_G) - \frac{b^2}{n}$$

Again, for embankments, figure 13.16 is inverted and the same formulae apply.

The greater and lesser *side widths* (W_G and W_L) are used to show the extent of the embankments and cuttings on the road drawings as discussed in section 13.11.

13.8 Three-level Cross-sections

A cutting with a transverse slope which changes gradient at the centre line is shown in figure 13.17, where W_1 and W_2 are the side widths, h = depth of cut on the centre line from the existing to the proposed levels, 1 in n = side slope, 1 in s_1 and 1 in s_2 = transverse slopes.

Cross-sections of this type are best considered as consisting of two separate half sections on either side of the centre line. There are eight possible types of half section as shown in figure 13.18.

Using techniques similar to those used to derive the formulae for the two-level section in figure 13.16, it is possible to derive the following formulae for three-level sections.

For the half sections shown in figure 13.18 *a*, *b*, *c* and *d*

$$W = \frac{s(b + nh)}{(s - n)}$$

For the half sections shown in figure 13.18 *e*, *f*, *g* and *h*

$$W = \frac{s(b + nh)}{(s + n)}$$

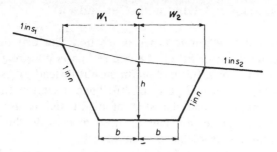

Figure 13.17 *Three-level section*

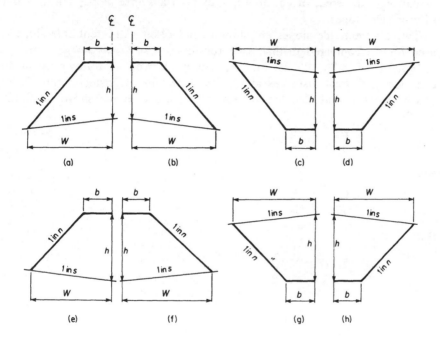

Figure 13.18 *Half sections*

The cross-sectional area (*A*) of any combination of any two of the eight types of half section is given by

$$A = \frac{1}{2}\left(h + \frac{b}{n} \right) \text{(sum of side widths)} - \frac{b^2}{n}$$

The *side widths* (*W* values) are used to show the extent of the embankments and cuttings on the road drawings as discussed in section 13.11.

13.9 Cross-sections Involving both Cut and Fill

Figure 13.19 shows the four types of section that can occur in practice where the depth of cut or fill on the centre line is not great enough to give either a full cutting or a full embankment but instead gives a cross-section consisting partly of cut and party of fill. Such a section can occur when a road is being built around the side of a hill and is used for economic reasons since the cut section can be used to provide the fill section and very little earth-moving distance is involved.

With reference to figure 13.19, h = depth of cut or fill on the centre line from the existing to the proposed levels, W_1 and W_2 = the side widths, A_1 and A_2 = the areas of cut or fill, 1 in s = transverse slope, 1 in n and 1 in m = side slopes.

Two different side slopes are shown since often a different side slope is used for cut compared to that used for fill.

Formulae for W_1, W_2, A_1 and A_2 can be derived as follows by again considering vertical distances along the centre line. Consider figure 13.20 which shows a more detailed version of the cross-section shown in figure 13.19a.

$$\frac{W_1}{n} = \frac{b}{n} + h_1$$

and

$$\frac{W_2}{m} = \frac{b}{m} + h_2$$

But

$$h_1 = \frac{W_1}{s} - h$$

and

$$h_2 = \frac{W_2}{s} + h$$

Substituting for h_1 and h_2 and multiplying by sn and sm respectively gives

$$W_1 s = bs + W_1 n - hsn$$

and

$$W_2 s = bs + W_2 m + hsm$$

Hence

$$W_1 = s\,\frac{b - nh}{s - n}$$

and

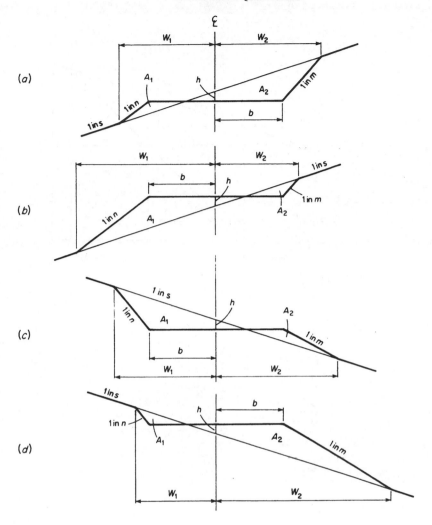

Figure 13.19 *Sections involving cut and fill*

$$W_2 = s \ \frac{b + mh}{s - m}$$

The *plan width* is given by $(W_1 + W_2)$.

The *cross-sectional areas* are obtained as follows

$$A_1 = \tfrac{1}{2} (b - sh)h_1$$

and

$$A_2 = \tfrac{1}{2} (b + sh)h_2$$

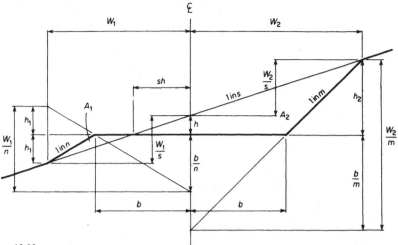

Figure 13.20

but

$$h_1 = \frac{W_1}{s} - h$$

$$= \frac{b - nh}{s - n} - h$$

$$= \frac{b - sh}{s - n}$$

and

$$h_2 = \frac{W_2}{s} + h$$

$$= \frac{b + mh}{s - m} + h$$

$$= \frac{b + sh}{s - m}$$

Hence

$$A_1 = \frac{(b - sh)^2}{2(s - n)}$$

and

$$A_2 = \frac{(b + sh)^2}{2(s - m)}$$

The formulae derived above for W_1, W_2, A_1 and A_2 apply to cross-sections similar to those shown in figure 13.19a and d.

For cross-sections similar to those shown in figure 13.19b and c, the formulae are amended slightly as follows

$$W_1 = s\,\frac{b + nh}{s - n} \qquad W_2 = s\,\frac{b - mh}{s - m}$$

$$A_1 = \frac{(b + sh)^2}{2(s - n)} \qquad A_2 = \frac{(b - sh)^2}{2(s - m)}$$

The *side widths* (W_1 and W_2) are again used to show the extent of the embankments and cuttings on the road drawings as discussed in section 13.11.

With any cross-section partly in cut and party in fill, it is essential that a drawing be produced so that the correct formulae can be used.

The worked example in section 13.13 illustrates the application of these formulae.

13.10 Irregular Cross-sections

Figure 13.21 shows a cutting, the ground surface of which has been surveyed using the levelling methods discussed in section 2.20. For each point surveyed, the RL and offset distance (x) will be known.

Although the area of such a section could be found using a planimeter or by a mathematical method, the *cross coordinate method* (see section 13.2) can also be used. In order to apply this method, a coordinate system which has its origin at the intersection of the formation level and the centre line is used. Offset distances (x values) to the right of the centre line are taken as positive and those to the left of the centre line are taken as negative; heights (y values) above the formation level are considered to be positive and those below the formation level are considered to be negative. For figure 13.21, the points defining the section will have the coordinates given

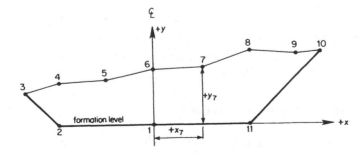

Figure 13.21 *Irregular section*

T A B L E 13.2

Point n =	1	2	3	4	5	6	7	8	9	10	11
E_n	0	$-b$	$-x_3$	$-x_4$	$-x_5$	0	x_7	x_8	x_9	x_{10}	b
N_n	0	0	y_3	y_4	y_5	y_6	y_7	y_8	y_9	y_{10}	0

in table 13.2. These coordinates are used to obtain the area of this section by substituting them into the following cross coordinate formula

$$2 \times \text{Area} = (N_1E_2 + N_2E_3 + N_3E_4 + \ldots + N_{11}E_1)$$
$$- (E_1N_2 + E_2N_3 + E_3N_4 + \ldots + E_{11}N_1)$$

since a clockwise order is given in figure 13.21 for the points.

A similar process can be applied to embankments and also to those sections involving both cut and fill. For the cut and fill sections, separate calculations are required.

In this type of area calculation, careful attention must be paid to the algebraic signs involved.

13.11 Using the Cross-sectional Areas and Side Widths

The cross-sectional areas are used to calculate volumes and this is discussed in section 13.12. However, before they can be used to obtain volumes it is necessary to modify their calculated values by making allowance for the depth of the road construction. The cross-sectional area of the road construction should be calculated or scaled from a plan of the construction and added to cross-sectional areas in cut and subtracted from cross-sectional areas in fill. Cross-sections partly in cut and partly in fill should be inspected on the cross-sectional drawings and their areas modified accordingly.

The side widths, either separately or together in the form of the plan width, are used to mark the extent of the embankments and cuttings on the working drawings.

This is shown in figure 13.22. They help firstly in the calculation of the area of land which must be obtained for the construction and secondly in the calculation of the area of the site to be cleared and stripped of topsoil before construction can begin.

13.12 Volumes from Cross-sections

The cross-sections calculated in sections 13.6 to 13.10 are used to calculate the volume contained between them. Two methods are considered, both comparable to the trapezoidal rule and Simpson's rule for areas.

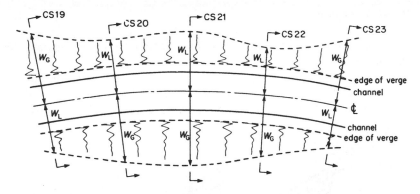

Figure 13.22 *Side widths*

End Areas Method

This is comparable to the trapezoidal rule for areas. If two cross-sectional areas A_1 and A_2 are a horizontal distance d_1 apart, the volume contained between them (V_1) is given by

$$V_1 = d_1 \; \frac{(A_1 + A_2)}{2}$$

This leads to the general formula for a series of N cross-sections

$$
\begin{aligned}
V_{\text{total}} &= V_1 + V_2 + V_3 + \ldots + V_{N\text{-}1} \\
&= d_1 \; \frac{(A_1 + A_2)}{2} + d_2 \; \frac{(A_2 + A_3)}{2} \\
&\quad + d_3 \; \frac{(A_3 + A_4)}{2} + \ldots \\
&\quad + d_{N-1} \; \frac{(A_{N-1} + A_N)}{2}
\end{aligned}
$$

and, if $d_1 = d_2 = d_3 = d_{N-1} = d$

$$\text{total volume} = \frac{d}{2} \; [A_1 + A_N + 2(A_2 + A_3 + \ldots A_{N-1})]$$

The end areas method will given accurate results if the cross-sectional areas are of the same order of magnitude. Worked example 13.5 at the end of this section shows how the end areas formula is applied.

Prismoidal Formula

This is comparable to Simpson's rule for areas and is more accurate than the end areas method.

The volume contained between a series of cross-sections a constant distance apart can be approximated to the volume of a *prismoid* which is a solid figure with plane parallel ends and plane sides. This is shown in figure 13.23.

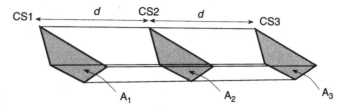

Figure 13.23   *Prismoid for volume calculation*

It can be shown that for a series of three cross-sections the volume, V_{1-3}, contained between them is given by

$$V_{1-3} = \frac{d}{3} (A_1 + 4A_2 + A_3)$$

This is the *prismoidal formula* and is used for earthwork calculations of cuttings and embankments and gives a true volume if *either*

(1) the transverse slopes at right angles to the centre line are straight and the longitudinal profile on the centre line is parabolic, *or*
(2) the transverse slopes are parabolic and the longitudinal profile is a straight line.

Hence, unless the ground profile is regular both transversely and longitudinally it is likely that errors will be introduced in assuming that the figure is prismoidal over its entire length. These errors, however, are small and the volume obtained is a good approximation to the true volume.

If figure 13.23 is extended to include cross-section 4 (A_4) and cross-section 5 (A_5), the volume from CS3 to CS5 (V_{3-5}) is given by

$$V_{3-5} = \frac{d}{3} (A_3 + 4A_4 + A_5)$$

Therefore, the total volume from CS1 to CS5 (V) is

$$V = \frac{d}{3} (A_1 + 4A_2 + 2A_3 + 4A_4 + A_5)$$

This leads to a general formula for N cross-sections, where N MUST be ODD, as follows

$$V = \frac{d}{3} (A_1 + A_N + 4 \Sigma \text{ even areas} + 2 \Sigma \text{ remaining}$$

$$\text{odd areas)}$$

This is often referred to as Simpson's rule for volumes and should be used wherever possible. The worked example given in section 13.13 illustrates the application of the prismoidal formula.

Effect of Curvature on Volume

The foregoing discussion has assumed that the cross-sections are taken on a straight road or similar. However, where a horizontal curve occurs the cross-sections will no longer be parallel to each other and errors will result in volumes calculated from either the end areas method or the prismoidal formula.

To overcome this, *Pappus' theorem* must be used and this states that a volume swept out by a plane constant area revolving about a fixed axis is given by the product of the cross-sectional area and the distance moved by the centre of gravity of the section.

Hence the volumes of cuttings which occur on circular curves can be calculated with a better degree of accuracy. Figure 13.24 shows an asymmetrical cross-section in which the centroid is situated at a horizontal distance c from the centre line, where c is referred to as the *eccentricity*. The centroid may be on either side of the centre line according to the transverse slope.

Figure 13.25 shows this cross-section occurring on a circular curve of radius R.

$$\text{Length of path of centroid} = (R + c) \ \theta \text{ rad}$$

From Pappus' theorem, the volume swept out is $V = A(R + c) \ \theta$ rad. But

$$\theta \text{ rad} = L/R, \text{ hence } V = LA(R + c)/R.$$

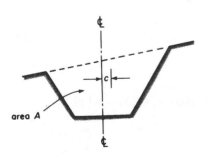

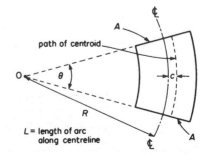

Figure 13.24 *Asymmetrical cross-section with eccentricity c*

Figure 13.25 *Effect of eccentricity on volume calculation*

Therefore

$$V = L\left(A + \frac{Ac}{R}\right) = LA\left(1 + \frac{c}{R}\right) = LA'$$

In the expression for V, LA is the volume of a prismoid of length L and the term Ac/R can be regarded as the correction to be made to the cross-sectional area before calculating the volume as that of a normal prismoid. The corrected area can be expressed as

$$A' = A\left(1 \pm \frac{c}{R}\right)$$

The $\pm$ sign is necessary since the centroid can lie on either side of the centre line. The negative sign is adopted if the centroid lies on the same side of the centre line as the centre of curvature and the positive sign if on the other side.

In practice, the shape of the cross-section will not be constant so that neither A nor c will be constant. However, the ratio c/R will usually be small and it is generally sufficiently accurate to calculate the correction for each cross-section and to use either the end areas method or the prismoidal formula to determine the volume.

Worked Example 13.5: Volume Calculation from a Series of Cross-sections

Question
Figure 13.26 shows a longitudinal section along the proposed centre line of a road together with a series of six cross-sections taken at 20 m intervals. The areas of cut and/or fill at each section are indicated.

Calculate the volumes of cut and fill contained between CS1 and C6.

Solution
The volumes of cut and fill are calculated using either the end areas method or the prismoidal formula. If the cross-sections are all of the same type, one of the formulae, preferably the prismoidal, can be used. However, if the cross-sections are changing as in figure 13.26 it is best to work from one cross-section to the next as follows.
CS1 to CS2

$$\text{Volume of cut} = \frac{20}{2}(110.6 + 64.3) = \mathbf{1749}\ \mathbf{m}^3$$

$$\text{Volume of fill} = \mathbf{zero}$$

CS2 to CS3

$$\text{Volume of cut} = \frac{20}{2}(64.3 + 36.2) = \mathbf{1005}\ \mathbf{m}^3$$

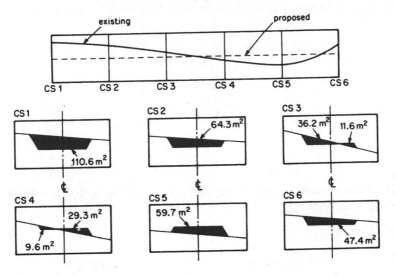

Figure 13.26 *Data for worked example 13.5*

The volume of fill presents a problem. Between CS2 and CS3 there is a point at which the fill begins. A good estimate of the position of this point must be made to enable accurate volume figures to be obtained. This is best done by assuming that the rate of increase of fill between CS2 and CS3 is the same as that between CS3 and CS4.

Between CS3 and CS4, the area of fill increases from 11.6 m² to 29.3 m² in a distance of 20 m. If this is extrapolated back as shown in figure 13.27, the point at which the fill begins between CS2 and CS3 can be found.

In figure 13.27

$$\frac{d_1}{11.6} = \frac{20}{(29.3 - 11.6)}$$

hence

$$d_1 = 13.1 \text{ m}$$

Therefore

$$\text{volume of fill} = \frac{13.1}{2} (0 + 11.6) = \textbf{76 m}^3$$

This extrapolation method will work only if the area of fill at CS4 is greater than twice the area of fill at CS3 otherwise a meaningless result will be obtained, that is, $d_1 > d$. The only solution to such an occurrence is to inspect the cross-sectional drawings and the longitudinal section and to make a reasoned estimate of the position at which the fill begins.

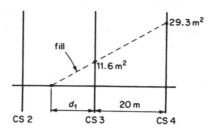

Figure 13.27

The extrapolation method is suitable for both cut and fill, whether increasing or decreasing.

CS3 to CS4

$$\text{volume of cut} = \frac{20}{2} \ (36.2 + 9.6) = \textbf{458 m}^3$$

$$\text{volume of fill} = \frac{20}{2} \ (11.6 + 29.3) = \textbf{409 m}^3$$

CS4 to CS5   The volume of cut must be calculated using the extrapolation method. From CS3 to CS4 the area of cut decreases from 36.2 m² to 9.6 m² in a distance of 20 m. Hence, the distance from CS4 towards CS5 at which the cut decreases (d_2) is given by

$$d_2 = \frac{20 \times 9.6}{36.2 - 9.6} = 7.2 \text{ m}$$

Therefore

$$\text{volume of cut} = \frac{7.2}{2} \ (9.6 + 0) = \textbf{35 m}^3$$

$$\text{volume of fill} = \frac{20}{2} \ (29.3 + 59.7) = \textbf{890 m}^3$$

CS5 to CS6   In this case the cross-section changes from all fill to all cut. Again, for accuracy, it is necessary to estimate the position where the fill section ends and the cut section begins. The relevant part of the longitudinal section is shown in figure 13.28.

A linear relationship involving the cross-sectional areas can be used

$$\frac{d_1}{A_s} = \frac{d}{A_s + A_6}$$

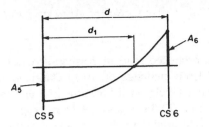

Figure 13.28

hence

$$d_1 = \frac{d\,A_5}{A_5 + A_6}$$

The volume of fill is obtained from $[d_1/2]\,(A_5 + 0)$ and the volume of cut is obtained from $[(d - d_1)/2]\,(0 + A_6)$. In this case

$$d_1 = \frac{20 \times 59.7}{59.7 + 47.4} = 11.1 \text{ m}$$

Therefore

$$\text{volume of fill} = \frac{11.1}{2}\,(59.7 + 0) = \mathbf{331 \text{ m}^3}$$

$$\text{volume of cut} = \frac{(20 - 11.1)}{2}\,(0 + 47.4) = \mathbf{211 \text{ m}^3}$$

This linear method applies equally when the cross-section changes from all cut to all fill.

13.13 Combined Cross-sectional Area and Volume Calculations

In sections 13.6 to 13.10, several methods by which cross-sectional areas can be calculated were discussed and in section 13.12, techniques by which these can be used to obtain volumes were described. The worked examples given in these sections concentrate either on areas or on volumes in order to illustrate the various principles involved. However, although it is necessary to calculate the cross-sectional areas before the volumes of cut and/or fill can be obtained, one follows naturally from the other and, in practice, the two processes are considered together as shown in the following worked example.

Worked Example 13.6: Cross-sectional Area and Volume Calculations

Question

The centre line of a proposed road of formation width 12.00 m is to fall at a slope of 1 in 100 from chainage 50 m to chainage 150 m.

The existing ground levels on the centre line at chainages 50 m, 100 m and 150 m are 71.62 m, 72.34 m and 69.31 m respectively and the ground slopes at 1 in 3 at right angles to the proposed centre line.

If the centre line formation level at chainage 50 m is 71.22 m and side slopes are to be 1 in 1 in cut and 1 in 2 in fill, calculate the volumes of cut and fill between chainages 50 m and 150 m.

Solution

Figure 13.29 shows the longitudinal section from chainage 50 m to chainage 150 m. Hence

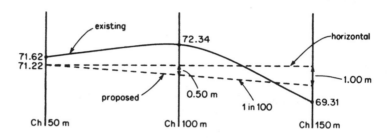

Figure 13.29

the centre line formation level at chainage 100 m = 71.22 − 0.50

$$= 70.72 \text{ m}$$

the centre line formation level at chainage 150 m = 71.22 − 1.00

$$= 70.22 \text{ m}$$

Figure 13.30 shows the three cross-sections.

Since all the cross-sections are part in cut and part in fill the formulae derived in section 13.9 apply

Cross-section 50 m $s = 3$, $b = 6$ m, $n = 2$, $m = 1$

$$h = 71.62 - 71.22 = + 0.40 \text{ m, that is, cut at the}$$
$$\text{centre line}$$

This cross-section is similar to that shown in figure 13.19*a*, hence

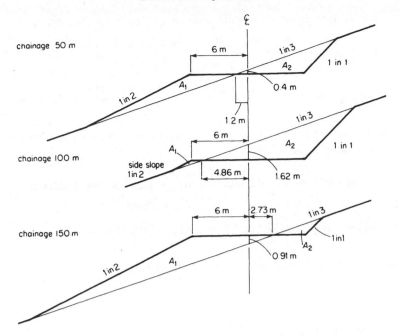

Figure 13.30

$$area\ of\ cut = A_2 = \frac{(b + sh)^2}{2(s - m)} = \frac{(6 + 3 \times 0.40)^2}{2(3 - 1)} = \textbf{12.96 m}^2$$

$$area\ of\ fill = A_1 = \frac{(b - sh)^2}{2(s - n)} = \frac{(6 - 3 \times 0.40)^2}{2(3 - 2)} = \textbf{11.52 m}^2$$

Cross-section 100 m $s = 3, b = 6$ m, $n = 2, m = 1$

$$h = 72.34 - 70.72 = + 1.62 \text{ m, that is, cut at the}$$
centre line

This cross-section is similar to cross-section 50 m, hence again

$$area\ of\ cut = A_2 = \frac{(6 + 3 \times 1.62)^2}{2(3 - 1)} = \textbf{29.48 m}^2$$

$$area\ of\ fill = A_1 = \frac{(6 - 3 \times 1.62)^2}{2(3 - 2)} = \textbf{0.65 m}^2$$

Cross-section 150 m $s = 3, b = 6$ m, $n = 2, m = 1$

$$h = 69.31 - 70.22 = -0.91 \text{ m, that is, fill at the}$$
centre line

This cross-section is similar to that shown in figure 13.19b, hence

$$area\ of\ cut = A_2 = \frac{(b - sh)^2}{2(s - m)} = \frac{(6 - 3 \times 0.91)^2}{2(3 - 1)} = \textbf{2.67 m}^2$$

$$area\ of\ fill = A_1 = \frac{(b + sh)^2}{2(s - n)} = \frac{(6 + 3 \times 0.91)^2}{2(3 - 2)} = \textbf{38.11 m}^2$$

The prismoidal formula can be used to calculate the volumes since the number of cross-sections is odd, hence

$$volume\ of\ cut = \frac{50}{3} [12.96 + 2.67 + 4(29.48)] = \textbf{2225.8 m}^3$$

$$volume\ of\ fill = \frac{50}{3} [11.52 + 38.11 + 4(0.65)] = \textbf{870.5 m}^3$$

These figures would normally be rounded to at least the nearest cubic metre.

13.14 Volumes from Spot Heights

This method is used to obtain the volume of large deep excavations such as basements, underground tanks and so on where the formation level can be sloping, horizontal or terraced.

A square, rectangular or triangular grid is established on the ground and spot levels are taken at each grid intersection as described in section 2.24. The smaller the grid the greater will be the accuracy of the volume calculated but the amount of fieldwork increases so a compromise is usually reached.

The formation level at each grid point must be known and hence the depth of cut from the existing to the proposed level at each grid intersection can be calculated.

Figure 13.31 shows a 10 m square grid with the depths of cut marked at each grid intersection. Consider the volume contained in grid square $h_1 h_2 h_6 h_5$; this is shown in figure 13.32.

It is assumed that the surface slope is constant between grid intersections, hence the volume is given by

$$volume = mean\ height \times plan\ area$$

$$= \tfrac{1}{4}(4.76 + 5.14 + 4.77 + 3.21) \times 100 = \textbf{447 m}^3$$

A similar method can be applied to each individual grid square and this leads to the following general formula for square or rectangular grids

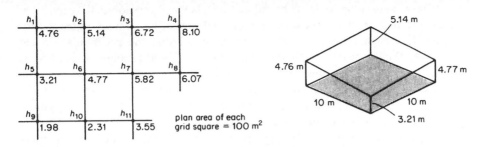

Figure 13.31 *Grid heights for volume calculation*

Figure 13.32 *Volume calculation for a grid square*

$$\text{total volume} = \frac{A}{4} \, (\Sigma \text{ single depths} + 2 \, \Sigma \text{ double depths}$$

$$+ \, 3 \, \Sigma \text{ triple depths} + 4 \, \Sigma \text{ quadruple depths)}$$

$$+ \, \delta V$$

where A = plan area of each grid square; single depths = depths such as h_1 and h_4 which are used once; double depths = depths such as h_2 and h_3 which are used twice; triple depths = depths such as h_7 which are used three times; quadruple depths = depths such as h_6 which are used four times; δV = the total volume outside the grid which is calculated separately.

Hence, in the example shown in figure 13.31

$$\text{volume contained within the grid area} = \frac{100}{4} \, [4.76 + 8.10$$

$$+ \, 6.07 + 1.98 + 3.55 + 2(5.14 + 6.72 + 3.21 + 2.31)$$

$$+ \, 3(5.82) + 4(4.77)]$$

$$= 25(24.46 + 34.76 + 17.46 + 19.08) = \textbf{2394 m}^3$$

The result is only an approximation since it has been assumed that the surface slope is constant between spot heights.

If a triangular grid is used, the general formula must be modified as follows

(1) $A'/3$ must replace $A/4$ where A' = plan area of each triangle and
(2) depths appearing in five and six triangles must be included.

13.15 Volumes from Contours

This method is particularly suitable for calculating very large volumes such as those of reservoirs, earth dams, spoil heaps and so on.

The system adopted is to calculate the plan area enclosed by each contour and then treat this as a cross-sectional area. The contour interval provides the distance between cross-sections and either the prismoidal or end areas method is used to calculate the volume. If the prismoidal method is used, the number of contours must be odd.

The plan area contained by each contour can be calculated using a planimeter or one of the methods discussed in section 13.3. The graphical method is particularly suitable in this case.

The accuracy of the result depends to a large extent on the contour interval but normally great accuracy is not required, for example in reservoir capacity calculations, volumes to the nearest 1000 m³ are more than adequate. Consider the following example.

Figure 13.33 shows a plan of a proposed reservoir and dam wall. The vertical interval is 5 m and the water level of the reservoir is to be 148 m. The capacity of the reservoir is required.

The volume of water that can be stored between the contours can be found by reference to figure 13.34, which shows a cross-section through the reservoir and the plan areas enclosed by each contour and the dam wall.

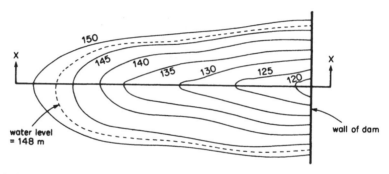

Figure 13.33 *Plan of proposed reservoir*

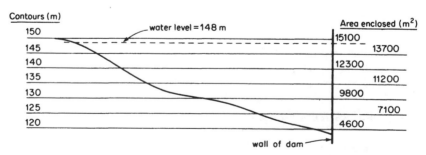

Figure 13.34 *Contour areas for proposed reservoir*

total volume = volume between 148 m and 145 m contours +

volume between 145 m and 120 m contours +

small volume below 120 m contour.

Volume between 148 m and 145 m contours is found by the end areas method to be

$$= \left(\frac{15\ 100\ +\ 13\ 700}{2}\right) \times 3 = \mathbf{43\ 200\ m^3}$$

Volume between 145 m and 120 m contours can also be found by the end areas method to be

$$= \frac{5}{2}\ (13\ 700\ +\ 4600\ +\ 2(12\ 300\ +\ 11\ 200\ +\ 9800\ +\ 7100))$$

$$= \mathbf{247\ 750\ m^3}$$

The small volume below the 120 m contour can be found by decreasing the contour interval to say, 1 m and using the end areas method or the prismoidal formula. Alternatively, if it is very small, it may be neglected. Let this volume = δV. Therefore

$$\text{total volume} = 43\ 200\ +\ 247\ 750\ +\ \delta V$$

$$= \mathbf{(290\ 950\ +\ \delta V\ )\,m^3} \tag{13.4}$$

(this would usually be rounded to the nearest 1000 m³). The second term in equation (13.4) was obtained by the end areas method applied between contours 145 m and 120 m. Alternatively, the prismoidal formula could have been used between the 145 m and 125 m contours (to keep the number of contours ODD) and the end areas method between the 125 m and 120 m contours. If this is done, the volume between the 145 m and 120 m contours is calculated to be **248 583 m³**.

13.16 Introduction to Mass Haul Diagrams

During the construction of long engineering projects such as roads, railways, pipelines and canals there may be a considerable quantity of earth required to be brought on to the site to form embankments and to be removed from the site during the formation of cuttings.

The earth brought to form embankments may come from another section of the site such as a tip formed from excavated material (known as a *spoil heap*) or may be imported on to the site from a nearby quarry. Any earth brought on to the site is said to have been *borrowed*.

The earth excavated to form cuttings may be deposited in tips at regular intervals along the project to form spoil heaps for later use in embankment formation or may be *wasted* either by spreading the earth at right angles to the centre line to form verges or by carting it away from the site area and depositing it in suitable local areas.

This movement of earth throughout the site can be very expensive and, since the majority of the cost of such projects is usually given over to the earth-moving, it is essential that considerable care is taken when planning the way in which material is handled during the construction. The *mass haul diagram* is a graph of volume against chainage which greatly helps in planning such earth-moving.

The *x* axis represents the chainage along the project from the position of zero chainage.

The *y* axis represents the aggregate volume of material up to any chainage from the position of zero chainage.

When constructing the mass haul diagram, volumes of cut are considered positive and volumes of fill are considered negative. The vertical and horizontal axes of the mass haul diagram are usually drawn at different scales to exaggerate the diagram and thereby facilitate its use.

The mass haul diagram considers only earth moved in a direction longitudinal to the direction of the centre line of the project and does not take into account any volume of material moved at right angles to the centre line.

13.17 Formation Level and the Mass Haul Diagram

Since the mass haul diagram is simply a graph of aggregate volume against chainage it will be noted that if the volume is continually decreasing with chainage, the project is all embankment and all the material will have to be imported on to the site since there will be no fill material available for use. Such an occurrence will involve a great deal of earth-moving and is obviously not an ideal solution.

If a better attempt had been made in the selection of a suitable formation level such that some areas of cut were balanced out by some areas of fill, a more economical solution would result. Because of this vital connection between the formation level and the mass haul diagram the two are usually drawn together as shown in figure 13.35 and section 13.18 describes its method of construction.

13.18 Drawing the Diagram

Figure 13.35 was drawn as follows.

(1) The cross-sectional areas are calculated at regular intervals along the project, in this case every 50 m.

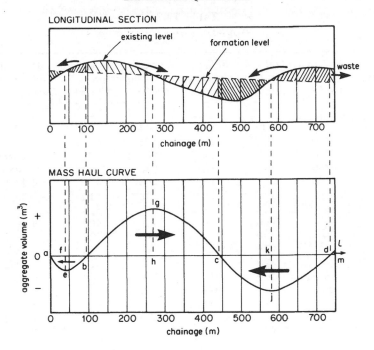

Figure 13.35 *Mass haul diagram*

(2) The volumes between consecutive areas and the aggregate volume along the site are calculated, cut being positive and fill negative.
(3) Before plotting, a table is drawn up as shown in table 13.3. One of the columns in table 13.3 shows bulking and shrinkage factors. These are necessary owing to the fact that material usually occupies a different volume when it is used in a man-made construction from that which it occupied in natural conditions. Very few soils can be compacted back to their original volume.

If 100 m³ of rock are excavated and then used for filling, they may occupy 110 m³ even after careful compaction and the rock is said to have undergone *bulking* and has a *bulking factor* of 1.1.

If 100 m³ of clay are excavated and then used for filling they may occupy only 80 m³ after compaction and the clay is said to have undergone *shrinkage* and has a *shrinkage factor* of 0.8.

Owing to the variable nature of the same material when found in different parts of the country, it is impossible to standardise bulking and shrinkage factors for different soil and rock types. Therefore, a list of such factors has deliberately not been included since it would indicate a uniformity that, in practice, does not exist. Instead, it is recommended that local knowledge of the materials in question should be considered together with tests on soil and rock samples from the area so that reli-

Table 13.3

Mass Haul Diagram Calculations

Chainage (metres)	Individual volume (m³) Cut(+)	Fill(−)	Bulking/ Shrinkage factors	Corrected individual volumes (m³) Cut(+)	Fill(−)	Aggregate volume (m³) Cut(+) Fill(−)
0	—	—	—	—	—	0
50	40	800	1.1	44	800	− 756
100	730	—	1.1	803	—	+ 47
150	910	—	1.1	1001	—	+1048
200	760	—	1.1	836	—	+1884
250	450	—	1.1	495	—	+2379
300	80	110	1.1	88	110	+2357
350	—	520	—	—	520	+1837
400	—	900	—	—	900	+ 937
450	—	1120	—	—	1120	− 183
500	—	970	—	—	970	−1153
550	—	620	—	—	620	−1773
600	200	200	0.8	160	200	−1813
650	590	—	0.8	472	—	−1341
700	850	—	0.8	680	—	− 661
750	1120	—	0.8	896	—	+ 235

able bulking and shrinkage factors (which apply only to that particular site) can be determined.

As far as the mass haul diagram is concerned, it is the volumes of fill that are critical, for example, if the hole in the ground is 1000 m³, the required volume is that amount of cut which will fill the hole. There are two methods which can be used to allow for such bulking and shrinkage. Either the calculated volumes of fill can be amended by dividing them by the factors applying to the type of material available for fill or the calculated volumes of cut can be amended by multiplying them by the factors applying to the type of material in the cut. In table 13.3 the latter has been done.

(4) The longitudinal section along the proposed centre line is plotted, the proposed formation level being included.

(5) The axes of the mass haul diagram are drawn underneath the longitudinal profile such that chainage zero on the profile coincides with chainage zero on the diagram.

(6) The aggregate volume up to chainage 50 is plotted at $x = 50$ m. The aggregate volume up to chainage 100 is plotted at $x = 100$ m and so on for the rest of the diagram.

(7) The points are joined by curves or straight lines to obtain the finished mass haul diagram.

13.19 Terminology of Mass Haul Diagrams

(1) *Haul distance* is the distance from the point of excavation to the point where the material is to be tipped.

(2) *Average haul distance* is the distance from the centre of gravity of the excavation to the centre of gravity of the tip.

(3) *Free haul distance* is that distance, usually specified in the contract, over which a charge is levied only for the volume of earth excavated and not its movement. This is discussed further in section 13.21.

(4) *Free haul volume* is that volume of material which is moved through the free haul distance.

(5) *Overhaul distance* is that distance, in *excess* of the free haul distance, over which it may be necessary to transport material. See section 13.21.

(6) *Overhaul volume* is that volume of material which is moved in excess of the free haul distance.

(7) *Haul*. This is the term used when calculating the costs involved in the earth-moving and is equal to the sum of the products of each volume of material and the distance through which it is moved. It is equal to the total volume of the excavation multiplied by the average haul distance and on the mass haul diagram is equal to the area contained between the curve and *balancing line* (see section 13.20).

(8) *Freehaul* is that part of the haul which is contained within the free haul distance.

(9) *Overhaul* is that part of the haul which remains after the freehaul has been removed. It is equal to the product of the overhaul volume and the overhaul distance.

(10) *Waste* is that volume of material which must be exported from a section of the site owing to a surplus or unsuitability.

(11) *Borrow* is that volume of material which must be imported into a section of the site owing to a deficiency of suitable material.

13.20 Properties of the Mass Haul Curve

Consider figure 13.35.

(1) When the curve rises the project is in cut since the aggregate volume is increasing, for example section ebg. When the curve falls the project is in fill since the aggregate volume is decreasing, for example section gcj. Hence, the end of a section in cut is shown by a maximum point on the curve, for example point g, and the end of a section in fill is shown by a minimum point on the curve, for example point j.

The vertical distance between a maximum point and the next forward minimum represents the volume of an embankment, for example (gh + kj), and the vertical distance between a minimum point and the next forward maximum represents the volume of a cutting, for example (ef + gh).

(2) Any horizontal line which cuts the mass haul curve at two or more points balances cut and fill between those points and because of this is known as a *balancing line*.

In figure 13.35 the *x* axis is a balancing line and the volumes between chainages a and b, b and c, and c and d are balanced out, that is, as long as the material is suitable, all the cut material between a and d can be used to provide the exact amount of fill required between a and d. The *x* axis, however, does not always provide the best balancing line and this is discussed further in section 13.22.

When a balancing line has been drawn on the curve, any area lying above the balancing line signifies that the material must be moved to the right and any area lying below the balancing line signifies that the material must be moved to the left. In figure 13.35, the arrows on the longitudinal section and the mass haul diagram indicate these directions of haul.

The length of balancing line between intersection points is the maximum haul distance in that section, for example the maximum haul distance in section bc is (chainage c − chainage b).

(3) The area of the mass haul diagram contained between the curve and the balancing line is equal to the *haul* in that section, for example afbea, bgchb and ckdjc.

If the horizontal scale is 1 mm = R m and the vertical scale is 1 mm = S m^3, then an area of T mm^2 represents a haul of TRS m^3 m. This area could be measured using one of the methods discussed in sections 13.2 to 13.4. Note that the units of haul are m^3 m (one cubic metre moved through one metre).

Instead of calculating centres of gravity of excavations and tips, which can be a difficult task, the *average haul distance* in each section can be easily found by dividing the *haul* in that section by the *volume* in that section, for example

$$\text{the } \textit{average haul distance} \text{ between b and c} = \frac{\text{area bgchb m}^3 \text{ m}}{\text{gh} \quad \text{m}^3}$$

(4) If a surplus volume remains, this is *waste* and must be removed from the site, for example lm; if a deficiency of earth is found at the end of the project this is *borrow* and must be imported on to the site. It is possible for waste and borrow to occur at any point along the site and this is discussed in section 13.22.

13.21 Economics of Mass Haul Diagrams

When costing the earth-moving, there are four basic costs which are usually included in the contract for the project.

(1) *Cost of freehaul*
   Any earth moved over distances not greater than the free haul distance is costed only on the excavation of its volume, that is £A per m³.

(2) *Cost of overhaul*
   Any earth moved over distances greater than the free haul distance is charged both for its volume and for the distance in excess of the free haul distance over which it is moved. This charge can be specified either for units of haul, that is, £B per m³ m, or for units of volume, that is £C per m³.

(3) *Cost of waste*
   Any surplus or unsuitable material which must be removed from the site and deposited in a tip is usually charged on units of volume, that is, £D per m³. This charge can vary from one section of the site to another depending on the nearness of tips.

(4) *Cost of borrow*
   Any extra material which must be brought on to the site to make up a deficiency is also usually charged on units of volume, that is, £E per m³. This charge can also vary from one section of the site to another depending on the nearness of borrow pits or spoil heaps.

The following worked example illustrates how the costs of freehaul and overhaul can be calculated. Worked example 13.8 given in section 13.22 illustrates how the costs of borrowing and wasting can affect the final decision as to how the earth should be moved around the site.

Worked Example 13.7: Costing Using the Mass Haul Diagram

Question
In a project for which a section of the mass haul diagram is shown in figure 13.36, the free haul distance is specified as 100 m. Calculate the cost of earth-moving in the section between chainages 100 m and 400 m if the charge for moving material within the free haul distance is £A per m³ and that for moving any overhaul is £B per m³ m.

The x axis should be taken as the balancing line and the areas between the curve and the balancing line in figure 13.36 were measured with a planimeter and, on conversion, found to be as follows

$$\text{area of } (J + K + L + M) = 396\,000 \text{ m}^3 \text{ m}$$

$$\text{area } J = 181\,300 \text{ m}^3 \text{ m}$$

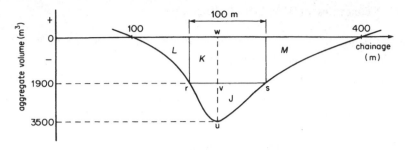

Figure 13.36

Solution
This type of problem can be solved in one of two ways.

Solution 1 – Using planimeter areas only
Between chainages 100 m and 400 m, the x axis balances cut and fill and the total volume to be moved in that section is given in figure 13.36 as uw $= 3500$ m^3.

The free haul distance of 100 m is fitted to figure 13.36 so that it touches the curve at two points r and s. This means that the volume uv is the free haul volume and is, therefore, only charged for volume.

$$uv = (3500 - 1900)m^3 = 1600 \ m^3$$

Therefore, area J can be removed since it is costed as **£(1600 A)**.

This leaves volume vw, which is equal to 1900 m^3, to be considered. This volume is the overhaul volume and has to be moved over a distance greater than the free haul distance. This distance through which it is moved has two components, the free haul distance and the overhaul distance, and this leads to two costs.

(1) The overhaul volume moved through the free haul distance is costed on its volume only. This is area K in figure 13.36. The cost = **£(1900A)**.
(2) The overhaul volume moved through the overhaul distance is the over-haul and is shown in figure 13.36 as areas L and M. The cost is that involved in moving area M to area L and is obtained as follows.

$$\text{area contained in } L \text{ and } M = (J + K + L + M) - (J + K)$$

$$= 396\ 000 - (181\ 300 + (1900{\times}100))$$

$$= 24\ 700 \ m^3 \ m$$

Hence

$$\text{cost of this overhaul} = \textbf{£(24 700B)}$$

Therefore

$$total\ cost\ =\ \text{free haul volume cost} + \text{overhaul volume costs}$$

$$=\ \pounds(1600A)\ +\ (\pounds(1900A)\ +\ \pounds(24\ 700B))$$

$$=\ \pounds(3500A)\ +\ \pounds(24\ 700B)$$

Solution 2 – Using average haul distance and overhaul distance
Average haul distance between chainages 100 m and 400 m

$$=\ \frac{\text{haul between chainages 100 m and 400 m}}{\text{total volume between chainages 100 m and 400 m}}$$

$$=\ \frac{396\ 000}{3500}\ =\ 113\ \text{m}$$

but the free haul distance = 100 m, hence

$$\text{overhaul distance} = 113\ -\ 100\ =\ 13\ \text{m}$$

therefore

$$\text{overhaul} = \text{overhaul volume} \times \text{overhaul distance}$$

$$=\ 1900\ \times\ 13\ =\ 24\ 700\ \text{m}^3\ \text{m}$$

As for solution 1, the cost of moving material over the free haul distance

$$=\ (\text{free haul volume} + \text{overhaul volume}) \times \pounds A$$

$$=\ \pounds 3500A\ (\text{areas } J \text{ and } K)$$

$$\text{cost of overhaul} = \pounds 24\ 700B\ (\text{moving area } M \text{ to area } L)$$

therefore

$$\textbf{total cost} = \textbf{\pounds 3500A} + \textbf{\pounds 24\ 700B}$$

13.22 Choice of Balancing Line

In worked example 13.7 given in section 13.21, the x axis was used as the balancing line. This is not always ideal. Figure 13.37 shows three possible balancing lines for the same mass haul diagram. In figure 13.37*a* the x axis has been used and this results in waste near chainage 230 m.

In figure 13.37*b* a balancing line is shown which gives wastage near chainage 0 m. This may be better and cheaper if local conditions provide a suitable wasting point near chainage zero.

In figure 13.37*c* two different balancing lines have been used, bc and de. This results in waste near chainage 0 m where the curve is rising from a to b, borrow near chainage 125 m where the curve is falling from c to d and waste near chainage 210 m where the curve is rising from e to f. The two waste sections may be used to satisfy the central borrow requirement if economically viable.

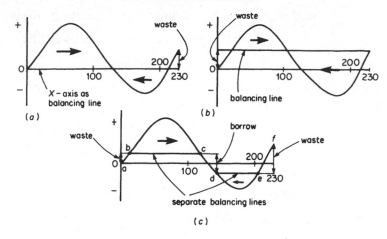

Figure 13.37 *Balancing lines*

Which choice is best depends on local conditions, for example, proximity of borrow pits, quarries and suitable tipping sites. However, the following factors should be considered before a final choice is made.

(1) The use of more than one balancing line results in waste and borrow at intermediate points along the project which will involve extra excavation and transportation of material.

(2) Short, unconnected, balancing lines are often more economical than one long continuous balancing line, especially where the balancing lines are shorter than the free haul distance since no overhaul costs will be involved.

(3) The direction of haul can be important. It is better to haul downhill to save power and, if long uphill hauls are involved, it may be better to waste at the lower points and borrow at the higher points.

(4) The main criterion should be one of economy. The free haul limit should be exceeded as little as possible in order that the amount of overhaul can be minimised.

(5) The haul is given by the area contained between the mass haul curve and the balancing line. Since the haul consists of freehaul and overhaul, if the haul area on the diagram can be minimised, the majority of it will be freehaul and hence overhaul will also be minimised. Therefore, the most economical solution from the haul aspect is to minimise the area between the curve and the balancing line. However, as shown in figure 13.37c, this can result in large amounts of waste and borrow at intermediate points along the project. The true economics can only be found by considering all the probable costs, that is, those of hauling, wasting and borrowing.

(6) Where long haul distances are involved, it may be more economical to waste material from the excavation at some point within the free haul limit at one end of the site and to borrow material from a location within the free haul limit at the other end of the site rather than cart the material a great distance from one end of the site to the other.

This possibility will become economical when the cost of excavating and hauling one cubic metre to fill from one end of the site to the other equals the cost of excavating and hauling one cubic metre to waste at one end of the site plus the cost of excavating and hauling one cubic metre to fill from a borrow pit at the other end of the site.

In practice, several different sets of balancing lines are tried and each costed separately with reference to the costs of wasting, borrowing and hauling. The most economical solution is usually adopted. The following worked example illustrates how this can be done.

Worked Example 13.8: The Use of Balancing Lines in Costing

Question
In a project for which a section of the mass haul diagram is shown in figure 13.38, the free haul distance is specified as 200 m. The earth-moving charges are as follows

$$\text{cost of free haul volume} = \pounds A \text{ per m}^3$$

$$\text{cost of overhaul volume} = \pounds B \text{ per m}^3$$

$$\text{cost of borrowing} = \pounds E \text{ per m}^3$$

Calculate the costs of each of the following alternatives

(1) borrowing at chainage 1000 m only
(2) borrowing at chainage 0 m only
(3) borrowing at chainage 300 m only.

Solution
The 200 m free haul distance is added to figure 13.38 as shown, that is, rs = tu = 200 m. The volumes corresponding to the horizontal lines rs and tu are interpolated from the curve to be +433 m³ and −2007 m³ respectively.

(1) Borrowing at chainage 1000 m only
In this case, acg is used as a balancing line and borrow is required at g (chainage 1000 m) to close the loop cefgc.

$$\text{free haul volume in section ac} = 1017 - 433 = 584 \text{ m}^3$$

$$\text{free haul volume in section cg} = 2553 - 2007 = 546 \text{ m}^3$$

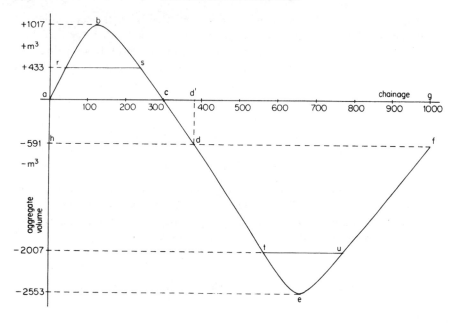

Figure 13.38

hence

$$total\ free\ haul\ volume = 584 + 546 = 1130\ m^3$$

$$overhaul\ volume\ in\ section\ ac = 433\ m^3$$

$$overhaul\ volume\ in\ section\ cg = 2007\ m^3$$

hence

$$total\ overhaul\ volume = 2440\ m^3$$

$$borrow\ at\ g = 591\ m^3$$

therefore

cost of borrowing at chainage 1000 m only

$$= £1130A + £2440B + £591E$$

(2) Borrowing at chainage 0 m only
In this case, hdf is used as a balancing line and borrow is required at h (chainage 0 m) to close the loop habdh.

The *total free haul volume* again equals *1130 m³*

$$overhaul\ volume\ in\ section\ hd = 433 + 591 = 1024\ m^3$$

$$overhaul\ volume\ in\ section\ df = 2007 - 591 = 1416\ m^3$$

hence

$$total\ overhaul\ volume\ =\ 2440\ m^3$$

$$borrow\ at\ h\ =\ 591\ m^3$$

therefore

cost of borrowing at chainage 0 m only

$$=\ £1130A\ +\ £2440B\ +\ £591E$$

This is the same as the cost of borrowing at chainage 1000 m provided that the cost of borrow is the same at chainages 0 m and 1000 m.

(3) Borrowing at chainage 300 m only
In this case, two separate balancing lines ac and df are used and borrow is required at c (chainage 300 m) to fill the gap between c and d.
As before, *total free haul volume = 1130 m³*, however

$$overhaul\ volume\ in\ section\ ac\ =\ 433\ m^3$$

$$overhaul\ volume\ in\ section\ df\ =\ 2007\ -\ 591\ =\ 1416\ m^3$$

hence

$$total\ overhaul\ volume\ =\ 1849\ m^3$$

$$borrow\ at\ c\ =\ 591\ m^3$$

therefore

cost of borrowing at chainage 300 m only

$$=\ £1130A\ +\ £1849B\ +\ £591E$$

This is the cheapest alternative assuming that the costs of borrow at chainages 0 m, 300 m and 1000 m are all equal. Considerably less overhaul is required when borrowing at chainage 300 m only.

13.23 Uses of Mass Haul Diagrams

Mass haul diagrams can be used in several ways.

In Design

In section 13.17 the close link between the mass haul diagram and the formation level was discussed. If several formation levels are tried and a mass haul diagram constructed for each, that formation which gives the most economical result and maintains any stipulated standards, for example, gradient restrictions in vertical curve design, can be used.

Nowadays, mass haul diagrams tend to be produced using computer software packages and these greatly reduce the time required to obtain several different possible mass haul diagrams for comparison purposes. This is discussed further in section 13.24.

In Financing

Once the formation level has been designed, the mass haul diagram can be used to indicate the most economical method of moving the earth around the project and a good estimate of the overall cost of the earth-moving can be calculated.

In Construction

The required volumes of material are known before construction begins, enabling suitable plant and machinery to be chosen, sites for spoil heaps and borrow pits to be located and directions of haul to be established.

In Planning Ahead

The mass haul diagram can be used to indicate the effect that other engineering works, for example tunnels and bridges, within the overall project will have on the earth-moving. Such constructions upset the pattern of the mass haul diagram by restricting the directions of haul but, since the volumes and hence the quantities of any waste and borrow will be known, suitable areas for spoil heaps and borrow pits can be located in advance of construction, enabling work to proceed smoothly.

13.24 Computer-Aided Earthwork Calculations

In simple terms, the main purpose of earthwork calculations is to determine the size of some geometrical figure as accurately as time and costs will allow. The figure can be two- or three-dimensional and its size can be measured as either an area or a volume. These can be stated as numerical quantities for costing purposes or, in the case of volumes, they can be presented graphically in the form of a mass haul diagram for planning purposes. With odd exceptions, such as the planimeter, the numerical values of area and volume are normally obtained from specific formulae and the mass haul diagram is basically a graph. Hence, the whole nature of earthwork calculations can be broken down into two related elements: calculating and plotting. While both of these can be easily performed using hand calculators and manual drawing techniques, they are also ideal subjects for analysis by computer. Not surprisingly, therefore, there are now a large number of computer software packages available which can be used for earthwork computations. Some are general graphics packages which incorporate area and volume calculation routines. Others, however, have been produced specifically for land and engineering surveying purposes and many of the packages mentioned in sections 9.11 and 11.23 include software modules for the computation of various earthwork quantities.

It is beyond the scope of this book to discuss individual suites of programs. However, the majority involve the use of a *digital terrain model* (DTM) in conjunction with a *database* of all the points surveyed and computed. DTMs and databases are discussed in section 9.12 and, once they have been established, they can be used together with some of the techniques covered in the previous sections of this chapter to compute any required earthwork quantities.

The basic method by which the various packages compute areas and volumes relies on the fact that every point surveyed and computed is referenced by a unique number and has its own set of three-dimensional coordinates. In addition, if a DTM is formed, its triangular structure provides the computer with an interlocking network which shows the interrelationship between the points. This enables the operator to interact with the computer and establish the geometrical figures which need to be measured. For example, areas can be defined by running straight lines and/or arcs between points which define the boundary. The computer can then either apply the relevant formula to the points in question or it can break down the defined figure into a series of geometrical shapes (triangles, squares, segments and so on) from which it can calculate the area. These principles can be applied to obtain a wide range of area and volume information as outlined below.

- *Surface areas* and *plan areas* can be obtained by summing the individual surface areas and plan areas, respectively, of specific triangles in the DTM. The individual areas are calculated using one of the formulae given in section 13.2. Alternatively, for plan areas, if no DTM has been created, the cross coordinate method as also described in section 13.2 can be applied to the E, N values of the appropriate points contained in the database.

- *Cross-sectional areas* are represented in the computer by a series of linked points each being referenced by its elevation above a datum and its offset distance from the centre line. These can be used to establish a coordinate system from which the individual areas of cut and fill can be computed. This is similar to the technique described in section 13.10. Alternatively, the cross-section can be broken down into a series of figures made up of straight lines and arcs, the individual areas of which can be summed.

- *Volumes of road works* can be computed from the cross-sectional areas using the end areas method or the prismoidal formula. Bulking and shrinkage factors can be included as necessary.
- *Volumes of stockpiles* can be computed by specifying their base levels and summing the individual triangular prisms (known as *isopachytes*) defined by each triangle in the DTM. This is identical to the method discussed in section 13.14 for volumes from spot heights. The *volume between two contours* and the *volume of a void* (such as a hole in the ground or a lake) can be computed in a similar manner although in these cases two levels are specified and the final result is obtained by subtracting the volumes above each level.
- *Mass haul diagrams* and *longitudinal* and *cross-sections* can easily be computed and plotted once the highway design and the volume computations have been completed. Again, the computer uses techniques similar to those described in sections 13.18 and 13.5, respectively.

The above examples by no means represent all the possibilities associated with computer-aided earthworks software systems. They are only intended to show some of their capabilities. However, the real advantage of such systems is their ability to speed up the calculations to such an extent that many different designs can be compared and an optimum solution found.

In addition, complete interaction with the software is possible and earthwork quantities can not only be calculated but also specified in order that the software can design the most appropriate geometrical figure. An example of this would be when subdividing an area into a number of lots each having a different area. Virtually any type of area or volumetric calculation can be performed by these packages and, should the reader wish to know more about the capabilities of specific modules, the articles by Mike Fort referenced in the following section are strongly recommended.

Further Reading

M.J. Fort, 'Software for Surveyors', in *Civil Engineering Surveyor*, Vol. 18, No. 3, Electronic Surveying Supplement, pp. 19–27, April 1993.

M.J. Fort, 'Surveying by Computer', in *Engineering Surveying Showcase '93*, pp. 24, 27–31 (PV Publications, 101 Bancroft, Hitchin, Hertfordshire, January 1993).

G. Petrie and T.J.M. Kennie, *Terrain Modelling in Surveying and Civil Engineering* (Whittles Publishing, in association with Thomas Telford, London 1990).

14

Setting Out

A definition often used for setting out is that it is the reverse of surveying. What is meant by this is that whereas surveying is the process of producing a plan or map of a particular area, setting out begins with the plan and ends with the various elements of a particular engineering project correctly positioned in the area. This definition can be misleading since it implies that setting out and surveying are opposites. This is not true. Most of the techniques and equipment used in surveying are also used in setting out and it is important to realise that setting out is simply one application of surveying.

A better definition of setting out is provided by the International Organisation for Standardization (ISO) in their publication ISO/DP 7078 Building Construction which states that

'Setting Out is the establishment of the marks and lines to define the position and level of the elements for the construction work so that works may proceed with reference to them. This process may be contrasted with the purpose of Surveying which is to determine by measurement the positions of existing features.'

Attitudes to setting out vary enormously from site to site, but, frequently, insufficient importance is attached to the process and it tends to be rushed, often in an effort to keep ahead of the contractor's workforce. This can lead to errors which in turn require costly corrections.

Fortunately, in recent years, greater emphasis has been placed on the need for good working practices in setting out. There are now a number of national and international standards specifically dealing with the accuracy requirements of setting out and the techniques that should be employed in order to minimise errors and ensure that the construction process proceeds smoothly. Further information on these is given in section 14.18.

However, although progress is being made in the production of standards, the main problems of the lack of education in and the poor knowledge of suitable setting-out procedures still remain. Good knowledge is vital since, despite the lack of importance often placed upon it, setting out is one of the most important stages in any civil engineering construction. Mistakes in setting out cause abortive work and delays which leave personnel, machinery and plant idle, all of which results in additional costs.

The aim of this chapter, therefore, is to discuss some of the equipment and techniques that should be used in setting-out operations. It begins with a review of the personnel involved in this type of work and then recommends good working practices that should be adopted. The types of plans used on site are discussed and the need for accurate horizontal and vertical control is emphasised. Various positioning techniques are described together with practical examples of their use. Sections are included on the methods by which verticality can be controlled and on the use of laser instruments in setting-out operations. The role of Quality Assurance in surveying and setting out is assessed and details of the relevant British Standards are given. Since the setting out of road alignments is dealt with in chapters 10, 11 and 12, emphasis in this chapter is placed on the setting-out procedures used for other engineering schemes.

14.1 Personnel Involved in Setting Out and Construction

The *Client*, *Employer* or *Promoter* is the person, company or government department who requires the particular scheme (the *Works*) to be undertaken and finances the project. Often, the Employer has no engineering knowledge and therefore commissions an *Engineer* (possibly a firm of Consulting Engineers or the City Engineer of a Local Authority) to provide the professional expertise. A formal contract is normally established between these two parties.

It is the responsibility of the Engineer to investigate the feasibility of the proposed project, to undertake site investigation and prepare various solutions for the Employer's consideration. Ultimately, the Engineer undertakes the necessary calculations and prepares the drawings, specifications and quantities for the chosen scheme. The Engineer also investigates the likely costs and programme for the project.

The *calculations* and *drawings* give the form and nature of construction of the Works. The *quantities* are used as a means of estimating the value of the project, for inviting competitive tenders for the project and, ultimately, as a basis for payment as the job is executed. The *specifications* describe the minimum acceptable standards of materials and workmanship included in the project. The *programme* identifies the overall time for completion of the project.

When these documents are complete, the project is put out to tender and contractors are invited to submit a price for which they will carry out the

Works described. A *Contractor* is chosen from the tenders submitted and a contract is formed between the Employer and the Contractor.

Hence, three parties are now involved, the Employer, the Engineer and the Contractor.

Although the Engineer is not legally a party to the contract between the Employer and the Contractor, the duties of the Engineer are described in the contract. The job of the Engineer is to act as an independent arbiter and ensure that the Works are carried out in accordance with the drawings, specifications and the other conditions as laid out in the contract. The Employer is rarely, if ever, seen on site. The Engineer is represented on site by the *Resident Engineer* (RE). The Contractor is represented on site by the *Agent*.

The responsibilities of the Resident Engineer, Engineer, Agent, Contractor and Employer are described in the document known as the *Conditions of Contract*. Every scheme has such a contract and a number of different ones are available. Two of the most commonly used are the *ICE Conditions of Contract* which is sponsored by the Institution of Civil Engineers, the Association of Consulting Engineers and the Federation of Civil Engineering Contractors, and the *JCT (Joint Contracts Tribunal) Standard Form of Building Contract* which is sponsored by the Royal Institute of British Architects, the Building Employers Federation, the Royal Institution of Chartered Surveyors, the Association of County Councils, the Association of Metropolitan Authorities, the Association of District Councils, the Confederation of Associations of Specialist Engineering Contractors, the Federation of Associations of Specialist and Sub-Contractors, the Association of Consulting Engineers, the British Property Federation and the Scottish Building Contract Committee.

The RE is in the employ of the Engineer who has overall responsibility for the contract. Many responsibilities are vested in the RE by the Engineer. The RE is helped on site by a staff which can include assistant resident engineers and clerks of works.

The Agent, being in the employ of the Contractor, is responsible for the actual construction of the Works. The Agent is a combination of engineer, manager and administrator who supervises assistant agents and site foremen who are involved in the day-to-day construction of the Works.

Many large organisations employ a Contracts Manager who mainly supervises financial dealings on several contracts and is a link between head office and site.

As regards setting out, the Resident Engineer and the Agent usually work in close cooperation and they have to meet frequently to discuss the work. The Agent undertakes the setting out and it is checked by the Resident Engineer. Good communication is essential since, although the Resident Engineer checks the work, the setting out is the responsibility of the Contractor and the cost of correcting any errors in the setting out has to be paid for by the

Contractor, provided the Resident Engineer has supplied reliable information in writing. If unreliable written information is given, the responsibility
for correcting any errors in setting out reverts to the Employer. The whole
question of responsibility for setting out will be covered by the formal contract used in the scheme. This will contain a definitive section on setting
out, for example, Clause 17 of the Sixth Edition of the ICE Conditions of
Contract takes the form of three statements. These are reproduced below by
kind permission of the Institution of Civil Engineers.

*(1) The Contractor shall be responsible for the true and proper setting out
of the Works and for the correctness of the position levels dimensions and
alignment of all parts of the Works and for the provision of all necessary
instruments appliances and labour in connection therewith.*
*(2) If at any time during the progress of the Works any error shall appear
or arise in the position levels dimensions or alignment of any part of the
Works the Contractor on being required so to do by the Engineer shall at
his own cost rectify such error to the satisfaction of the Engineer unless
such error is based on incorrect data supplied in writing by the Engineer or
the Engineer's Representative in which case the cost of rectifying the same
shall be borne by the Employer.*
*(3) The checking of any setting-out or of any line or level by the Engineer
or the Engineer's Representative shall not in any way relieve the Contractor of his responsibility for the correctness thereof and the Contractor shall
carefully protect and preserve all bench-marks sight rails pegs and other
things used in setting out the Works.*

It is essential, therefore, that setting-out records, to monitor the progress,
accuracy and any changes from the original design, are kept by both the
Engineer and the Contractor as the scheme proceeds. These can be used to
settle claims, to provide the basis for amending the working drawings and
to help in costing the various stages of the project.

Further detailed information on the topics discussed in this section can
be found in some of the publications listed in Further Reading at the end of
the chapter.

14.2 Aims of Setting Out

There are two main aims when undertaking setting-out operations.

(1) The various elements of the scheme must be correct in all three dimen-
    sions both relatively and absolutely, that is, each must be its correct
    size, in its correct plan position and at its correct reduced level.
(2) Once setting out begins it must proceed quickly and with little or no
    delay in order that the Works can proceed smoothly and the costs can

be minimised. It must always be remembered that the Contractor's main commercial purpose is to make a profit. Efficient setting-out procedures will help this to be realised.

In practice, there are many techniques which can be used to achieve these aims. However, they are all based on three general principles.

(1) Points of known E, N coordinates must be established within or near the site from which the design points can be set out in their correct plan positions. This involves *horizontal control* techniques and is discussed in section 14.6.
(2) Points of known elevation relative to an agreed datum are required within or near the site from which the design points can be set out at their correct reduced levels. This involves *vertical control* techniques and is discussed in section 14.7.
(3) Accurate methods must be adopted to establish design points from this horizontal and vertical control. This involves *positioning techniques* and is discussed in section 14.8.

In addition, the chances of achieving the aims and minimising errors will be greatly increased if the setting-out operations are planned well in advance. This requires a careful study of the drawings for the project and the formulation of a set of good working practices. These are discussed in sections 14.3 and 14.4, respectively.

14.3 Plans and Drawings Associated with Setting Out

Before any form of construction can begin, a preliminary survey is required. This may be undertaken by the Engineer or a specialist team of land surveyors and the result will be a contoured plan of the area at a suitable scale (usually 1:500 or larger) showing all the existing detail. As discussed in chapter 9, it is usually prepared from a network of control stations established around the site. These stations are often left in position to provide a series of horizontal and vertical control points which may be used to help with any subsequent setting out. This first plan is known as the *site* or *survey* plan.

The Engineer takes this site plan and uses it for the design of the project. The proposed scheme is drawn on the site plan and this becomes the *layout* or *working* drawings. All relevant dimensions are shown and a set of documents relating to the project and the drawings is included. These form part of the scheme when it is put out to tender. The Contractor who is awarded the job will be given these drawings.

The Contractor uses these layout drawings to decide on the location of the horizontal and vertical control points in the area from which the project is to be set out and on the positions of site offices, stores, access points, spoil heaps and so on. All this information together with the angles and lengths necessary to relocate the control points should they become disturbed is recorded on a copy of the original site plan and forms what is known as the *setting-out plan*.

As work proceeds, it may be necessary to make amendments to the original design to overcome unforeseen problems. These will be agreed between the Resident Engineer and the Agent. Any such alterations are recorded on a copy of the working drawings. This copy becomes the *latest amended drawing* and should be carefully filed for easy access. It is essential that the latest version of any drawing is always used, particularly if setting-out operations are to be undertaken. It is also important to keep the drawings which show the earlier amendments; they may be needed to resolve a dispute or for costing purposes. When the scheme is finally completed, the drawing which shows all the alterations that have taken place during the course of the Works becomes the *as-built drawing* or *record drawing*.

14.4 Good Working Practices when Setting Out

The basic procedures involved in setting out utilise conventional surveying instruments and techniques and, given sufficient practice, a young engineer can become highly proficient at undertaking setting-out activities. Unfortunately, this is not sufficient if the aims stated in section 14.2 are to be achieved. There is more to setting out than simply using equipment and many of the problems that occur are often due to lack of thought rather than lack of technical competence.

During the setting out and construction of a scheme a number of difficulties will inevitably arise. These can be concerned with such diverse matters as site personnel, equipment, ground and weather conditions, changes in materials, design amendments and financial constraints. Most will be unforeseen but an experienced engineer will always expect some unplanned events to occur and will take steps to minimise their effects as and when they happen. Setting out is no exception to the vagaries of construction work and anyone given the task of undertaking such operations should be equally prepared for the unexpected. Of course, it is not possible to be ready for every eventuality but, by adopting a professional approach and a series of good working practices, most problems can be overcome.

Emphasis has, so far, been placed on having to deal with difficulties. Although these do arise, the majority of the setting-out activities on site would normally be expected to proceed without any problems. This, of course, is one of the aims as stated in section 14.2. However, such trouble-free progress does not happen by accident and the chances of mistakes and errors occurring will be greatly diminished if, as before, a series of good working practices is carefully followed.

While it is impossible to discuss all the procedures that should be adopted when setting out, the working practices given below cover most of the important considerations and it is strongly recommended that they be followed.

Keep Careful Records

Always record any activities in writing and date the entries made. Get into the habit of carrying a notebook and/or a diary and record each day's work at the end of each day (not a few days later) when it is still fresh in the mind. Anything which has an influence on the Works should be noted including the names of all the personnel involved. If requested to carry out a particular piece of work, note it in the diary and ask the person who gave the instruction to sign to confirm what has been agreed. Once it has been completed to the agreed specification, a second signature should be requested.

Try to be neat when keeping records and using field books and sheets. A dispute may not arise until months after the work was done and recorded. It is essential that the records can be fully understood in such a case, not only by the person who did the recording (who may have been transferred to another project) but also by someone who has no previous knowledge of the work. A good test is to allow a colleague to review your notebook/diary to see if it is clear.

Adopt Sensible Filing Procedures

As work proceeds, the quantity of level and offset books, booking forms and other setting-out documents will grow quickly. These are often the only record of a particular activity and, as such, could be called upon to provide evidence in the case of a dispute. They are also used to monitor the progress of the Works and to help the Quantity Surveyors with their costings. Consequently, they are extremely important and should be carefully stored in such a way that they are not only kept safe but also are easily retrieved when requested. If the records of a number of different jobs are being stored in the same site office, great care must be taken to ensure that they are not mixed up. Different filing cabinets and plan chests clearly marked with the relevant job name should be used for each. Once a file or drawing has been consulted, it should be returned *immediately* to its correct place in order that it can be easily located the next time it is required.

Look After the Instruments

Surveying instruments are the engineer's tools of the trade when setting out. Modern equipment is very well manufactured and can achieve very accurate results if used properly and regularly checked. No instrument will perform well, however, if it is neglected or treated badly. If an automatic level is allowed to roll about in the back of a Land Rover, for example, even if it is in its box, it should not come as too much of a surprise if the compensator is found to be out of adjustment. Similarly, if a theodolite is carried from one station to another while still on its tripod any jarring will be transferred to the instrument which could affect its performance. Should the person carrying it trip and fall over, of course the instrument could be ruined altogether.

All instruments must be treated with respect and should be inspected and checked both before work commences and at regular intervals during the work, ideally, once per week when used daily and at least once every month if used occasionally. In the case of total stations, EDM instruments, theodolites and levels, the permanent tests should be carried out and the instruments checked to ensure that all the screws, clamps and so on are functioning correctly. The purpose of testing equipment is to find out if it is in correct adjustment. Instruments which are found to be out of adjustment should, normally, be returned to the company surveying store or to the hirer for repair. There is not usually time or adequate facilities on site actually to carry out any adjustments. Other equipment, such as chains, tapes, ranging rods and tripods should be kept clean and oiled where necessary. Any damaged equipment should be sent for repair or returned to the hirer. If the end comes off a tape, the whole tape should be thrown away; the small cost of a new one is nothing to the costs that may arise from errors caused by incorrectly allowing for its shortened length.

If equipment gets wet it should be dried as soon as possible. Optical instruments should be left out of their boxes in a warm room to prevent condensation forming on their internal lenses. Electronic equipment, although water resistant, is not waterproof. Should it become wet, it can start to behave unpredictably and give false readings. In such a case it is advisable to dry it off and also leave it overnight in a warm dry place out of its box. It should then be checked to ensure that it is working properly before it is returned to site.

Common sense should be shown when using equipment on site. A tripod can be left set up above a station but its instrument should be detached and put back into its box if it is not to be used for a while. This will prevent accidental damage should one of the tripod legs be knocked by a passing vehicle. Ranging rods are not javelins and should not be thrown. Levelling staves should only be used fully extended if absolutely necessary as their centre of gravity is much higher in such circumstances, making them difficult to handle, particularly in windy conditions. Great care should also be exercised when using levelling staves near overhead power lines.

The consequences of loss of time due to badly adjusted or damaged equipment can be extremely serious. Expensive plant and personnel will be kept idle, the programme will be delayed and material such as ready mixed concrete may be wasted. The expression *time is money* is one of the overriding considerations on site. It is essential that no time is wasted as a result of poor equipment.

Further information on the care of instruments can be found in the standards referenced in section 14.18.

Check the Drawings

Before beginning any setting-out operations, care must be taken to ensure that the correct information is at hand. Much of this will be obtained from the drawings for the scheme and it is essential that these are checked for consistency and completeness. It is not unusual for errors to be present in the dimensions quoted or for critical dimensions to be omitted. The first step, therefore, is to study the plans very carefully, abstracting all relevant information that will be needed for the setting-out operations. Should any errors or omissions be found, these must be reported immediately in writing in order that corrections can be made. It is also essential to ensure that the latest versions of the drawings are being used. A logical plan storage system must be adopted to ensure that previous versions are not used by mistake. The various different types of plans and drawings that are associated with setting out are discussed in section 14.3.

Walk the Site

Even if all the drawings are correct and the relevant setting-out information can be obtained, the topography and nature of the site may hamper construction. Initially, therefore, it is essential to walk over the whole of the site and carry out a reconnaissance. The engineer must become very familiar with the area and this cannot be done from inside the site office. Any irregularities or faults in the ground surface which may cause problems should be noted and any discrepancies between the site and the drawings should be reported in writing.

Fix the Control Points

During the reconnaissance, any existing horizontal and vertical control points should be inspected and suitable positions for any new points temporarily marked with ranging rods or wooden pegs. Once they have been finalised they can be permanently marked. Many different types of marker can be used, for example, an iron bar set in concrete at ground level is ideal. In addition, there are a number of commercially available ground markers ranging from plastic discs to elaborate ground anchors.

Ideally, all control points should be placed well away from any traffic

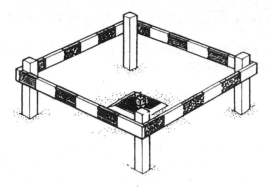

Figure 14.1 *Protecting a control point (from the CIRIA/Butterworth–Heinemann book* Setting Out Procedures, *1988)*

routes on site and all must be carefully protected as shown in figure 14.1. The protection takes the form of a small wooden barrier completely surrounding the point and painted in very bright colours, for example, red and white stripes. Such barriers are not meant to prevent points from being disturbed but to serve as a warning to let site personnel know where the points are in order that they can be avoided. There is nothing more frustrating to an engineer than to spend several days establishing control only to find that half the points have been accidentally disturbed. Careful planning coupled with a thorough knowledge of the site will help to avoid such occurrences.

Inspect the Site Regularly

As work progresses, the engineer should inspect the site daily for signs of moved or missing control points. A peg, for example, may be disturbed and replaced without the engineer being informed. Points of known reduced level should be checked at regular intervals, preferably at least once a week, and points of known plan position should be checked from similar points nearby.

Work to the Programme

The detailed programme for the Works should be posted in the form of a bar chart on the wall of the site office. Using this, the engineer should plan the various setting-out operations well in advance and execute them on time to prevent delays. It is not always advisable to work too far in advance of the programme since points established at an early stage may be disturbed before they are required. Any agreed changes to the programme should be recorded immediately on the chart.

Work to the Specifications

In the contract documents, details will be given of the various tolerances which apply to the different setting-out operations. It is essential that the engineer becomes familiar with these and works to them throughout the project. Suitable techniques and equipment must be adopted to ensure that all specified tolerances are met.

Maintain Accuracy

Once the control framework of plan and level points has been established, all design points must be set out from these and not from other design points which have already been set out. This is another example of *working from the whole to the part* (see section 14.6) and it avoids any errors in the setting out of one design point being passed to another.

Check the Work

Each setting-out operation should incorporate a checking procedure. A golden rule is that *work is not completed until it has been checked.* However, it is not advisable simply to carry out the same operation in exactly the same manner on two separate occasions. The same errors could be made a second time. Instead, any check should be designed to be completely *independent* from the initial method used, for example:

- Points fixed from one position should be checked from another and, if possible, from a third.
- If the four corners of a building have been established, the two diagonals should be measured as a check.
- All levelling runs should start and finish at points of known reduced level.
- Once a distance has been set out it should be measured twice as a check, once in each direction.
- Points set out by intersection should be checked by measuring the appropriate distances.

Communicate

Lack of communication is one of the main causes of errors on construction sites. The engineer must understand exactly what has to be done *before* going ahead and doing it.

In many cases, verbal communication will be perfectly acceptable. However, for matters which may be disputed such as an agreed change in working procedures, the discovery of a discrepancy and the acceptance of a decision, it is advisable to obtain confirmation *in writing*. Signatures should also be obtained whenever possible.

Any errors in setting out should be reported as soon as they are discovered. Prompt action may save a considerable amount of money. There is nothing to be gained from trying to hide errors. This does not remove them and they will only reappear at a later stage when dealing with them will be that much more difficult and expensive.

14.5 Stages in Setting Out

As the Works proceed, the setting out falls into two broad stages. Initially, techniques are required to define the site, to set out the foundations and to monitor their construction. Once this has been done, emphasis changes to the above-ground elements of the scheme and methods must be adopted which will ensure that they are fixed at their correct levels and positions. These two stages are discussed below but the division between the two is not easily defined and a certain amount of overlap is inevitable.

First Stage Setting Out

The first stage when setting out any scheme is to locate the boundaries of the Works in their correct position on the ground surface and to define the major elements. In order to do this, horizontal and vertical control points must be established on or near to the site. These are then used not only to define the perimeter of the site which enables fences to be erected and site clearance to begin but also to set out critical design points on the scheme and to define slopes, directions and so on. For example, in a structural project, the main corners and sides of the buildings will be located and the required depths of dig to foundation level will be defined. In a road project, the centre line and the extent of the embankments and cuttings will be established together with their required slopes.

When the boundaries and major elements have been pegged out, the top soil is stripped and excavation work begins. During this period, it may be necessary to relocate any pegs that are accidentally disturbed by the site plant and equipment. Once the formation level is reached, the foundations are laid in accordance with the drawings and the critical design points located earlier. Setting-out techniques are used to check that the foundations are in their correct three-dimensional position. The first stage ends once construction to ground floor level, sub-base level or similar has been completed. The relevant sections of this chapter are 14.6 to 14.10, inclusive.

Second Stage Setting Out

This continues on from the first stage, beginning at the ground floor slab, road sub-base level or similar. Up to this point, all the control will still be outside the main construction, for example, the pegs defining building corners, centre lines and so on will have been knocked out during the earth-moving work and only the original control will be undisturbed. Some offset pegs (see section 14.6) may remain but these too will be set back from the actual construction itself.

The purpose of second stage setting out, therefore, is to transfer the horizontal and vertical control used in the first stage into the actual construction in order that it can be used to establish the various elements of the scheme. The relevant sections of this chapter are 14.11 to 14.15.

14.6 Methods of Horizontal Control

In order that the design points of the scheme can be correctly fixed in plan position, it is necessary to establish points on site for which the E, N coordinates are known. These are *horizontal control points* and, once they have been located, they can be used with one of the methods discussed in section 14.8 to set out the design points.

In general terms, the process of establishing horizontal control is one of *working from the whole to the part*. This involves starting with a small number of very accurately measured control points (first level) which enclose the area in question and using these to set out a second level of control points near the site. This second level can then be used either to set out the design points of the scheme and/or to establish a third level of control points, as necessary. The process is one of extending control throughout the site until all the design points have been fixed. Inevitably, the accuracy of the control will decrease slightly as each new level is established and great care must be taken to ensure that the tolerances stated in the contract specifications are met. The working practices discussed in section 14.4 will help to maintain the required accuracy.

An example of working from the whole to the part which could be adopted in Great Britain is given in figure 14.2. In this, the first level of control is provided by four National Grid stations. These are incorporated in a traverse which is run through the site in question to provide the second level which takes the form of main site control points. These, in turn, are used to establish a third level of control, in this case secondary site points at each end of a series of baselines which define important elements of the scheme.

If National Grid stations are used in this way, any distances calculated from their coordinates which are used to establish further control points will

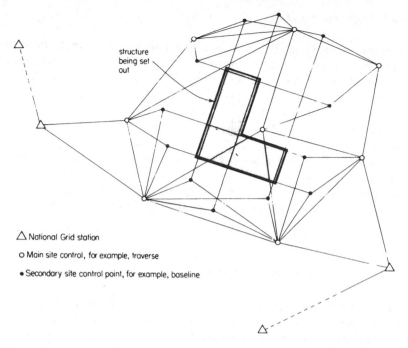

structure
being set
out

△ National Grid station

○ Main site control, for example, traverse

● Secondary site control point, for example, baseline

Figure 14.2 *Site control*

have to be corrected using the appropriate scale factor. This is discussed in section 5.23.

On some schemes, the main site control itself provides the first level of control and the same ground points are chosen as those which were used in the production of the site plan prior to design work. If this is the case and they are to be used in the setting out operations, they must be re-surveyed before setting out commences. They may have altered their position owing to settlement, heave or vandalism in the time period between the original survey and the start of the setting out operations.

Horizontal control points should be located as near as possible to the site in open positions for ease of working, but well away (up to 100 m if necessary since this is easily accommodated by modern EDM equipment) from the construction areas and traffic routes on site to avoid them being disturbed. Since design points are to be established from them, they must be clearly visible and as many proposed design points as possible should be capable of being set out from each control point.

The construction and protection of control points is very important. Wooden pegs are often used for nonpermanent stations but they are not recommended owing to their vulnerability. Should they be the only means available, figure 7.3 shows suitable dimensions.

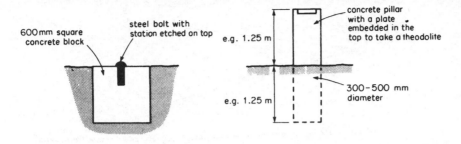

Figure 14.3 *Permanent control stations*

For longer life the wooden peg can be surrounded in concrete but, preferably, permanent stations similar to those illustrated in figure 14.3 should
be built.

All points must be clearly marked with their reference numbers or letters
and painted so that they can be easily found. They should also be surrounded
by a brightly painted protective barrier to make them clearly visible to site
traffic. Figure 14.1 shows a suitable arrangement.

Once established and coordinated, the main site control points are used
to set out design points of the proposed structure. They are, generally, used
in one of the following ways.

Baselines

Main site control points, such as traverse stations, can be used to establish
baselines from which setting out can be undertaken. Examples are shown in
figures 14.2 and 14.4.

Subsidiary lines can be set off from the baseline to establish design corner points.

The baseline may be specified by the designer and included in the contract between the Promoter and the Contractor.

Baselines can take many forms: they can run between existing buildings;
mark the boundary of an existing development; be the direction of a proposed pipeline or the centre line of a new road.

The accuracy is increased if two baselines at right angles to each other
are used on site. Design points can be established by offsetting from both
lines or a grid system can be set up to provide additional control points in
the area enclosed by the baselines.

The use of baselines to form grids leads to the use of reference grids on
site.

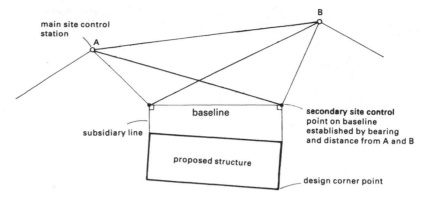

Figure 14.4 *Baseline*

Reference Grids

A control grid enables points to be set up over a large area. Several different grids can be used in setting out.

(1) The *survey grid* is drawn on the survey plan from the original traverse or triangulation scheme. The grid points have known eastings and northings related either to some arbitrary origin or to the National Grid. Control points on this grid are represented by the original control stations.

(2) The *site grid* is used by the designer. It is usually related in some way to the survey grid and should, if possible, actually be the survey grid, the advantage of this being that if the original control stations have been permanently marked then the designed points will be on the same coordinate system and setting out is greatly simplified. If no original control stations remain, the designer usually specifies the positions of several points in the site grid which are then set out on site prior to any construction. These form the site grid on the ground.

Since all design positions will be in terms of the site grid coordinates, the setting out is easily achieved as shown in figure 14.5.

The grid itself may be marked with wooden pegs set in concrete, the interval between points being small enough to enable every design point to be set out from at least two and preferably three grid points but large enough to ensure that movement on site is not restricted.

(3) The *structural grid* is established around a particular building or structure which contains much detail, such as columns, which cannot be set out with sufficient accuracy from the site grid. An example of its use is in the location of column centres (section 14.13).

The structural grid is usually established from the site grid points and uses the same coordinate system.

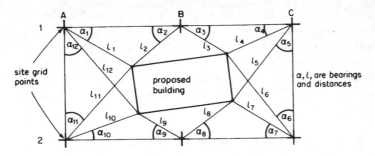

Figure 14.5 *Site grid*

(4) The *secondary grid* is established inside the structure from the structural grid when it is no longer possible to use the structural grid to establish internal features of the building owing to vision becoming obscured.

Note: Errors can be introduced in the setting out each time one grid system is established from another hence, wherever possible, only one grid system should be used to set out the design points.

Offset Pegs

Whether used in the form of a baseline or a grid, the horizontal reference marks are used to establish points on the proposed structure. For example, in figure 14.5, the corners of a building have been established by polar coordinates from a site grid.

However, as soon as excavations for the foundations begin, the corner pegs will be lost. To avoid having to re-establish these from reference points, extra pegs are located on the lines of the sides of the building but offset back from the true corner positions. Figure 14.6 shows these *offset pegs* in use.

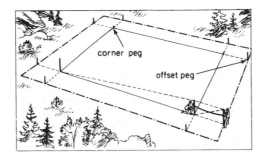

Figure 14.6 *Offset pegs*

The offset distance should be great enough to avoid the offset pegs being disturbed during excavation.

These pegs enable the corners to be re-established at a later date and are often used with *profile boards* in the construction of buildings; this is further discussed in section 14.7. Offset pegs can be used in all forms of engineering construction to aid in the relocation of points after excavation.

14.7 Methods of Vertical Control

In order that design points on the Works can be positioned at their correct levels, *vertical control points* of known elevation relative to some specified datum must be established on the site. In Great Britain, a datum commonly used is *ordnance datum* (see section 2.2) and all the levels on a site will normally be reduced to a nearby *ordnance bench mark* (OBM). The actual OBM used will be agreed in writing between the Engineer and the Contractor. The bench mark chosen is known as the *master bench mark* (MBM) and it is used for two main purposes.

First, to establish points of known reduced level near to and on the elements of the proposed scheme. These are known as *transferred bench marks* (TBMs). Although TBMs are often located in new positions on the scheme, any existing horizontal control stations can be used as TBMs providing that they have been permanently marked.

Second, if there are other OBMs nearby, their reduced levels are checked with reference to the MBM and in the case of any discrepancy, their amended values are used. This ensures that the overall vertical control remains with the MBM.

Once they have been established, the vertical control points are used to define reference planes in space, parallel to and usually offset from selected planes of the proposed construction. These planes may be horizontal, for example, a floor level inside a building, or inclined, for example, an embankment slope in earthwork construction.

As with horizontal control, it is essential that the principle of *working from the whole to the part* is adopted. In practice this means ensuring that all vertical design points are set out either from the MBM or from a nearby TBM, and not from another vertical design point which has been established earlier. This prevents an error in the reduced level of one design point being carried forward into that of another.

Transferred or Temporary Bench Marks (TBMs)

The positions of TBMs should be fixed during the initial site reconnaissance so that their construction can be completed in good time and they can

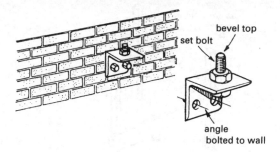

Figure 14.7 *TBM on the side of a wall (from the CIRIA/Butterworth–Heinemann book* Setting Out Procedures, *1988)*

be allowed to settle before levelling them in. For this reason, permanent, existing features should be used wherever possible. In practice, 20 mm diameter steel bolts 100 mm long driven into existing door steps, ledges, footpaths, low walls and so on are ideal.

Any TBM constructed on the side of a wall should be such that the base of a levelling staff will always be at the same reduced level every time it is placed on the mark. For this reason an etched or scribed horizontal line is not recommended since it can be difficult always to return the base of the staff to exactly the same position. Instead, a bolt fitted to a piece of angle iron should be attached to the wall as shown in figure 14.7. This provides an excellent permanent point on which to rest the staff. Where TBMs are constructed at ground level on site, a design similar to that shown in figure 14.8 is recommended.

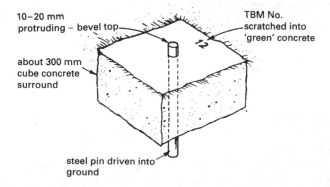

Figure 14.8 *TBM on solid ground (from the CIRIA/Butterworth–Heinemann book* Setting Out Procedures, *1988)*

Each TBM is referenced by a number or letter on the site plan and the setting-out plan and should be protected since re-establishment can be time consuming. A suitable method of protection is shown in figure 14.9.

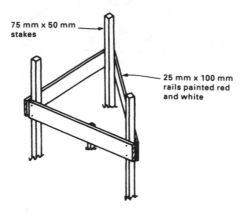

Figure 14.9 *TBM protection*

Any TBMs set up on site must be levelled with reference to the agreed MBM or some other agreed datum. It is vital that the agreed datum is used since the design levels are usually based on this.

There should never be more than 80 m between TBMs on site and the accuracy of levelling should be within the following limits

> site TBM relative to the MBM ± 0.010 m
> spot levels on soft surfaces relative to a TBM ± 0.010 m
> spot levels on hard surfaces relative to a TBM ± 0.005 m

Because TBMs are vulnerable, they must be checked by relevelling at regular intervals and, as soon as the project has reached a suitable stage, TBMs should be established on permanent points on the new construction. To avoid confusion, all the TBMs should be clearly marked on a copy of the site plan, together with their reduced levels, and this should be displayed in the site office.

Sight Rails

These consist of a horizontal timber cross piece nailed to a single upright or a pair of uprights driven into the ground. Figure 14.10 shows several different types of sight rail.

The upper edge of the cross piece is set to a convenient height above the required plane of the structure, usually to the nearest half metre, and should be at a height above ground to ensure convenient alignment by eye with the upper edge. The level of the top edge of the cross piece is usually written on the sight rail together with the length of traveller required. *Travellers* are discussed in the following section. *Double sight rails* are discussed in section 14.9.

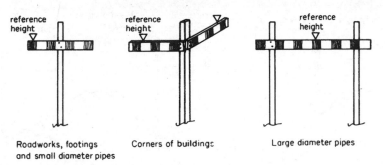

reference
height

Roadworks, footings
and small diameter pipes

reference
height

Corners of buildings

reference
height

Large diameter pipes

Figure 14.10 *Sight rails*

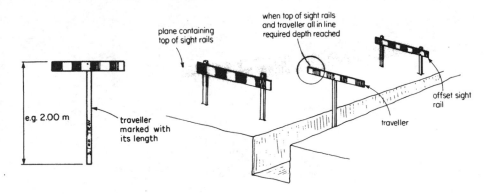

plane containing
top of sight rails

when top of sight rails
and traveller all in line
required depth reached

e.g. 2.00 m

traveller
marked with
its length

offset sight
rail

traveller

Figure 14.11 *Sight rails and traveller used for excavation of trench*

Sight rails are usually offset 2 or 3 metres at right angles to construction lines to avoid them being damaged as excavation proceeds. This is shown in figure 14.11.

Travellers and Boning Rods

A *traveller* is similar in appearance to a sight rail on a single support and is portable. The length from upper edge to base should be a convenient dimension to the nearest half metre.

Travellers are used in conjunction with sight rails. The sight rails are set some convenient value above the required plane and the travellers are constructed so that their length is equal to this value. As excavation proceeds, the traveller is sighted in between the sight rails and used to monitor the cutting or filling. Excavation or compaction stops when the tops of the sight rails and the traveller are all in line.

Figure 14.11 shows a traveller and sight rails in use in the excavation of

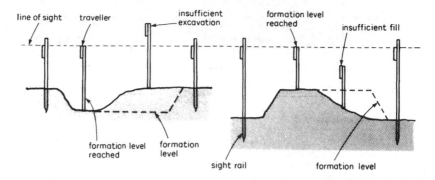

Figure 14.12 *Sight rails and traveller used for forming cutting and embankment*

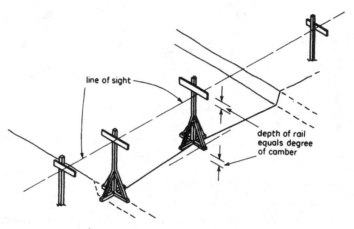

Figure 14.13 *Free standing traveller*

a trench and figure 14.12 shows the ways in which travellers and sight rails can be used to monitor cutting and filling in earthwork construction.

Boning rods are discussed in section 14.9.

There are several different types of traveller. *Free-standing travellers* are frequently used in the control of superelevation on roads, a suitable foot being added to the normal traveller as shown in figure 14.13. *Pipelaying travellers* are discussed in section 14.9.

Slope Rails or Batter Boards

For controlling side slopes in embankments and cuttings, sloping rails are used. These are known as *slope rails* or *batter boards*.

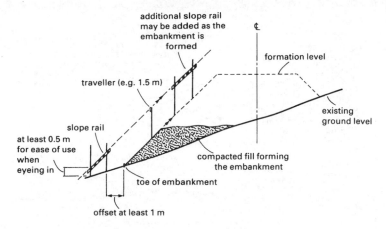

Figure 14.14 *Slope rail for an embankment*

For an *embankment*, the slope rails usually define planes parallel to but offset some vertical distance from the proposed embankment slopes as shown in figure 14.14. In addition, they are usually offset a horizontal distance of at least 1 m from the toe of the embankment to prevent them from being covered during the filling operations. Travellers are always used in conjunction with the slope rails to monitor the formation of the embankments.

For a *cutting*, the slope rails can be set to define either the actual slope of the cutting as shown in figure 14.15 or a parallel offset slope for use in conjunction with a short traveller as shown in figure 14.16. Both methods are satisfactory but each has its limitations. If the exact slope is defined, it is not possible to erect an additional slope rail as the excavation proceeds. If a parallel offset slope is defined, the height at which the slope rail must be fixed will increase by an amount equal to the length of the traveller used

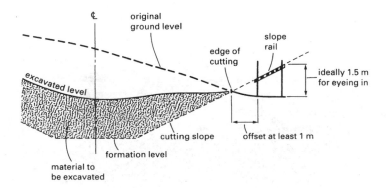

Figure 14.15 *Slope rail defining a cutting slope without a traveller*

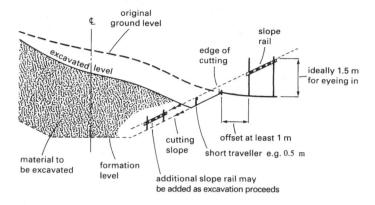

Figure 14.16   *Slope rail defining a cutting slope with a traveller*

and this may make the operation of viewing along the slope rail very difficult to accomplish. In both methods, the wooden stakes supporting the slope
rail are usually offset a horizontal distance of at least 1 m from the edge of
the proposed cutting to prevent them being disturbed during excavation.

    All relevant information is usually marked on the slope rails, for example,
chainage of centre line, distance from wooden stakes to centre line, length
of traveller, side slopes and so on.

    During the setting out, the positions of the toes of the embankments and
the edges of the cuttings must be fixed in order that the wooden stakes onto
which the slope rails are to be attached can be located in their correct positions. One method of doing this is as follows.

    Consider figure 14.17 in which the toe, T, of an embankment is to be
fixed. The top of the embankment is to run from point A to point B and is
to have a width of 12 m. Point C is on the existing ground directly below

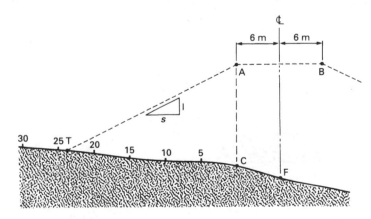

Figure 14.17   *Locating the toe of an embankment*

point A and the centre line is defined by point F which is also on the existing ground. The sides of the embankment are to slope at 1 in s. The procedure is as follows.

(1) From the road design, obtain the reduced level of point A.
(2) Peg out point C by measuring a distance of 6 m horizontally from F at right angles to the centre line.
(3) Peg out points at 5 m horizontal distance intervals from point C along the line FC produced. Locate sufficient points to ensure that the toe T will fall between two of them.
(4) Measure the reduced level on the ground surface at the first 5 m peg.
(5) Calculate the proposed reduced level on the embankment slope directly above this point from

$$\text{RL at 5 m point} = RL_A - (5/s)$$

(6) Compare the values of the reduced levels obtained in (4) and (5).
(7) If the ground level is lower than the proposed level then the toe of the embankment is further than 5 m from point C. Move to the 10 m peg and measure the RL of the ground surface. Calculate the proposed reduced level on the embankment slope directly above this point from

$$\text{RL of 10 m point} = RL_A - (10/s)$$

Compare the two RLs. If the ground level is still lower than the proposed level repeat step (7) for the 15 m peg. Continue moving from one peg to the next until the ground level is higher than the proposed level.
(8) Once the ground level at a 5 m peg is measured to be higher than the proposed level, the toe of the embankment has been passed and its position is somewhere between this 5 m peg and the previous one.
(9) To locate the exact position of the toe, return to the previous peg and repeat step (7) but advancing forward in 1 m intervals.
(10) Once the ground level is equal to the proposed level within 50 mm, point T has been located and a peg should be hammered into the ground at this point. As a precaution, the distance along the ground surface from point T to the centre line peg F should be measured and recorded in case the toe peg is disturbed.

On first reading, the above procedure appears to be rather slow and laborious. In practice, however, this is not the case and an experienced engineer can very quickly locate embankment toes by this method. Often, it is not necessary to work from point C in 5 m intervals. If the cross-sectional drawings are available, good estimates of the positions of toes can be obtained and these will indicate the best location for the pegs set out in step (3), for example, the first at 25 m and then every 2 metres.

Although the above procedure has concentrated on locating an embank-

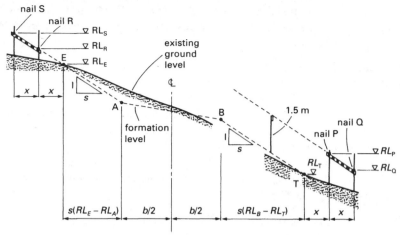

Figure 14.18 *Positioning slope rails*

ment toe, a similar technique can be used to locate the edge of a cutting.

Once the toes and edges have been located, the wooden stakes which are to carry the slope rails can be hammered into the ground at offset horizontal distances from these as shown in figure 14.18. The next stage is to calculate the required reduced levels at which the top edges of the slope rails must be fixed on the wooden stakes. In practice, nails are hammered into the stakes at the required levels and the rails are attached with their top edges butted up against them.

For an *embankment*, assuming that a 1.5 m traveller is to be used as shown on the right-hand side of figure 14.18, the reduced levels at which two nails P and Q should be placed on the wooden stakes is obtained as follows. It is assumed that the RL at the toe of the embankment (RL_T) is known.

$$RL_Q = RL_T - 2x/s + 1.5$$

$$RL_P = RL_T - x/s + 1.5 = RL_Q + x/s$$

Once it has been calculated, RL_Q should be compared with the reduced level of the ground directly below point Q to ensure that the difference is at least 0.5 m to enable the slope rail to be sighted along without too much difficulty. If it is less than 0.5 m, a longer traveller should be used.

For a *cutting*, as shown on the left-hand side of figure 14.18, the reduced levels at which two nails R and S should be placed on the wooden stakes is obtained as follows. It is assumed that the RL at the edge of the embankment (RL_E) is known and that a traveller is not being used.

$$RL_S = RL_E + 2x/s$$

$$RL_R = RL_E + x/s = RL_S - x/s$$

Finally, the tops of the stakes are levelled and the values obtained are compared with the reduced levels calculated above for the nails. This gives the required distances to be measured down from the tops of the stakes and nails P, Q, R and S are hammered into them at these levels. The slope rails are then attached with their top edges butted up against these nails. If it is found that some of the wooden stakes are not long enough, it will be necessary to add extension pieces to them and then attach the slope rails to these.

Profile Boards

These are very similar to sight rails but are used to define corners or sides of buildings.

In section 14.6 it was shown that offset pegs are used to enable building corners to be relocated after foundation excavation.

Normally a *profile board* is erected near each offset peg and used in exactly the same way as a sight rail, a traveller being used between profile boards to monitor excavation.

Figure 14.19 shows profile boards and offset pegs at the four corners of a proposed building.

The arrangement shown in figure 14.19 is quite an elaborate one and a simpler, more often used type of corner arrangement is shown in figure 14.20. Nails or sawcuts are placed in the tops of the profile boards to define the width of the foundations and the line of the outside face of the wall. String or piano wire is stretched between opposite profile boards to guide the width of cut while a traveller is used to control the depth of cut.

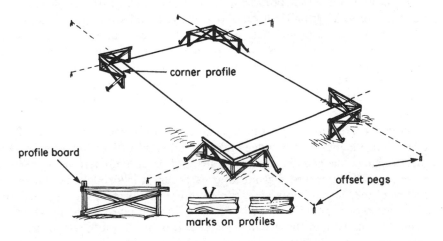

Figure 14.19 *Profile boards*

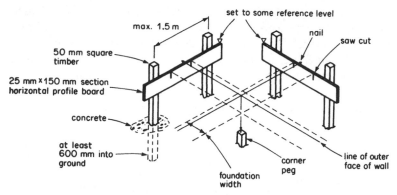

Figure 14.20 *Profile boards*

Figure 14.21 *Continuous profile*

A variation on corner profiles is to use a *continuous profile* all round the building set to a particular level above the required structural plane. Figure 14.21 shows such a profile with a gap left for access into the building area.

The advantage of a continuous profile is that the lines of the internal walls can be marked on the profile and strung across to guide construction.

Another type of profile is a *transverse profile* and this is used together with a traveller to monitor the excavation of deep trenches as shown in figure 14.22.

Profile boards and their supports are normally made from timber. However, sledge-hammering wooden stakes into the ground and nailing on cross pieces can be dangerous, especially in hard or difficult terrain. Great care must be taken to ensure that no injuries occur. It is strongly recommended that steel-toe-capped boots are worn by those involved in this type of work and that no one should be asked to hold a stake in place by hand while it is being hammered into the ground. Instead, it is not too difficult to manufacture a simple grip holder out of a piece of reinforcing rod and to use this to support the stake while it is being hit.

Recently, an alternative to such traditional wooden materials has been

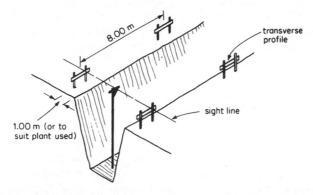

Figure 14.22   *Transverse profile*

Figure 14.23   *Pro-Set profiles*

developed by Pro-Set Profiles Ltd. who are based in Manchester, England. They have produced a system of portable and reusable profiles each consisting of two 1500 mm long anti-rust treated 60 mm diameter tubes made from 2.5 mm thick steel plus a 1500 mm long cross bar having a 25 mm square section. Replaceable high impact strength nylon points and solid steel caps are provided for the two uprights.

Each set of two uprights and a cross bar is an integral unit. The three pieces always remain fastened together but can be folded flat for easy storage and carriage. In use, the uprights are hammered into the ground, the cross bar is set at the required level and held in place by large adjustable clamps. Nylon sliders are supplied which attach to the cross bar and can be moved and clamped as necessary to define the required reference directions. Stringlines can then be run between these as shown in figure 14.23. The sliding markers also have sight lines for use with theodolites.

The manufacturers claim that, in addition to having the advantages over conventional methods of being easy to use, easy to store, extremely portable and fully reusable, this new system also has considerable cost savings when compared to traditional timber profiles.

14.8 Positioning Techniques

As discussed in section 14.2, one of the main aims of setting out is to ensure that the design points of the scheme are located in their correct plan positions. Depending on the equipment available, there are a number of different methods which can be adopted to ensure that this aim is achieved. Those commonly used in engineering surveying are described in this section.

From Existing Detail

On small sites or for single buildings, the location of the new structure may have to be fixed by running a line between corners of existing buildings and offsetting from this. The offset dimensions have to be scaled from the plan but this can be inaccurate and it is not recommended. However, if there is no alternative, such a method can be carried out successfully if great care is taken, particularly with the scaling of dimensions.

From Coordinates

These are undoubtedly the best methods. Design points will be coordinated in terms of the site grid or referenced to a baseline and they can be established by one of the following techniques.

(1) By calculation of the *bearing* and *distance* from at least three horizontal control points to each design point (this is known as setting out by polar coordinates) as shown in figure 14.24a.

The angle α in figure 14.24a can be set out by one of two methods. In one method, α is the angle to be set out after being calculated from α = WCB(ST) − WCB(SA). The length l and the WCBs of ST and SA are calculated from the coordinates of S, T and A as described in section 1.5. In the alternative method, the horizontal circle of the theodolite is set to read the WCB of ST and the telescope aligned on point T with the instrument at station S. The telescope is then rotated towards point A until the WCB of SA is read on the horizontal circle. In both methods, l is the horizontal length to be set out from S to A and its value is calculated using methods described in section 1.5.

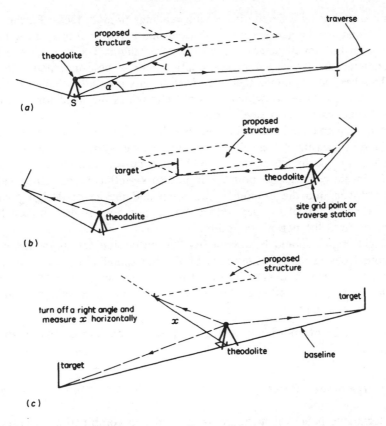

Figure 14.24 *Positioning techniques using coordinates*

(2) By *intersection*, with two theodolites, from two of the control points using bearings only and checking from a third. Intersection is shown in figure 14.24*b*.

(3) By *offsetting* from one or more baselines as shown in figure 14.24*c*, the offsets being calculated from the coordinates of the ends of the baselines and the design point coordinates. If only one baseline is used, extra care should be taken since there is very little check on the set-out points.

Whichever method is used, the following points must be taken into consideration.

• All angles must be set off using a correctly adjusted theodolite otherwise both faces should be used and the mean position taken.

• Since the design dimensions will be in the horizontal plane, any distance set out with a steel tape should be stepped to a plumb line or computation of the slope distance will be necessary. The slope can be measured using a theodolite, taking readings on both faces. Further corrections may be necessary if high precision is required (see section 4.4). If possible, for distances greater than the length of a steel tape, a total station should be used in conjunction with a pole-mounted reflector. This has the advantage that such instruments normally display horizontal distances directly which not only eliminates the need for additional calculations but also saves time.

• It is recommended that, wherever possible, each design point be set out from at least three control points. This increases the accuracy since the effect of one of the control points being out of position is reduced.

• To locate each design point, a large cross-section wooden peg should be driven into the ground at the point and the exact design position marked on top of the peg with a fine tipped pen. A nail is then hammered into the peg at this point.

• Right angles should be set out by theodolite and the angle turned on both faces using opposite sides of the horizontal circle to remove eccentricity and graduation errors, for example, on face right use 0° to 90° and on face left use 180° to 270°. The mean of two pointings is the correct angle.

Applications of coordinate methods of setting out are discussed in section 14.17.

From Free Station Points

This technique is shown in figure 14.25 and is a combination of resection (see section 7.30) and setting out from coordinates. It is particularly applicable to large sites where the coordinates of prominent features and targets on nearby buildings or parts of the construction are known. The procedure is as follows

(1) A total station instrument is set up at some suitable place in the vicinity of the points which are to be set out. This gives rise to the term *free station* since the choice of instrument position is arbitrary.

(2) A distance or angular resection is carried out to fix the position of the free station. Preferably, observations should be taken to more site control points than the minimum for checking purposes.

(3) The coordinates of the free station are calculated (see sections 7.31 and 7.32).

(4) Using the method of polar coordinates described earlier in this section, the required design points are set out using the total station instrument set at the free station point.

If free station points are to be used widely on a particular site, it is essential that there is a sufficient number of well-established control points

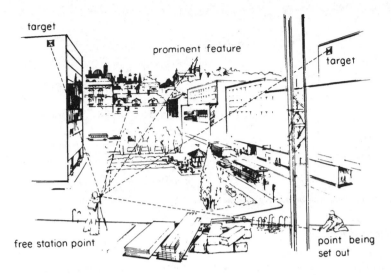

Figure 14.25 *Free station point*

around the site to enable enough obstruction free sightings to be achieved while construction proceeds.

14.9 Setting Out a Pipeline

The foregoing principles are now considered in relation to the setting out of a gravity sewer pipeline. The whole operation falls within the category of first stage setting out.

General Considerations

Sewers normally follow the natural fall in the land and are laid at gradients which will induce a self-cleansing velocity. Such gradients vary according to the material and diameter of the pipe. Figure 14.26 shows a sight rail offset at right angles to a pipeline laid in granular bedding in a trench.

Depth of cover is, normally, kept to a minimum but the sewer pipe must have a concrete surround at least 150 mm in thickness where cover is less than 1 m or greater than 7 m. This is to avoid cracking of pipes owing to surface or earth pressures.

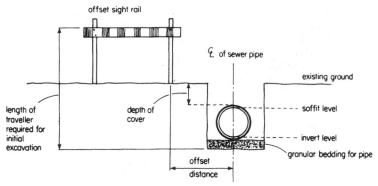

Figure 14.26 *Sight rail for a sewer pipeline*

Horizontal Control

The working drawings will show the directions of the sewer pipes and the positions of manholes.

The line of the sewer is normally pegged at 20 to 30 m intervals using coordinate methods of positioning from reference points or in relation to existing detail. Alternatively, the direction of the line can be set out by theodolite and pegs sighted in.

Manholes are set out at least every 100 m and also at pipe branches and changes of gradient.

Vertical Control

This involves the erection of sight rails some convenient height above the invert level of the pipe (see figure 14.26).

The method of excavation should be known in advance such that the sight rails will not be covered by the excavated material (the *spoil*).

A suitable scheme for both horizontal and vertical control is shown in figure 14.27.

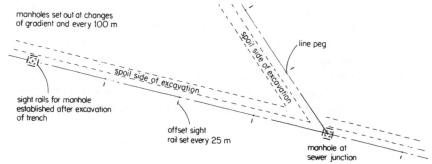

Figure 14.27 *Layout of horizontal and vertical control for a sewer pipe system*

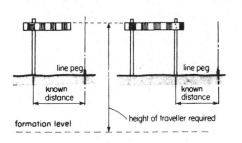

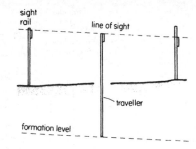

Figure 14.28 *Sight rail positions*   Figure 14.29 *Lining in traveller*

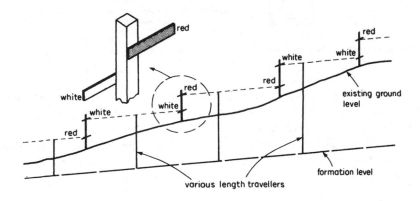

Figure 14.30 *Double sight rails*

Erection and Use of Sight Rails

The sight rail upright or uprights are hammered firmly into the ground, usually offset from the line rather than straddling it. Using a nearby TBM and levelling equipment, the reduced levels of the tops of the uprights are determined. Knowing the proposed depth of excavation, a suitable traveller is chosen and the difference between the level of the top of each upright and the level at which the top edge of the cross piece is to be set is calculated (see the first worked example in section 14.19). Figure 14.28 shows examples of sight rails fixed in position. The excavation is monitored by lining in the traveller as shown in figure 14.29.

Where the natural slope of the ground is not approximately parallel to the proposed pipe gradient, *double sight rails* can be used as shown in figure 14.30.

Often, it is required to lay storm water and foul water sewers in adjacent trenches. Since the storm water pipe is usually at a higher level than the foul water pipe (to avoid the foul water overflowing into the storm water),

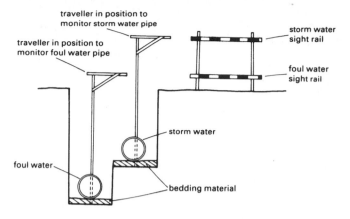

Figure 14.31   *Setting out storm water and foul water pipes in same trench*

it is common to dig one trench to two different invert levels as shown in figure 14.31. Both pipe runs are then controlled using different sight rails nailed to the same uprights. To avoid confusion, the storm water sight rails are painted in different colours from the foul water ones. The same traveller is used for each pipe run. It is made with only one cross piece and is used in conjunction with the storm water or foul water sight rails as appropriate.

Manholes

Control for manholes is usually established after the trench has been excavated and can be done by using sight rails as shown in plan view in figure 14.32 or by using an offset peg as shown in section in figure 14.33.

Pipelaying

On completion of the excavation, the sight rail control is transferred to pegs in the bottom of the trench as shown in figure 14.34. The top of each peg is set at the invert level of the pipe.

Pipes are usually laid in some form of bedding and a *pipelaying traveller* is useful for this purpose. Figure 14.35 shows such a traveller and its method of use.

Pipes are laid from the lower end with sockets facing uphill. They can be bedded in using a straight edge inside each pipe until the projecting edge just touches the next forward peg or the pipelaying traveller can be used. Alternatively, three travellers can be used together as shown in figure 14.36. When used like this the travellers are known as *boning rods*.

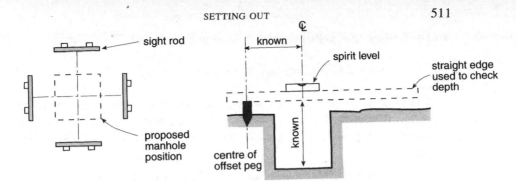

Figure 14.32 *Control for manholes: sight rails*   Figure 14.33 *Control for manholes: offset pegs*

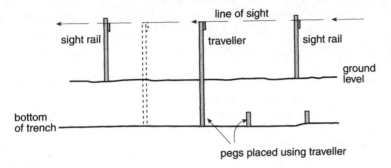

Figure 14.34 *Setting invert pegs in trench with traveller*

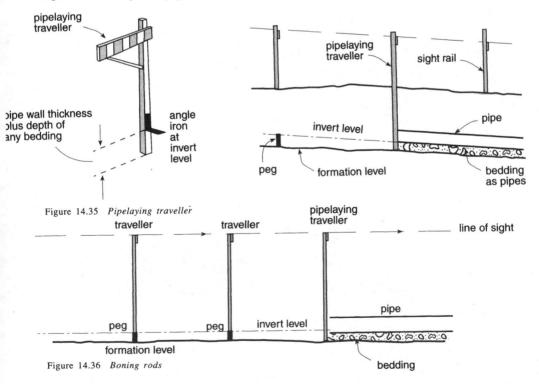

Figure 14.35 *Pipelaying traveller*

Figure 14.36 *Boning rods*

14.10 Setting Out a Building to Ground-floor Level

This also comes into the category of first stage setting out. It is summarised below.

It is vital to remember when setting out that, since dimensions, whether scaled or designed, are almost always horizontal, slope must be allowed for in surface taping on sloping ground. The slope correction is additive when setting out.

(1) Two corners of the building are set out from the baseline, site grid or traverse stations using one of the methods shown in figure 14.24.

(2) From these two corners, the sides are set out using a theodolite to turn off right angles as shown in figure 14.37. The exact positions of each corner are then marked in the top of wooden pegs by nails and offset pegs are established at the same time as the corner pegs (see figure 14.6).

(3) The diagonals are checked as shown in figure 14.38 and the nails repositioned on the tops of the pegs as necessary.

(4) Profile boards are erected at each corner or a continuous profile is used (see figures 14.19, 14.20 and 14.21) and excavation begins. The next step is to construct the foundations; these can take several forms but for the purposes of the remainder of the chapter it will be assumed that concrete foundations have been used and a concrete ground floor slab

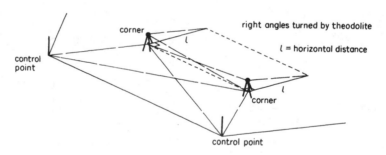

Figure 14.37 *Setting out building sides by right angles*

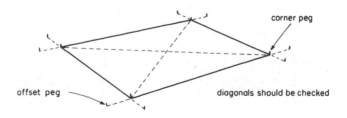

Figure 14.38 *Checking diagonals*

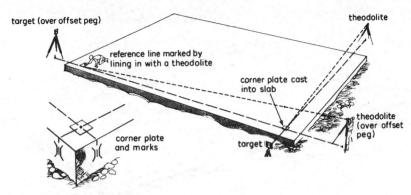

Figure 14.39 *Transfer of horizontal control to ground floor slab*

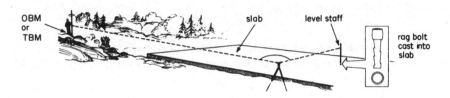

Figure 14.40 *Transfer of vertical control to ground floor slab*

laid. This would have required formwork to contain the wet concrete and this could have been set out by aligning the shuttering with string lines strung between the profiles.

14.11 Transfer of Control to Ground-floor Slab

This is done for horizontal control by setting a theodolite and target over opposite pairs of offset pegs as shown in figure 14.39 and for vertical control as shown in figure 14.40.

14.12 Setting Out Formwork

The points required for formwork can be set out with reference to the control plates by marking the lines between these plates as shown in figure 14.41.

One method of marking these lines on the slab is by means of chalked string held taut and fixed at each corner position. The string is pulled vertically away from the slab and released. It hits the surface of the slab, marking it with the chalk.

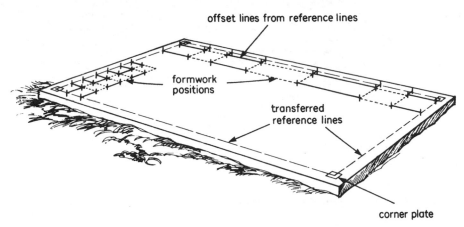

Figure 14.41 *Setting out formwork lines*

These slab markings are used as guidelines for positioning the formwork and should be extended to check the positioning as shown in figure 14.42.

14.13 Setting Out Column Positions

Where columns are used, they can be set out with the aid of a structural grid as discussed in section 14.6. Column centres should be positioned to within ± 2 to 5 mm of their design position. The structural grid enables this to be achieved.

Figure 14.43 shows a structural grid of wooden pegs set out to coincide with the lines of columns. The pegs can either be level with the ground floor slab or profile boards can be used.

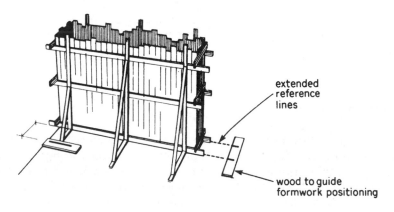

Figure 14.42 *Setting out formwork*

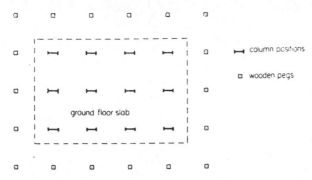

Figure 14.43 *Setting out column positions*

Lines are strung across the slab between the pegs or profiles to define the column centres. If the pegs are at slab level the column positions are marked directly. If profiles are used, a theodolite can be used to transfer the lines to the slab surface. The intersections of the lines define the column centres.

Once the centres are marked, the bolt positions for steel columns can be accurately established with a template, equal in size to the column base, placed exactly at the marked point. For reinforced concrete columns, the centres are established in exactly the same way but usually prior to the slab being laid so that the reinforcing starter bars can be placed in position.

14.14 Controlling Verticality

One of the most important setting out operations is to ensure that those elements of the scheme which are designed to be vertical are actually constructed to be so and there are a number of techniques available by which this can be achieved. Several are discussed in this section and particular emphasis is placed on the control of verticality in multi-storey structures. In order to avoid repeating information given earlier in this chapter, the following assumptions have been made

(i)   Offset pegs have been established to enable the sides of the building to be re-located as necessary.
(ii) The structure being controlled has already had its ground floor slab constructed and the horizontal control lines have already been transferred to it as shown in figure 14.39.

The principle behind verticality control is very straightforward: if the horizontal control on the ground floor slab can be accurately transferred to each higher floor as construction proceeds, then verticality will be maintained.

Depending on the heights involved, there are several different ways of achieving verticality. For single-storey structures, long spirit levels can be

plumbing a column by
using a spirit level

Figure 14.44

used quite effectively as shown in figure 14.44. For multi-storey structures, however, one of the following techniques is preferable:

(1) Plumb-bob methods
(2) Theodolite methods
(3) Optical plumbing methods
(4) Laser methods.

The basis behind all these methods is the same. They each provide a means of transferring points vertically. Once four suitable points have been transferred, they can be used to establish a square or rectangular grid network on the floor in question which can be used to set out formwork, column centres, internal walls and so on at that level. Plumb-bobs, theodolite methods and optical plummets are discussed in the following sections. Laser methods are discussed in section 14.16.

Plumb-bob Methods

The traditional method of controlling verticality is to use *plumb-bobs*, suspended on piano wire or nylon. A range of weights is available, from 3 kg to 20 kg and two plumb-bobs are needed in order to provide a reference line from which the upper floors may be controlled.

In an ideal situation, the bob is suspended from an upper floor and moved until it hangs over a datum reference mark on the ground floor slab. If it is impossible or inconvenient to hang the plumb-bob down the outside of the structure then holes and openings must be provided in the floors to allow the plumb-bob to hang through and some form of centring frame will be necessary to cover the opening to enable the exact point to be fixed. Service ducts can be used but often these are not conveniently placed to provide a suitable baseline for control measurements. It is also not always possible to use a plumb-bob over the full height of a building owing to the need to 'finish' each floor as work progresses, for example, the laying of a concrete screed would obliterate the datum reference mark.

Unfortunately, the problem of wind currents in the structure usually causes the bob to oscillate and the technique can be time consuming if great accuracy is required. To overcome this, two theodolites, set up on lines at right angles to each other, could be used to check the position of the wire and to estimate the mean oscillation position. However, limited space or restricted lines of sight may not allow for the setting up of theodolites and their use tends to defeat the object of the *simple* plumb-bob.

One partial solution to dampening the oscillations is to suspend the bob in a transparent drum of oil or water. However, this tends to obscure the ground control mark being used and, if this occurs it becomes necessary either to reference the plumbline to some form of staging built around or above the drum or to measure offsets to the suspended line. This is shown in figure 14.45, in which a freshly concreted wall is being checked for verticality. The plumb-bob is suspended from a piece of timber nailed to the top of the formwork and immersed in a tank of oil or water. Offsets from the back of the formwork are measured at top and bottom with due allowance for any steps or tapers in the wall. Any necessary adjustments are made with a push–pull prop.

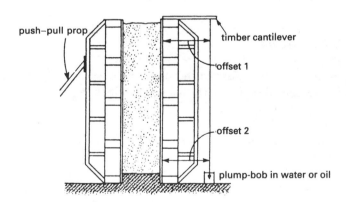

Figure 14.45 *Use of plumb-bob to control verticality in multi-storey structure (from the CIRIA/ Butterworth-Heinemann book* Setting Out Procedures, *1988)*

Increasing the weight of the bob reduces its susceptibility to oscillations but these are rarely eliminated completely. Plumb-bobs do have their uses, however, for example, they are very useful when constructing lift shafts and they are ideal for heights of one or two storeys. Figure 14.46 shows a plumb-bob being used to check the verticality of a single-storey structure in which offsets A and B will be equal when the structure is in a vertical position.

The advantages of plumb-bobs are that they are relatively inexpensive and straightforward in use. They are particularly useful for monitoring verticality over short distances, for example, when erecting triangular timber frames, a wire can be stetched across the base of the frame and a plumb-bob attached to the apex of the triangle. To ensure that it is erected in a vertical position, the frame is simply pivoted about its base until the suspended plumb-bob touches the stretched wire.

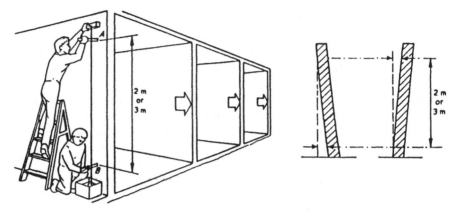

Figure 14.46 *Use of plumb-bob to control verticality in single-storey structure (reproduced from BS5606:1990 with permission of the BSI)*

Theodolite Methods

These methods assume that the theodolite is in perfect adjustment so that its line of sight will describe a vertical plane when rotated about its trunnion axis.

Controlling a multi-storey structure using a theodolite only

The theodolite is set up on extensions of each reference line marked on the ground floor slab in turn and the telescope is sighted on to the particular line being transferred. The telescope is elevated to the required floor and the point at which the line of sight meets the floor is marked. This is repeated at all four corners and eight points in all are transferred as shown in figure 14.47.

Once the eight marks have been transferred, they are joined and the distances between them and their diagonal lengths are measured as checks.

If the centre lines of a building have been established, a variation of this method is to set up a theodolite on each in turn and transfer four points instead of eight as shown in figure 14.48. This establishes two lines at right angles on each floor from which measurements can be taken.

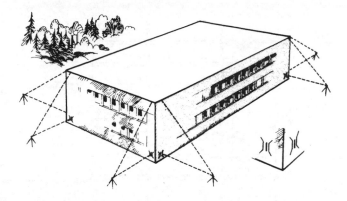

Figure 14.47   *Transfer of control in a multi-storey structure*

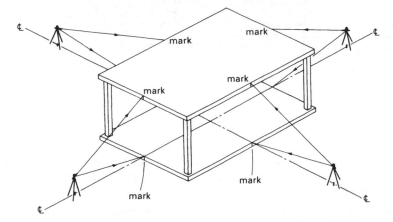

Figure 14.48   *Transfer of centre lines*

If the theodolite is not in perfect adjustment, the points must be transferred using both faces and the mean position used. In addition, because of
the large angles of elevation involved, the theodolite must be carefully levelled and a diagonal eyepiece attachment may be required to enable the
operator to look through the telescope (see figure 14.52).

Controlling a multi-storey structure using a theodolite and targets

In figure 14.49, A and B are offset pegs. The procedure is as follows.

(a) The theodolite is set over reference mark A, carefully levelled and aligned
     on the reference line marked on the side of the slab (see figure 14.39).
(b) The line of sight is transferred to the higher floor and a target accu-
     rately positioned.

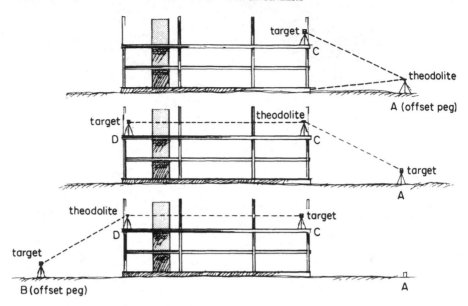

Figure 14.49 *Transfer of control by three-tripod traversing*

(c) A three-tripod traverse system is used with the target replacing the theodolite and vice versa.

(d) The theodolite, now at C, is sighted onto the target at A, transitted and used to line in a second target at D. Both faces must be used and the mean position adopted for D.

(e) A three-tripod traverse system is again used and the theodolite checks the line by sighting down to the reference mark at B, again using both faces.

(f) It may be necessary to repeat the process if a slight discrepancy is found.

(g) The procedure is repeated along the other sides of the building.

Again, the two centre lines can be transferred instead of the four reference lines if this is more convenient.

Controlling column verticality using theodolites only

Although short columns can be checked by means of a long spirit level held up against them as shown earlier in figure 14.44, long columns are best checked with two theodolites as shown in figure 14.50*a* and *b*. Either the edges or, preferably, the centre lines of each column are plumbed with the vertical hairs of two theodolites by elevating and depressing the telescopes. The theodolites are set up directly over the necessary control lines and because of the potentially high angles of elevation, they must be very carefully levelled. In addition, because they may not be in perfect adjustment,

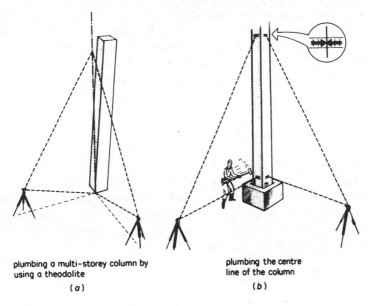

plumbing a multi-storey column by
using a theodolite

(a)

plumbing the centre
line of the column

(b)

Figure 14.50 *Control of column verticality using two theodolites*

the verticality of the column must be checked using both faces of the
theodolites.

It is not always necessary to use two theodolites, it is possible to use just
one as shown in figure 14.51 where the formwork for a tall column is being
plumbed. The theodolite is set up on a plane parallel to but offset from one
face of the formwork and sighted on suitable offset marks at the top and
bottom. Ideally the theodolite should be some distance away to avoid very
steep angles of elevation but this is not always possible. Observations should

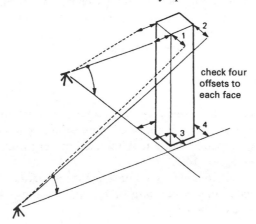

check four
offsets to
each face

Figure 14.51 *Control of column verticality using one theodolite (from the CIRIA/Butterworth-
Heinemann book* Setting Out Procedures, *1988)*

be taken to both edges of the face as a check on twisting, that is, offsets 1, 2, 3 and 4 should all be checked and, as a further precaution, both faces of the theodolite should be used. Any discrepancy between offsets 1, 2, 3 and 4 should be adjusted and the column face rechecked. Once this face has been plumbed, the whole procedure is repeated for the adjacent column face.

Optical Plumbing Methods

Optical plumbing can be undertaken in several ways. Either the optical plummet of a theodolite can be used or the theodolite can be fitted with a diagonal eyepiece or, preferably, an optical plumbing device specially manufactured for the purpose can be employed. When carrying out optical plumbing, holes and openings must be provided in the floors and a centring frame must be used to establish the exact position.

Optical plummet of a theodolite

The optical plummet of a theodolite provides a vertical line of sight in a downwards direction which enables the instrument to be centred over a ground mark. Optical plummets are usually incorporated into all modern theodolites but there are also special attachments which fit into a standard tribrach and enable high-accuracy centring to be obtained not only to reference marks below the instrument but also to control points above the instrument, for example, in the roof of a tunnel. These are optical roof and ground point plummets which enable centring to be achieved to ±0.3 mm over a distance of 1.5 m. On some instruments, a switch-over knob permits a selection between ground or roof point plumbings. After centring has been achieved, the plummet is replaced by the instrument or target which, by virtue of the system of controlled centring, is now correctly centred. These devices are meant for short-range work only and do not provide a vertical line of sight of sufficient accuracy to control a high-rise structure.

Diagonal eyepiece

Diagonal eyepiece attachments are available for most theodolites. These are interchanged with the conventional eyepiece and enable the operator to look through the telescope while it is inclined at very high angles of elevation as shown in figure 14.52.

They can be used to transfer control points upwards to special targets, either up the outside of the building or through openings left in the floors. The procedure is as follows.

(a) The theodolite with the diagonal eyepiece attached is centred and lev-

Figure 14.52 *Diagonal eyepiece*

elled over the point to be transferred as normal using its built-in optical plummet.

(b) The telescope is rotated horizontally until the horizontal circle reads 0°.

(c) The telescope is transitted until it is pointing vertically upwards. If an electronic reading instrument is being used, the display will indicate when the telescope is vertical. In the case of an optical reading instrument, an additional diagonal eyepiece must be fitted to the optical reading telescope to enable it to be read.

(d) A perspex target, as shown in figure 14.53, is placed over the hole on the upper floor and an assistant is directed by the theodolite operator to mark a line on the perspex which coincides with the image of the horizontal cross hair in the telescope.

(e) The telescope is rotated horizontally until the horizontal circle reads 180°.

(f) The telescope is set to give a vertical line of sight.

(g) The perspex target is again viewed. If the instrument is in correct adjustment, the horizontal hair of the telescope will coincide with the line drawn on the target at step (d). If this is not the case, the assistant is directed to mark another line on the target corresponding to the new position of the horizontal hair. A line mid-way between the two will be the correct one as shown in figure 14.53.

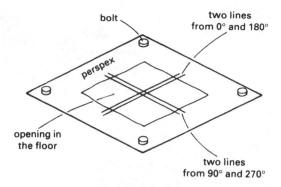

Figure 14.53 *Perspex target used with diagonal eyepiece*

(h) The whole procedure is now repeated twice with the horizontal circle of
the theodolite reading first 90° and then 270°. The mean of the two
lines obtained for these values will be the correct one. The transferred
point lies at the intersection of this mean line and that obtained in the
0° and 180° positions as shown in figure 14.53.

If care is taken with this method, the accuracy can be high, with precisions
of ±1 mm in 30 m being readily attainable. However, it is essential that the
horizontal hair is used for alignment because its mean position in this procedure is unaffected by any non-verticality of the vertical axis, which is not
the case with the vertical hair.

Special optical plumbing devices

There are several variations of these purpose-built optical plummets. Some
can only plumb upwards, some only downwards and some can do both. The
accuracy attainable is extremely high, for example the Wild ZL zenith plummet
and the Wild NL nadir plummet shown in figure 14.54 are each capable of
achieving precisions of 1:200 000 (0.5 mm at 100 m). Such high accuracy is
due to these instruments being fitted with compensator devices similar to
those used in automatic levels. They are first approximately levelled using
their small circular level and the compensator then takes over. This ensures
a much higher degree of accuracy in the vertical line of sight than could be
achieved with a theodolite fitted with a diagonal eyepiece.

In use, the plummet is first centred over the ground point to be transferred and then used with a special target in a similar way to that described
above for diagonal eyepieces. Some plummets have their own centring system to enable them to be set over the point while others are designed to fit
into a standard tribrach which has previously been centred over the ground
mark. As with diagonal eyepieces, the control should always be transferred
at the four major points of the circle, that is 0°, 90°, 180° and 270°, and the

Figure 14.54   *Wild ZL zenith and Wild NL nadir plummets (courtesy Leica UK Ltd)*

mean position used. Even if the plummet is in correct adjustment, this should still be carried out as a check. The handbook supplied with each plummet describes how it can be set to perfect adjustment.

Figures 14.55*a* and *b* show an optical plummet being used to plumb upwards to transfer control to a special centring device on the floor above. This device is fitted with an index mark which is moved as necessary by an assistant until it defines the transferred point. As it is difficult for the observer and assistant to keep in touch over a large number of storeys, the use of two-way radios is recommended.

Figure 14.55*c* shows an optical plummet being used to plumb downwards in order to transfer control up from the floor below. In this case, the optical plummet is adjusted until the reference mark on the floor below is bisected by its cross hairs. The centring device is then set in place beneath the tripod and moved until the line of sight passes through its centre.

Once at least three and preferably four points have been transferred, a grid can be established on the floor by offsetting from the transferred points. The offset points chosen for the grid should be sited away from any columns so that they can be used to fix the position of any formwork. When this has been completed, the centring devices can be removed and replaced by safety plates to avoid accidents.

14.15 Transferring Height from Floor to Floor

Height can be transferred by means of a weighted steel tape measuring each time from a datum in the base of the structure as shown in figure 14.56.

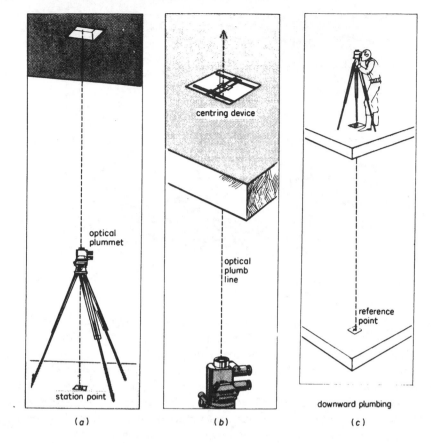

optical
plummet

centring device

optical
plumb
line

reference
point

station point

downward plumbing

(a)   (b)   (c)

Figure 14.55   *Use of optical plummet*

The base datum levels should be set in the bottom of lift wells, service
ducts and so on, such that an unrestricted taping line to roof level is provided. Worked example (3) in section 4.6 covers the calculations involved
in this method.

Each floor is then provided with TBMs in key positions from which
normal levelling methods can be used to transfer levels on each floor.

Alternatively, if there are cast-*in-situ* stairs present, a level and staff can
be used to level up and down the stairs as shown in figure 14.57. Note that
both up and down levelling must be done as a check.

14.16 Setting Out Using Laser Instruments

Although a detailed description of laser techniques and equipment is beyond
the scope of this book, laser instruments are now widely used in setting-out
operations and a few of the more common methods are discussed here.

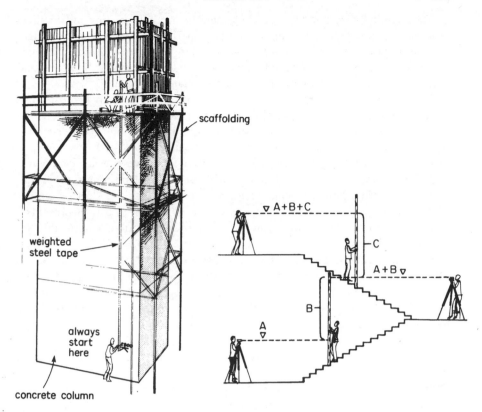

Figure 14.56 *Transfer of height from floor to floor using steel tape*

Figure 14.57 *Transfer of height from floor to floor using levelling*

The laser generates a beam of high intensity and of low angular divergence, hence it can be projected over long distance without spreading significantly. These characteristics are utilised in specially designed laser equipment and it is possible to carry out many alignment and levelling operations by laser.

There are two types of laser used in surveying equipment, those which generate a bright red *visible beam* and those which generate an *invisible beam*. Until 1988, all the visible beams used in surveying equipment were produced from a mixture of Helium and Neon (HeNe) gas housed inside a glass tube. In 1988, however, Toshiba produced the world's first commercially available visible laser diode using an Indium Gallium Aluminium Phosphorus (InGaA1P) diode source and, increasingly, visible laser diodes are being incorporated into surveying instruments. Such diode sources enable reductions in equipment size, weight and power consumption to be achieved when compared with HeNe sources. The invisible beams are produced from semiconductor diodes, commonly Gallium Arsenide (GaAs).

The visible beam instruments are generally used for setting-out applications such as pipelaying, tunnelling and any operation in which precise alignment is required and for levelling and grading purposes. Their beams can either be detected by eye or intercepted on translucent targets. Special hand-held or rod-mounted photoelectric detectors can also be used.

The invisible beam instruments are generally restricted to levelling and grading applications since special photoelectric detectors are always required to locate the beams. The fact that their beams are invisible makes such instruments difficult to use for alignment applications.

From a safety point of view, the lasers used in surveying are low power with outputs ranging from less than 1 mW to 5 mW. This represents absolutely no hazard when the beam or its reflection strikes the skin or clothes of anyone in the vicinity. However, an output of 1 mW to 5 mW presents a serious hazard to the eyes and on no account should anyone look directly into a laser beam.

Laser instruments are classified as Class 1, Class 2, Class 3A and Class 3B and every laser should be clearly marked with a label indicating its class. If this label is missing, the laser should not be used. Instead it should be returned to the hire company or the firm from which it was purchased to have its label replaced. The classes are as follows:

Class 1 lasers are completely safe but this category only applies to some invisible beam lasers.

Class 2 lasers are virtually harmless although staring into the beam should be avoided. Looking at the reflection of the beam on a wall through an optical instrument is perfectly safe.

Class 3A lasers are more powerful. Staring into the beam should again be avoided and approval must be obtained from a laser safety officer if a reflection of the beam is to be viewed through an optical instrument. Special safety notices are required.

Class 3B lasers are the most powerful used in surveying equipment. Staring at or viewing the beam optically must not be done. Suitable safety eyewear must be worn and protective clothing may be required. The area of operation must be roped off and special warning notices erected.

The majority of laser instruments used in surveying are of Safety Class 1, 2 or 3A and they fall into two main categories, either *alignment lasers* or *rotating lasers*. These are discussed in the following sections.

Alignment Lasers

This type of laser produces a single visible beam which, when used for alignment purposes, has the important advantage of producing a constantly

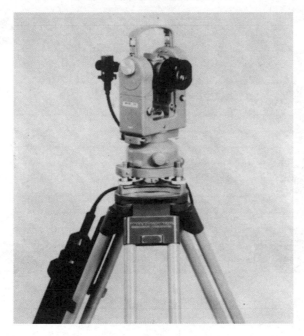

Figure 14.58 *Laser eyepiece attachment (courtesy Leica UK Ltd)*

present reference line which can be used without interrupting the construction works. A number of different instruments are manufactured for alignment purposes, examples include laser eyepiece attachments, laser theodolites,
pipe lasers and tunnel lasers. Invariably, a HeNe gas laser is used in these
instruments and they each require their own power source, either an external 12 V battery or an adapted mains supply.

Laser eyepiece attachments can be fitted to conventional theodolites, levels
and plummets to turn them into laser instruments. The instrument's eyepiece is simply unscrewed and replaced with the laser eyepiece in a few
seconds. Figure 14.58 shows a Wild GLO2 laser eyepiece fitted to a Wild
theodolite. The HeNe laser tube is fitted to one of the tripod legs and the
beam it generates is passed into the telescope of the instrument through a
fibre optic cable.

Laser theodolites are purpose built instruments which have the laser tube
permanently attached as shown in figure 14.59 for Sokkia's LDT5S electronic laser digital theodolite.

The laser beam generated by a laser eyepiece attachment or a laser theodolite coincides exactly with the line of collimation and is focused using
the telescope focusing screw to appear as a red dot in the centre of the cross
hairs as shown in figure 14.60. On looking into the eyepiece, the observer
sees a reflection of the beam which is perfectly safe. The beam can be

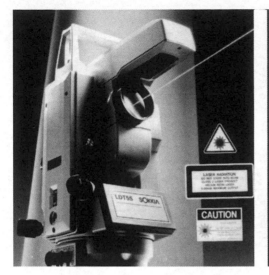
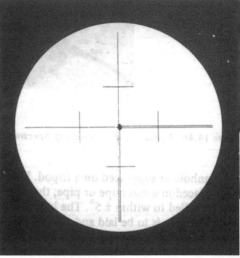

Figure 14.59 *Sokkia LDT5S electronic laser digital theodolite (courtesy Sokkia Ltd)*

Figure 14.60   *Laser eyepiece image*

intercepted with the aid of suitable targets over daylight ranges of 200–300 m and night ranges of 400–600 m. Such instruments can be used in place of a conventional theodolite in almost any alignment or intersection technique and, once set up, the theodolite can be left unattended. However, since the instrument could be accidentally knocked or vibration of nearby machinery could deflect the beam, it is essential that regular checks are taken to ensure that the beam is in its intended position.

For *controlling verticality*, a very narrow visible red reference line can be produced using either a laser theodolite or by fitting a laser eyepiece to an optical plummet. This is then set up on the ground floor slab directly over the ground point to be transferred and the beam is projected vertically either up the outside of the building or through special openings in the floors. The beam is intercepted as it passes the floor to be referenced by the use of plastic targets fitted in the openings or attached to the edge of the slab. The point at which the beam meets the target is marked to provide the reference.

When controlling floor-by-floor construction in multi-storey buildings, intermediate targets with holes in to enable the beam to pass through can be placed on completed floors, with a solid target being used on the floor in progress. This system has the advantage that should the laser instrument be moved accidentally, the beam will be cut out by the lower targets and the operator will immediately be aware that a problem has arisen.

With conventional optical plumbing, two people are required: one to look through the instrument and one to move the target into the correct position.

Figure 14.61 *Pipelaser and target (courtesy Leica Ltd)*

Such methods can cause problems of communication between the two operators. Laser plumbing has the great advantage that only the operator on the target is required since, once it has been set up, a constant visible beam will be projected for as long as is necessary without the need to look through the optics of the laser instrument. There is no danger of lack of communication and the spot can be clearly seen. However, care must be taken to ensure that the instrument is not disturbed while in use.

The essential requirement of the system is to ensure that the beam is truly vertical and, to check that this is the case, it is necessary to use four mutually perpendicular positions of the vertical telescope or optical plummet as described in section 14.14.

When *setting out pipelines*, the use of a laser system eliminates the need for sight rails and pipelaying travellers. A purpose-built *pipe laser* is shown in figure 14.61 together with its target. A pipe laser can be set up on a stable base positioned either within the pipe itself, in a manhole or on a variety of different supports. Figure 14.62 shows typical arrangements.

Pipe lasers are completely waterproof and made to be very robust. They are fully self-levelling over a wide range, typically $\pm5°$. In use, the laser is correctly aligned along the direction in which the pipe is to run and the gradient of the pipe is set on the grade indicator of the laser. During pipelaying, a plastic target is placed in the open end of the pipe length being laid which is then moved horizontally and/or vertically until the laser beam hits the centre of the target as shown in figure 14.63. The pipe is then carefully bedded in that position. The target is removed and the procedure repeated. If the laser is unintentionally moved off grade, the beam blinks on and off to provide a warning of this until the unit has re-levelled itself. If this should occur, the laser must be checked to ensure that the beam is still in the

Figure 14.62   *Various pipelaser supports and laser inside a pipe (courtesy AGL)*

required direction and at the required grade. Although pipe lasers will self-level, they will not re-level at the same height if they have been moved in a vertical direction.

Lasers have also been used very successfully for *tunnel alignment* and several *tunnel lasers* were used to control the tunnel boring machines used on the Channel Tunnel. Tunnel lasers are made to be shockproof, water-

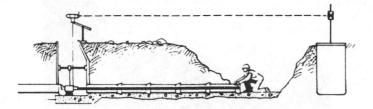

Figure 14.63 *Pipelaying with the aid of a laser*

Figure 14.64 *Controlling an automatically driven tunnel boring machine (courtesy AGL)*

proof and even flameproof, if required, owing to the need for them to function in the adverse conditions that can often prevail in tunnels. In use, a tunnel laser is fixed to the wall of the tunnel and aimed in the required direction which is defined by a series of intermediate targets with holes in to allow the beam to pass as shown in figure 14.64. The beam is detected by special targets on the tunnel boring machine and its position on the target indicates whether or not the machine is on line. In a manual system, a bull's-eye target is used and the operator adjusts the controls of the machine keeping the visible laser spot on the centre of the target. In an automatic system, the beam is detected by a photoelectric sensor which relays information about the boring machine to its control computer. The computer then automatically takes the necessary action to keep the machine on line.

Rotating Lasers

These instruments, which are also known as *laser levels*, generate a plane of laser light by passing the beam through a rotating pentaprism. Various mountings are available, from tripods to special column and wall brackets.

The beams generated by these instruments can be either visible or invis-

Figure 14.65   *Examples of rotating beam lasers*

ible, depending on the manufacturer. Either HeNe or laser diode sources are available. The HeNe instruments require either an external 12 V battery or an adapted mains supply whereas the diode instruments can function on built-in rechargeable or replaceable batteries. Each invisible beam laser comes complete with its special photoelectric detector which is essential to locate the beam. The rotating visible beam instruments are also normally supplied with photoelectric detectors since these enable the beams to be detected to a higher accuracy than by eye. Such high accuracy is essential when levelling and grading operations are being undertaken. Figure 14.65 shows examples of rotating beam lasers.

The majority of rotating beam instruments incorporate self-levelling devices and all can generate horizontal planes. Many can also generate vertical planes and some can generate sloping planes at known grades. In use, they are very simple to operate, for example, to generate a horizontal plane, the instrument is attached to a tripod or other suitable support, set approximately level using a circular level, if fitted, and turned on. After a few seconds, the laser plane will start to be generated.

On some instruments, the speed of rotation can be varied as required, typically from 0–720 rpm. The accuracy of the plane generated is normally better than ±10 seconds of arc (±5 mm in 100 m). If it is knocked accidentally out of its self-levelling range, the beam is cut off until it re-levels itself. Should this occur, the height of the plane must be remeasured to ensure that it is at the required value. Although the instrument will re-level, it will not necessarily re-level at the same height.

When the instrument is operating correctly, the height of the horizontal plane can be set to any required height by taking a reading on a levelling staff fitted with a photoelectric detector. The accuracy to which this can be done is in the order of ±2 mm, depending on the instrument.

On some instruments, known as *universal lasers*, the pentaprism which creates the laser plane can be removed to enable the laser to be used for alignment applications, such as pipelaying. These instruments are also ideal for use in *controlling verticality*. If set to generate a single beam in a vertical direction, the instrument's self-levelling system will ensure that the beam is truly vertical within, typically, ±10 seconds of arc. The only problem is in centring over a ground mark. Since no optical plummet is provided on the instrument, this must normally be done by mounting it on a tripod and suspending a conventional plumb-bob over the mark.

Another class of instruments can generate both a single beam and a rotating beam, simultaneously, at right angles to each other. These are the *interior lasers* which generate visible beams and have been specially developed for controlling the installation of mutually perpendicular internal fittings in a building such as raised floors, partition walls and suspended ceilings. They are designed for use without photoelectric detectors. Instead, targets such as graduated pieces of perspex with a magnetic edge are used. These are fitted to ceiling members allowing the operator to have both hands free during the installation. The visible beam is set to the required height by taking a reading on a levelling staff and the operator moves the ceiling member into position as shown in figure 14.66, finally fixing it when the beam passes through the correct graduations on the perspex target.

Figure 14.66 *Installing a suspending ceiling (courtesy Spectra-Physics Ltd)*

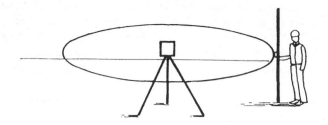

Figure 14.67 *Single-person levelling (courtesy Spectra-Physics Ltd)*

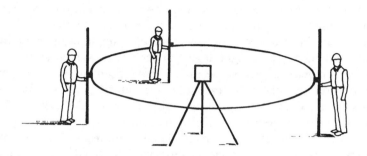

Figure 14.68 *Multi-user levelling (courtesy Spectra-Physics Ltd)*

For general *levelling* work, the laser can be set to generate a rotating plane at a known height and then left unattended. Single-person levelling can be undertaken as shown in figure 14.67 or several operators can use the same plane simultaneously as shown in figure 14.68. However, if more than one rotating laser is being used on a site, great care must be taken to ensure that the plane from one laser is not being sensed by the detector from another. When used for setting out foundations, floor levels and so on, the sensor can be fixed at some desired reading and the staff used as a form of traveller, the laser reference plane replacing sight rails as shown in figure 14.69.

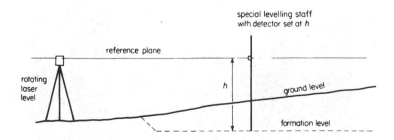

Figure 14.69 *Setting out with rotating laser level*

For *grading* work, the detectors can be fitted to the blades and arms of earth-moving machinery, as shown in figure 14.70, and the laser can be set to generate either a horizontal plane, a single-grade or a dual-grade. Both manual and fully automatic systems are available. In a manual system, the driver is presented with a series of lights which indicate whether the blade is too high or too low. If the middle light is kept illuminated, the operation is on line. In a fully automatic system, information about the position of the beam on the detector is sent to a controlling computer which automatically adjusts the machine's hydraulic systems to keep the operation on line. Long operating ranges are possible in this type of work with some of the rotating beams having working radii in excess of 300 m. Fast rotation speeds are used to ensure that the detector is not having to hunt for the beam and that the blade passes smoothly over the ground surface.

Concrete *floor screeding* can be very accurately controlled by laser. Figure 14.71 shows a system which involves two detectors mounted at each end of the screed carriage, simultaneously sensing the plane being generated by a single visible beam rotating laser. The information collected by the detectors is passed to an on-board control box which checks and adjusts the screed elevation five times per second through automatic control of the machine's hydraulics.

14.17 Applications of Setting Out from Coordinates

Coordinate methods of setting out are described in section 14.8 and their application in the setting out of horizontal curves is discussed in sections 11.16 and 11.17, where an appraisal of the advantages and disadvantages of such methods can be found.

The system used in coordinate setting out is that of establishing design points by either bearings and distances or by intersection using bearings only from nearby control points. The bearings and distances are calculated from the coordinates of the control and design points by the method described in section 1.5.

If the bearing and distance method is used, either a total station or a theodolite and some form of distance measuring system are required whereas if the intersection method is used, two theodolites are necessary. These two techniques are described in section 14.8.

The development of EDM equipment and total stations, in which the reflecting unit can be mounted on a detail pole, enables distances to be set out very accurately and quickly regardless of terrain and this is now widely used in bearing and distance methods. However, the use of EDM methods in setting out has its limitations. It is ideal for use with a theodolite or as part of a total station in bearing and distance methods for establishing site grids and other control points but such methods should not be used where

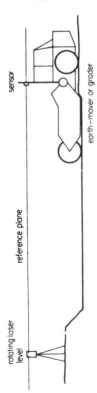

Figure 14.70 *Earthmoving and grading by laser (courtesy Spectra-Physics Ltd)*

Figure 14.71 *Laser-controlled floor screeding (courtesy John Kelly Lasers Ltd)*

alignment is critical, for example, setting out column centres, since the alignment obtained would not be satisfactory owing to slight angular and/or distance errors. In any alignment situation, it is best either to use one theodolite to establish the line and measure distances along it using a steel tape or EDM or, preferably, to use two theodolites positioned at right angles such that the intersection of their lines of sight establishes the point which can be located by lining in a suitable target.

In general, on site, EDM should not be employed if a steel tape could be used satisfactorily and usually all lengths less than one tape length (up to 50 m) would be set out and checked using steel tapes. EDM would be used in cases where distances in excess of one tape length were involved and over very uneven ground where steel taping would be difficult.

In all setting out, if the design is based on National Grid coordinates, the scale factor must be taken into account as described in section 5.23.

The great advantage of coordinate methods is that they can be used to set out virtually any civil engineering construction providing the points to be located and the control points are on the same rectangular coordinate system. The calculations can be undertaken on a computer and the results presented on a printout similar in format to that shown in table 11.2. With the increasing use of electronic data loggers on site, setting-out data generated by a computer can be transferred directly into a logger via a suitable interface for immediate use on site. The required bearing and distance information to establish the corners of a building, for example, can be recalled from the logger's memory at the press of a key. The following two examples demonstrate the versatility of coordinate based methods.

Setting Out and Controlling Piling Work

The equipment used in piling disturbs the ground, takes up a lot of space and obstructs sightings across the area. Hence, it is not possible to establish all the pile positions before setting out begins since they are very likely to be disturbed during construction. Coordinate methods can be used to overcome this difficulty as follows.

(1) Before piling begins a baseline is decided upon and the lengths and angles necessary to set out the pile positions from each end of the baseline are calculated from the coordinates of the ends of the baseline and the design coordinates of the pile positions.
   The position of the baseline must be carefully chosen so that taping and sighting from each end will not be hindered by the piling rig. Figure 14.72 shows a suitable scheme.
(2) Each bearing is set out by theodolite from one end of the baseline and checked from the other. The distances are measured using a steel tape or, if possible, EDM equipment with the reflecting unit mounted on a movable pole. A total station would be ideal.
(3) The initial two or three positions are set out and the piling rig follows the path shown in figure 14.72.
(4) The engineer goes on ahead and establishes the other pile positions as work proceeds.
(5) A variation is to use two baselines on opposite sides of the area and establish the pile positions from four positions instead of two.

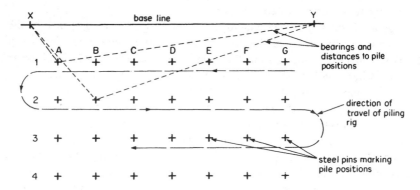

Figure 14.72 *Setting out pile positions*

Setting Out Bridges

Figure 14.73 shows the plan view of a bridge to carry one road over another.

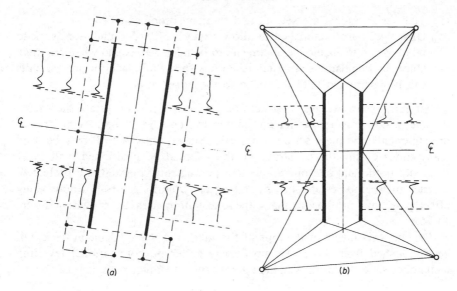

Figure 14.73 *Setting out bridges: (a) using structural grid; (b) by bearing and distance*

Procedure

(1) The centre lines of the two roads are set out by one of the methods discussed in chapters 10 and 11.

(2) The bridge is set out in advance of the road construction. Secondary site control points are established either in the form of a structural grid, which itself is set up from main site control stations by bearings and distances (see figure 14.73*a*), or, if this is not possible, in positions from which the bridge abutments can be set out by bearings and distances. These positions may be at traverse stations or site grid points (see figure 14.73*b*).

Whichever method is used, all the points must be permanently marked and protected to avoid their disturbance during construction, and positioned well away from the traffic routes on site.

(3) TBMs are set up. These can be separate levelled points or a control point can be levelled and used as a TBM.

(4) If the method shown in figure 14.73*a* is used, the distances from the secondary site control points to abutment design points are calculated and set out by steel tape or EDM equipment, the directions being established by theodolite. Alternatively, total stations could be used.

If the method shown in figure 14.73*b* is used, the bearings and distances from the secondary site control points are calculated from their respective coordinates such that each design point can be established from at least two and, preferably, three control points.

(5) The design points are set out and their positions checked.
(6) Offset pegs are established to allow excavation and foundation work to proceed and to enable the abutments to be relocated as and when required.
(7) Once the foundations are established, the formwork, steel or precast units can be positioned with reference to the offset pegs.

For multi-span bridges, a structural grid can again be established from the site grid or traverse stations as shown in figure 14.74 and the centres of the abutments and piers set out from this. Since points A to P may be used many times during the construction, they should be positioned well away from site traffic and site operations and permanently marked and protected.

Each pier can be established by setting out from its centre positions using offset pegs and profiles to mark the excavation area as shown in figure 14.75.

The required levels of the tops of the piers and the subsequent deck will be established from TBMs set up nearby either by conventional levelling techniques or by using a weighted steel tape as shown in figure 14.56.

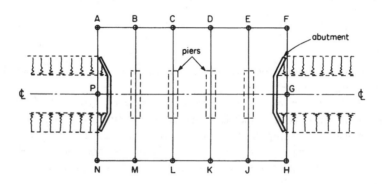

Figure 14.74 *Setting out a bridge from a site grid*

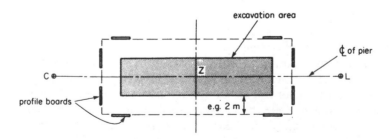

Figure 14.75 *Setting out a bridge pier*

14.18 Quality Assurance and Accuracy in Surveying and Setting Out

In recent years, there has been a marked increase in concern about the quality of work achieved on construction sites and the standards against which this quality should be assessed. Since 1987, a number of publications have been produced which deal specifically with the concept of *Quality Assurance* (QA) and several new British Standards have been released which are directly related to quality in the form of accuracy. This section discusses the role of Quality Assurance in surveying and setting-out operations and reviews those British Standards which are directly concerned with the accuracy attainable in these activities.

Quality Assurance

The term *Quality Assurance* (QA) is nowadays widely used on construction sites to indicate that certain standards of quality have been or are expected to be achieved. Unfortunately, as with many terms, it has become part of everyday jargon and those using it do not necessarily appreciate the concepts on which it is based. In general terms, QA is the creation of a fully competent management and operations structure which can consistently deliver a high-quality product to the complete satisfaction of the client. It has its origins in 1979 when, in order to provide guidelines on which a company could base a quality system, the British Standards Institute issued BS5750 *Quality Systems*. Since then, more than 9000 firms in the UK have been assessed and registered against this standard. Included in these are a number of firms involved in surveying and setting out operations.

In 1983, in response to a wider interest in quality standards, the International Organisation of Standards (ISO) was formed and in 1987 it published the ISO9000 series of five standards which were largely based on BS5750: 1979. These were adopted for publication in the UK by the British Standards Institute (which is the UK member of ISO) and were released as an updated version of BS5750. They became BS5750: 1987 which consisted of Parts 0, 1, 2 and 3. Part 4, which is a guide to using Parts 1, 2 and 3, was released in 1990.

Also in 1987, the European Committee for Standardisation (CEN) which is based upon members from Austria, Belgium, Denmark, Finland, France, Germany, Greece, Ireland, Italy, Netherlands, Norway, Portugal, Spain, Sweden, Switzerland and the United Kingdom issued their own quality standard, EN29000, which was entirely based on BS5750.

Hence, BS5750, ISO9000 and EN29000 can all be considered to be the same.

In order to achieve BS5750 accreditation, UK firms are required to submit three sets of documents for consideration by the National Accreditation

Council for Certification Bodies (NACCB) which acts on behalf of the Department of Trade and Industry (DTI). First, a quality manual must be prepared in which the details of the firm's policies, objectives, management structure and its methods of implementing quality control are given. Second, documents detailing each quality related management procedure must be produced. Third, each actual day-to-day working practice should be described to show how quality is maintained.

From a surveying and setting-out point of view, it could be argued that engineers and surveyors have always undertaken quality assurance procedures as part of their duties. Planning the work, following a programme, recording all relevant information, checking each setting-out operation and working to specified tolerances are just some of the many examples of quality procedures undertaken as a matter of course during engineering surveying work. However, BS5750 accreditation can only be obtained by submission of the correct documents for approval. The preparation of these represents a major task for any firm and those wishing to know more about the procedures involved are recommended to read the two articles written by Frank Shepherd referenced in Further Reading at the end of the chapter. In these, Frank gives the background to the concept of Quality Assurance, from which some of the information above has been taken, and then describes how a small survey company prepared for and obtained BS5750 accreditation.

Accuracy and British Standards

The question of accuracy in surveying and setting out arises at all stages of the construction process. When the initial site survey is being carried out, the survey team must use equipment and techniques which will ensure that the plan produced shows the required detail and is within the tolerances agreed with the client. At the design stage, the designer must specify suitable tolerances which will ensure not only that the construction will function properly but also that it can be built. At the construction stage, the contractor must choose equipment and adopt working procedures which will ensure that the scheme is located correctly in all three dimensions and that the tolerances used by the designer and specified in the contract are met.

Those involved in engineering surveying work must, therefore, become familiar with the limitations and capabilities of the various surveying instruments they use in their day-to-day operations and must adopt suitable working procedures that will ensure that the tolerances specified in the contract are achieved. These requirements are directly related to the question of Quality Assurance discussed in the previous section and, in a QA-accredited surveying firm, some of their QA documents will give details of the standard working procedures used by the company in its day-to-day activities.

For those actively involved in surveying and setting out, whether preparing for QA accreditation or not, the development of suitable working procedures and the choice of appropriate equipment are of paramount importance. Considerable guidance on these matters can be obtained by consulting those British Standards specifically produced for these requirements.

Currently, there are a number of British Standards which relate to surveying and setting out operations. The most important ones are BS5606, BS5964, BS7307, BS7308 and BS7334. These are briefly discussed below. Since many of them have also been adopted by the ISO, their equivalent ISO references are given where appropriate.

BS5606: 1990 *Guide to Accuracy in Building*

Originally published in 1978, BS5606 was most recently reviewed in 1990. Its main objective is to guide the construction industry on ways to avoid problems of inaccuracy or fit arising on site. It is based on work carried out by the Building Research Establishment (BRE). Accuracies which can be achieved in practice are given and used to stress the need for realistic tolerances at the design stage. Table 14.1 has been taken directly from BS5606 and shows the accuracy in use (A) of various pieces of equipment when used in engineering surveying by reasonably proficient operators. The values shown in the table should be used both by the designer when specifying the deviations allowed in the design in order that what is designed can actually be set out (see BS5964 below), and by the engineer undertaking the setting out in order that equipment can be chosen which will maintain the design standards and specifications.

BS5606 was one of the first documents produced by the British Standards Institution which dealt with the accuracy of surveying and setting out and it remains the basic reference for this type of work. It is deliberately broad and basic in coverage, leaving guidance on more detailed and complex work to other more specific standards.

BS5964: 1990 *Building Setting Out and Measurement* (ISO 4463)

This standard, which was first published in 1990, translates the requirements of BS5606 to relate to the processes of setting out and measurement. It is a wide ranging standard which applies to all types of construction and to the control from which the construction is set out.

One of the most important aspects of BS5964 is that it defines an accuracy measure known as a *permitted deviation* (P). This term represents the required tolerance specified in the contract documents. In practice, it is compared with the appropriate A values from table 14.1 or BS7334 (as discussed later) in order that suitable equipment is chosen to ensure that $A \leq P$. Since many building practices and instrument manufacturers express

Table 14.1 Accuracy in use of measuring instruments (Reproduced from BS5606 : 1990 with permission of the British Standards Institution. Complete copies can be obtained by post from BSI Sales, Linford Wood, Milton Keynes, MK14 6LE)

Measurement	Instrument	Range of deviations	Comment (see also NOTE)
T.3.1 Linear	30 m carbon steel tape for general use	± 5 mm up to an including 5 m ± 10 mm for over 5 m and up to and including 25 m ± 15 mm for over 25 m	With sag eliminated and slope correction applied
	30 m carbon steel tape for use in precise work	± 3 mm up to and including 10 m ± 6 mm for over 10 m and up to and including 30 m	At correct tension and with slope, sag and temperature corrections applied
	Electronic distance measuring (EDM) instruments (short range models) for general use	± 10 mm for distances over 30 m and up to 50 m ± (10 mm + 10 p.p.m.³) for distances greater than 50m	Accuracies of EDM instruments vary, depending on make and model of instruments. Distances measured by EDM should normally be greater than 30 m and measured from each end.
	Precise work	± (5 mm + 5 p.p.m.³)	
T.3.2 Angular	Opto-mechanical (eg. glass arc) theodolite 1) (with optical plummet or centering rod) reading directly to 20"	± 20" (± 5 mm in 50 m)	Scale readings estimated to the nearest 5". Mean of two sights, one on each face with readings in opposite quadrants of the horizontal circle.
	Opto-mechanical (eg. glass arc) theodolite (with optical plummet or centering rod) reading directly to 1"	± 5" (± 2 mm in 80 m)	Mean of two sights, one on each face with readings in opposite quadrants of the horizontal circle.
	1" opto-electronic theodolite/total station	± 3" (± 1 mm in 50 m)	Mean of two sights, one on each face with readings in opposite quadrants of the horizontal circle.
T.3.3 Verticality	Spirit level	± 10 mm in 3 m	For an instrument not less than 750 mm long
	Plumb-bob (3 kg) freely suspended	± 5 mm in 5 m	Should only be used in still conditions
	Plumb-bob (3 kg) immersed in oil to restrict movement	± 5 mm in 10 m	Should only be used in still conditions
	Theodolite (with optical plummet or centering rod) and diagonal eye-piece	± 5 mm in 30 m²)	Mean of at least four projected points, each one established at a 90° interval
	Optical plumbing device	± 5 mm in 100 m	Automatic plumbing device incorporating a pendulous prism instead of a levelling bubble
	Laser upwards or downwards alignment	± 7 mm in 100 m	Four readings should be taken in each quadrant of the horizontal circle and the mean value of readings in opposite quadrants accepted Appropriate safety precautions should be applied according to power of instrument used
T.3.4 Levels	Spirit level Water level Lightweight self-levelling level Optical level	± 5 mm in 5 m distance ± 5 mm in 15 m distance ± 5 mm in 25 m distance	Instrument rod not less than 750 mm long Sensitive to temperature variation
	(a) 'builders' class	± 5 mm per single sight of up to 60 m²)	Where possible sight lengths should be equal
	(b) 'engineers' class	± 3 mm per single sight of up to 60 m²) ± 10 mm per km	
	(c) 'precise' class	± 2 mm per single sight of up to 60 m ± 8 mm per km	If staff readings of less than 1 mm are required the use of a precise level incorporating a parallel plate micrometer is essential but the range per sight preferably should be about 15 m and should be not more than 20 m.
	Laser level (visible light source) (invisible light source)	± 7 mm per single sight up to 100 m ± 5 mm per single sight up to 100 m	Appropriate safety precautions should be applied according to power of instrument used

1) If a single sighting only is made when using a correctly adjusted theodolite to establish an angle the likely deviations will be increased by a factor of 3. Therefore a single sight should not be taken.
2) Value based on measured data
3) Parts per million of measured distance.
NOTE: Equipment should be checked periodically according to BS 7334

tolerances in terms of the statistical *standard deviation* (σ), *P* values have
been related to σ values by the following equation

$$P = 2.5 \times \sigma$$

Having defined *P* values, BS5964 goes on to break down setting out into its
component activities and assesses each in turn. Recommended practices are
given for various activities and permissible deviations are defined for primary control, secondary control, setting out, levelling and plumb.

BS7307: 1990 *Building Tolerances: Measurements of Buildings and Building Products* (ISO 7976)

In Part 1 of this standard, which first appeared in 1990, alternative methods
for determining shape, dimensions and deviations of building components
both in the factory and on site are given. Diagrams illustrating the procedures
that should be used are included as are accuracy tables which recommend
suitable equipment and its associated permitted deviations. Suitable positions
for measuring points on buildings and building components are covered in
Part 2 of the standard.

BS7308: 1990 *Presentation of Accuracy Data in Building Construction* (ISO 7737)

With the standardisation of the methods by which data is collected being
covered in BS7307: 1990, the need arises for standard formats for the presentation and processing of data. This is covered by BS7308: 1990 which
states how measured data used to check and assess accuracy should be presented. Initially, the standard defines which dimensions should be measured
and then goes on to give guidance on the correct presentation of dimensional accuracy data. Blank copies of standard booking forms and tables are
also included.

BS7334: 1990 *Measuring Instruments for Building Construction* (ISO 8322)

This is a very important British Standard which should be consulted by all
those involved in the use of surveying instruments for surveying and setting out. Its main purpose is to enable users of equipment to determine the
accuracy of the instruments they are using on site. It is divided into ten
parts, parts 1 to 3 appeared in 1990, parts 4 to 8 in 1992, and parts 9 and
10 are in preparation.

Part 1 gives the theory of how to determine the accuracy of measuring
instruments. Users can employ this theory when devising test procedures
for instruments not covered in other parts of the standard. Accuracy results
are expressed as accuracy in use (*A*) values as discussed earlier for BS5606.

Details are also provided on how to establish that the accuracy associated with a particular surveying technique using specific equipment is appropriate to the intended measuring task. This is done by comparing the calculated A values with P values specified using BS5964.

Part 2 gives a step-by-step guide to the observation procedures and calculation methods to be employed to determine the accuracy in use of measuring tapes.

Part 3 deals with optical levelling instruments.

Part 4 deals with theodolites.

Part 5 deals with optical plumbing instruments.

Part 6 deals with laser instruments.

Part 7 deals with instruments when used for setting out.

Part 8 deals with EDM instruments up to a range of 150 m.

Part 9 deals with EDM instruments up to a range of 500 m.

Part 10 deals with testing short-range reflectors.

Further Reading at the end of the chapter lists all the British Standards referred to above together with other publications which will be found useful by anyone concerned with the correct methods of setting out on construction sites.

14.19 Worked Examples

(1) Pipeline Example

Question

An existing sewer at P is to be continued to Q and R on a falling gradient of 1 in 150 for plan distances of 27.12 m and 54.11 m consecutively, where the positions of P, Q and R are defined by wooden uprights.

Given the following level observations, calculate the difference in level between the top of each upright and the position at which the top edge of each sight rail must be set at P, Q and R if a 2.5 m traveller is to be used.

Level reading to staff on TBM on wall (RL 89.52 m AOD)	0.39 m
Level reading to staff on top of upright at P	0.16 m
Level reading to staff on top of upright at Q	0.35 m
Level reading to staff on top of upright at R	1.17 m
Level reading to staff on invert of existing sewer at P	2.84 m

All readings were taken from the same instrument position.

Solution

Consider figure 14.76.

Height of collimation of instrument	$= 89.52 +$	$0.39 = 89.91$ m
Invert level of existing sewer at P	$= 89.91 -$	$2.84 = 87.07$ m

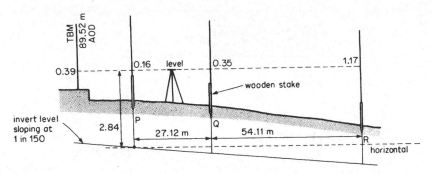

Figure 14.76

Hence, sight rail top edge level at P = 87.07 + 2.50 = 89.57 m
Level of top of upright at P   = 89.91 − 0.16 = 89.75 m
Hence, upright level − sight rail level = 89.75 − 89.57 = +0.18 m

That is, the top edge of the sight rail must be fixed **0.18 m** *below* the top of the upright at P.

Fall of sewer from P to Q   = −27.12 × (1/150) = −0.18 m
Hence, invert level at Q   = 87.07 − 0.18 = 86.89 m
Hence, sight rail top edge level at Q = 86.89 + 2.50 = 89.39 m
But, level of top of upright at Q   = 89.91 − 0.35 = 89.56 m

Hence, upright level − sight rail level = 89.56 − 89.39 = +0.17 m. That is, the top edge of the sight rail must be fixed **0.17 m** *below* the top of the upright at Q.

Fall of sewer from P to R   = −(27.12 + 54.11)/150 = −0.54 m
Hence, invert level at R   = 87.07 − 0.54 = 86.53 m
Hence, sight rail top edge level at R = 86.53 + 2.50 = 89.03 m
But, level of top of upright at R   = 89.91 − 1.17 = 88.74 m
Hence, upright level − sight rail level = 88.74 − 89.03 = −0.29 m

That is, the top edge of the sight rail must be fixed **0.29 m** *above* the top of the upright at R.

This is achieved by nailing the sight rail to an extension piece to form a short traveller and then nailing this to the upright such that it adds 0.29 m to its height.

(2) Coordinate Example

Question
A rectangular building having plan sides of 75.36 m and 23.24 m is to be set out with its major axis aligned precisely east–west on a coordinate sys-

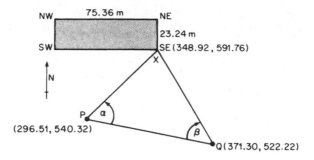

Figure 14.77

tem. Coordinates of the SE corner have been fixed as (348.92, 591.76) and this corner is to be fixed by theodolite intersections from two stations P and Q whose respective coordinates are (296.51, 540.32) and (371.30, 522.22). All dimensions are in metres.

Existing ground levels at the corners of the proposed structure were determined as follows

SE (156.82 m AOD), SW (149.73 m AOD), NE (151.45 m AOD), NW (146.53 m AOD)

Calculate

(1) The respective clockwise angles (to the nearest 20″) to be set off at P relative to PQ and at Q relative to QP in order to intersect the position of the SE corner.
(2) Surface setting-out measurements around the four sides of the building together with the two diagonals, assuming even gradients along all lines.

Solution
Consider figure 14.77.

(1) *Calculation of α and β*

Let the SE corner of the building be X.

easting of X	348.92	northing of X	591.76
easting of P	296.51	northing of P	540.32
ΔE_{PX}	+52.41	ΔN_{PX}	+51.44

Therefore from a rectangular/polar conversion

$$\text{bearing } PX = 45°32'07''$$

easting of X	348.92	northing of X	591.76
easting of Q	371.30	northing of Q	522.22
ΔE_{QX}	−22.38	ΔN_{QX}	+69.54

Therefore from a rectangular/polar conversion

bearing QX = 342°09′37″

easting of Q	371.30		northing of Q	522.22
easting of P	296.51		northing of P	540.32
ΔE_{PQ}	+74.79		ΔN_{PQ}	−18.10

Therefore from a rectangular/polar conversion

bearing PQ = 103°36′17″

Therefore

angle α = bearing PQ − bearing PX = 58°04′10″

Hence

clockwise angle to be set off at P relative to PQ = 360° − 58°04′10″
= **301°56′00″**

and

angle β = bearing QX − bearing QP = 58°33′20″

Hence

clockwise angle to be set off at Q relative to QP = **58°33′20″**

Both answers have been rounded to the nearest 20″.

(2) *Calculation of surface measurements*
Slope correction = $+(\Delta h^2/2L)$ (see section 4.4) where Δh is the height difference and L the slope distance (but horizontal distance may be used without significant error).

From SE to SW corners, Δh = 156.82 − 149.73 = 7.09; Δh^2 = 50.27
From NE to NW corners, Δh = 151.45 − 146.53 = 4.92; Δh^2 = 24.21
From SE to NE corners, Δh = 156.82 − 151.45 = 5.37; Δh^2 = 28.84
From SW to NW corners, Δh = 149.73 − 146.53 = 3.20; Δh^2 = 10.24

Slope distances are as follows

SE to SW corners = 75.36 + (50.27/(2 × 75.36)) = 75.36 + 0.33
= **75.69 m**

NE to NW corners = 75.36 + (24.21/(2 × 75.36)) = 75.36 + 0.16
= **75.52 m**

SE to NE corners = 23.24 + (28.84/(2 × 23.24)) = 23.24 + 0.62
= **23.86 m**

SW to NW corners = 23.24 + (10.24/(2 × 23.24)) = 23.24 + 0.22
= **23.46 m**

For the diagonals

Horizontal diagonals = $(75.36^2 + 23.24^2)^{\frac{1}{2}}$ = 78.86
From SE to NW corners, Δh = 156.82 − 146.53 = 10.29; Δh^2 = 105.88
From SW to NE corners, Δh = 151.45 − 149.73 = 1.72; Δh^2 = 2.96

Diagonal slope distances are as follows

SE to NW corners = 78.86 + (105.88/(2 × 78.86)) = 78.86 + 0.67
= **79.53 m**
SW to NE corners = 78.86 + (2.96/(2 × 78.86))   = 78.86 + 0.02
= **78.88 m**

Further Reading

S.G. Brightly, *Setting Out: A Guide for Site Engineers* (Granada, London, 1981).

P. Bryant, 'Quality and the Engineer – Do QA Systems Compromise the Engineer's Traditional Role?', in *Civil Engineering Surveyor*, Vol. 17, No. 8, p. 12, September 1992.

M.J. Fort, '+/− Permitted Deviation and all that', in *Civil Engineering Surveyor*, Vol. 16, No. 4, p. 22, May 1991.

W. Irvine, *Surveying for Construction*, 3rd Edition (McGraw-Hill, Maidenhead, 1988).

W.F. Price and J. Uren, *Laser Surveying* (Van Nostrand Reinhold, London, 1989).

B.M. Sadgrove , *Setting-out Procedures* (Butterworths, in conjunction with CIRIA, London, 1988).

F.A. Shepherd, 'Quality Assurance – What does it mean for you?', Part 1 in *Civil Engineering Surveyor*, Vol. 18, No. 3, pp. 17–18, April 1993. Part 2 in *Civil Engineering Surveyor*, Vol. 18, No. 4, pp. 10–11, May 1993.

I. Sinclair, 'Quality – Is it a Waste of Time?', in *Surveying World*, Vol. 1, No. 2, p. 17, January 1993.

R. Stirling, 'Quality – A Much Misunderstood Issue', in *Surveying World*, Vol. 1, No. 2, pp. 36–37, January 1993.

A.C. Twort, *Civil Engineering Supervision and Management* (Arnold, London, 1972).

BRE Digest 234, *Accuracy in Setting Out* (Building Research Establishment, 1980).

BS5606, *Guide to Accuracy in Building* (British Standards Institution, London, 1990).

BS5750, *Quality Systems*, (British Standards Institution, London, 1987).

BS5964, *Building Setting Out and Measurement* (British Standards Institution, London, 1990).

BS7307, *Building Tolerances: Measurement of buildings and building products* (British Standards Institution, London, 1990).

BS7308, *Presentation of Accuracy Data in Building Construction* (British Standards Institution, London, 1990).

BS7334, *Measuring Instruments for Building Construction* Parts 1, 2 and 3

– 1990. Parts 4, 5, 6, 7 and 8 – 1992 (British Standards Institution, London).

Bulletin M83: 16, *Measuring Practice on the Building Site*, CIB Report No. 69 (National Swedish Institute for Building Research, 1983).

ICE *Conditions of Contract, Sixth Edition* (Thomas Telford, London, 1991).

ISO 4463: *Measurement Methods for Building – Setting Out and Measurement – Permissible Measuring Deviations* (International Organisation for Standardisation, Geneva).

ISO/DP 8322 *Procedure for Determining the Accuracy in Use of Measuring Instruments* (International Organisation for Standardisation, Geneva).

JCT *Standard Form of Building Contract* (1980 Edition) (RIBA Publications, London).

15

Deformation Monitoring

Over the last decade, an interest has grown among the civil engineering and building professions in monitoring the movement of different types of structure both during and after completion of construction. There are many reasons why a structure may need to be monitored for movement. For example, it is well known that dam walls change shape with varying water pressure, that the foundations of large buildings are affected by changes in ground conditions and that landslips sometimes occur on embankments and cuttings. For all of these, deformation surveys can be used to measure the amount by which a structure moves both vertically and horizontally over regular time intervals. Although the principles of many of the techniques used to do this are recognisable as those used for site surveying and setting out, it is the taking of very precise periodic measurements that distinguishes a deformation survey from other types of survey.

In summary, the purposes of a deformation survey are to ascertain if movement is taking place and to assess whether a structure is stable and safe. In addition, movement may be analysed to assess whether it is due to some daily, seasonal or other factor and, most importantly, it may be used to predict the future behaviour of a structure.

The terms *relative* and *absolute* are often used for deformation surveys. The measurement of relative movement is generally much easier since movements are related to the structure itself or to some arbitrary point(s) nearby. These may move during a survey but this does not affect the results obtained. Absolute measurements, on the other hand, are related to datum points that are assumed not to move during a survey.

The accuracy required for a deformation survey depends on many factors including the type and size of building or structure, what is causing the movement (environmental factors or loading) and whether an understanding of the movement is needed. For many surveys, deformations of less than 1 mm are measured and precisions of this order require the best survey tech-

554

niques to be used throughout a deformation survey. This is to ensure that the systematic and random errors propagated in observations are less than the smallest displacement that is being measured or is expected.

Many different types of measurement can be taken during a deformation survey using a wide variety of equipment. This chapter will only consider methods for monitoring that involve some of the surveying techniques already described in this book and will not include details of monitoring using such methods as photogrammetry and automated systems which involve the use of transducers and sensors. Details of these can be found in the references given at the end of this chapter.

15.1 Vertical Movement

This type of movement can be measured by levelling or by measurement of vertical angles.

For nearly all applications where levelling is to be used, an ordinary level and staff are inadequate and special equipment is required so that sub-millimetre readings can be taken. This consists of a precise optical level fitted with a parallel plate micrometer and an invar staff (see figure 15.1). Alternatively, a precise digital level such as the Wild NA3000 with its bar-coded invar staff can be used (figure 15.2). Whatever equipment is used, the basic approach to fieldwork is similar to that described in chapter 2 but much more care is taken with the observations and with keeping sight lengths equal and much shorter than for ordinary levelling.

The stability of TBMs used in the measurement of absolute vertical movements is very important and one or more stable datum points must be established for the entire survey. Sometimes, when working on or near to large structures, it may be possible to locate a datum point on the structure itself, but very often it is necessary to locate TBMs well away from the building to be monitored in order to avoid any settlement affecting them. If, however, the TBMs are too far away from the monitoring area, errors may be introduced when transferring height over a large distance. Up to distances of about 1 km, this problem can be overcome by using intermediate bench marks and by checking the difference in height between these to monitor their stability.

At each point to be levelled in a deformation survey, a *levelling station* should be installed. A nail in a wall or some other crude mark is unacceptable since the prime function of a levelling station is that it achieves positional repeatability. In other words, each time the levelling staff is placed on it, a levelling station should ensure that the staff occupies exactly the same position. As well as this, a levelling station should be permanent and easy to install; it should also be vandal-proof and weather and corrosion resistant in addition to being unobtrusive. A levelling station made by the Building Research

(a)

(b)

Figure 15.1 *(a) Precise level; (b) invar staff and stand (courtesy Leica UK Ltd)*

Establishment (BRE) is shown in figure 15.3. This consists of a stainless steel socket which is permanently fixed to the structure to be monitored. When not in use, the socket is sealed with a plastic bung, the outer edge of which is flush with the structure, as shown in figure 15.3*a*. During a sur-

Figure 15.2   *Wild NA3000 precise digital level with bar-coded invar staff (courtesy Leica UK Ltd)*

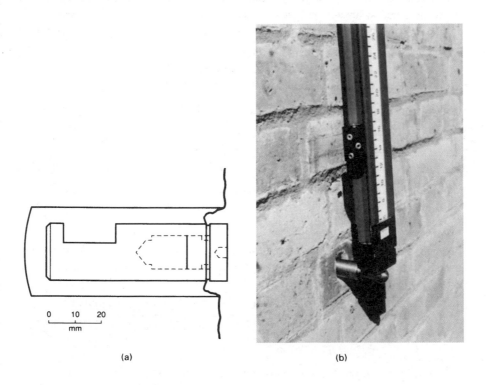

(a)   (b)

Figure 15.3   *BRE levelling station: (a) section through wall socket; (b) levelling staff on spherical plug*
(Building Research Establishment: Crown Copyright)

vey, a stainless steel plug that is threaded at its inner end is inserted into the socket. This ensures that the other end of the plug, which is spherical, accepts a levelling staff in the same position each time the levelling station is used (see figure 15.3b).

The principles of monitoring vertical movement by measurement of angles involve using a theodolite to measure horizontal and vertical angles from the ends of a known baseline. These are demonstrated in the worked example in section 15.2. The precision to which vertical movements can be determined is also of vital importance when monitoring what may be very small displacements. The determination of precision is demonstrated in the worked example in section 15.3.

15.2 Worked Example: Monitoring Vertical Settlement

Question
In order to check for any subsidence of a multi-storey block of flats, two pillars A and B were set up on solid foundations adjacent to the building. A target T was fixed near the top of the building and a series of angles and distances were measured at three-month intervals using the same equipment and methods.

The following values remained constant throughout the three-month period

$$\text{horizontal distance AB} = 76.987 \text{ m}$$
$$\text{horizontal angle TAB} = 52°34'21.1''$$
$$\text{horizontal angle TBA} = 64°09'12.3''$$

The vertical angle from A to T varied as follows

$$\text{initially} = 15°56'18.5''$$
$$\text{after three months} = 15°56'06.6''$$
$$\text{after six months} = 15°56'00.9''$$

Calculate the amounts of vertical movement over the six-month period, assuming that the angular changes are due solely to settlement of the building.

Solution
Figure 15.4 shows a plan view of the three points A, B and T. With reference to this, the horizontal distance D_{AT} is obtained by solving triangle ABT using the sine rule as follows

$$D_{AT} = \frac{D_{AB} \sin T\hat{B}A}{\sin(T\hat{A}B + T\hat{B}A)} = \frac{76.987 \sin 64°09'12.3''}{\sin(52°34'21.1'' + 64°09'12.3'')}$$

$$= 77.5729 \text{ m}$$

The height of target T relative to the theodolite collimation at pillar A is given by $h_i = D_{AT} \tan \theta_i$ where $i = 1, 2$ and 3 for each three-month interval. This gives

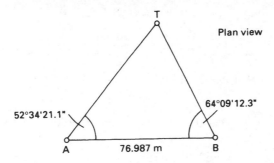

Figure 15.4

$$h_1 = 77.5729 \tan 15°56'18.5'' = 22.1535 \text{ m}$$
$$h_2 = 77.5729 \tan 15°56'06.6'' = 22.1487 \text{ m}$$
$$h_3 = 77.5729 \tan 15°56'00.9'' = 22.1464 \text{ m}$$

From these

vertical movement in three months = 22.1487 − 22.1535
                                   = **−4.8 mm**
vertical movement in six months   = 22.1464 − 22.1535
                                   = **−7.1 mm**

In this example, the effects of curvature and refraction have been ignored as they cancel (assuming refraction is the same) when the movements are calculated.

15.3   Worked Example: Propagation of Errors in Vertical Monitoring

Question
In the previous example, the vertical movements were measured using equipment and methods that gave the following standard errors

$$se_{D_{AB}} = \pm 1\text{mm}, \ se_{T\hat{A}B} = se_{T\hat{B}A} = \pm 1.5'', \ se_\theta = \pm 1.5''$$

For the previous example, calculate the precision of the calculated vertical movements.

Solution
In section 15.2, the height of target T above the theodolite collimation at pillar A is given by $h_i = D_{AT} \tan \theta_i$ with $i = 1, 2$ and 3. The movement during the first three months can be written as

$$m_1 = h_2 - h_1 = D_{AT} (\tan \theta_2 - \tan \theta_1)$$

Equation (6.5) gives

$$se^2_{m_1} = \left(\frac{\partial m_1}{\partial D_{AT}}\right)^2 se^2_{D_{AT}} + \left(\frac{\partial m_1}{\partial \theta_2}\right)^2 se^2_\theta + \left(\frac{\partial m_1}{\partial \theta_1}\right)^2 se^2_\theta$$

in which the partial differentials are

$$\frac{\partial m_1}{\partial D_{AT}} = \tan \theta_2 - \tan \theta_1 = \tan 15°56'06.6'' - \tan 15°56'18.5''$$

$$= -62.40 \times 10^{-6}$$

$$\frac{\partial m_1}{\partial \theta_2} = D_{AT} \sec^2 \theta_2 = 77.5729 \sec^2 15°56'06.6'' = 83.90$$

$$\frac{\partial m_1}{\partial \theta_1} = D_{AT} \sec^2 \theta_1 = 77.5729 \sec^2 15°56'18.5'' = 83.90$$

The magnitude of these differentials indicates that $(\partial m_1/\partial D_{AT})$ can be ignored since it is small in comparison to the $(\partial m_1/\partial \theta)$ values. In practical terms, this shows that the distance D_{AT} (and hence distance $\dot{D}_{AB}$ and angles TÂB and TB̂A) does not have to be determined with any high degree of precision and that the standard error of the vertical movements is governed by the precision of the measurement of the vertical angles θ.

Since $(\partial m_1/\partial \theta_2) = (\partial m_1/\partial \theta_1) = (\partial m_1/\partial \theta)$ (ignoring any small differences), the standard error in movement m_1 is given by

$$se^2_{m_1} = 2 \left(\frac{\partial m_1}{\partial \theta}\right)^2 se^2_\theta$$

from which

$$se_{m_1} = \sqrt{2} \left(\frac{\partial m_1}{\partial \theta}\right) se_\theta$$

For any movement m

$$se_m = \sqrt{2} D_{AT} \sec^2 \theta \, se_\theta$$

Substituting values from section 15.2 gives

$$se_m = \sqrt{2} (77.5729)\sec^2(15°56')(1.5''/206\ 265)$$

$$= \pm 0.86 \text{ mm}$$

If the vertical angle precision could be improved to, say, $\pm 1''$, the standard error of the vertical movements improves to ± 0.56 mm.

Sections 15.1 to 15.3 demonstrate how a basic deformation survey could be established for monitoring vertical movement. Compared with the examples given, however, many of the schemes that might be used on site would be quite complicated and would involve many redundant observations so that a least squares adjustment and calculation of movement could be carried out. Nevertheless, the examples show how a few simple measurements can produce movement data.

15.4   Horizontal Movement

This is measured using geometrical methods such as triangulation, trilateration, intersection (see chapter 7), or bearing and distance.

Triangulation and trilateration are seldom used in deformation surveys as it is necessary to occupy the point to be monitored. Where *triangulation* is used, a 1″ or 0.1″ reading theodolite is recommended for angle measurement and it may be necessary to measure a baseline with a precise EDM instrument in order to provide scale for the survey. Compared with triangulation, *trilateration* is quicker but distance measurement must be carried out with the best precision possible using such instruments as the Leica ME5000 (see section 5.14), the COM-RAD Geomensor CR234 and the Tellumat MA200. Trilateration is sometimes used in situations where it is difficult to observe angles as might be the case on long lines of sight or where heat haze persists. For all trilateration surveys, the measurement of meteorological conditions at all times is vital and atmospheric corrections must be rigorously applied to all the EDM measurements.

The most widely used methods for monitoring horizontal movement involve intersection or bearing and distance. Unlike triangulation and trilateration, both of these methods do not require the points being monitored to be occupied and, in most cases, remote targets and reflectors are set in place and left unattended for a complete survey.

15.5   Control Networks for Monitoring

One of the most important aspects of an intersection or bearing and distance survey is that a network of coordinated control points is established from which deformation measurements are taken. Because movements are usually small and often approach measuring errors, the specifications for control surveys in deformation monitoring demand that a rigorous network analysis is carried out and demand that great care is taken to record all field data with the best possible precision and reliability.

Two types of horizontal control network are used for deformation monitoring: absolute and relative networks. An *absolute network* is a network in which one or more points are considered to be stable so that a reference datum is provided against which coordinate changes can be assessed. One of the more difficult problems with an absolute network is to identify and then confirm the stability of fixed reference points. A *relative network* is one in which all surveyed points are assumed to be moving and provide no stable datum.

Several methods are available for analysing both types of network, the full details of which are outside the scope of this book. Whatever method is used, the first part of any network analysis involves design in which the precision of control point positions is examined before measurements are taken to ensure the tolerances for the survey are met. This is carried out using information gathered from a reconnaissance which includes approximate values for all the various angles and distances to be measured with their expected standard errors, these being deduced from instrument specifications. When completed, the network design will give the optimum positions for the control stations in relation to the structure to be monitored.

Following the design stage, stations are constructed and are often pillars or monuments of some type with a forced centring device (see section 3.2) of some description fixed to the top for instruments and targets. The OS pillar shown in figure 1.17 is a good example of a permanent control station (or monument) but less elaborate versions are possible. The location of monuments on stable ground is of the utmost importance, especially for absolute networks. Between control stations, a combination of angles and distances are measured using theodolites, EDMs and total stations with precisions dictated by the network design. These will usually be first-order instruments with measuring precisions of 1″ or less for angles and (1–2 mm ± 1–2 ppm) or better for distances, all of which are capable of recording data electronically. All measurements are subjected to a least squares adjustment (see section 7.20) to ensure that precise coordinates are obtained for the control stations as well as assessments of the precision for the position of each station.

As can be seen, the provision of a fully coordinated and adjusted control network for a deformation survey is major task and, as it is also expensive, the need for a survey in the first place must always be carefully examined.

15.6 Intersection

Having established a control network, suitable targets are put in place at selected points to be monitored on the structure or building. As already stated, these are usually permanent and are left undisturbed for the duration of a survey, often over long periods.

Intersection can be carried out by measuring angles to these targets from control stations using traditional 1″ optical theodolites and it is possible to obtain useful information with these. The limiting factor with their use is the speed at which observations are taken, bearing in mind that many points may need to be monitored at regular intervals. One of the reasons for this is the normal practice of taking two rounds of angles on two faces to achieve a reasonable precision and to check results. This does, of course, have to be repeated from at least one other station to complete the intersections and creates a situation where many readings have to be taken.

These problems are overcome by using electronic theodolites with dual-axis compensation (see section 3.5) that eliminates the need to take readings on both faces. By making use of electronic data recording (see section 5.15) and by linking the theodolites to a computer, a further saving in time is possible if coordinates can be calculated and checked instantly. This eliminates the need to take more than one round of intersection angles.

Apart from this saving in time, the advantage of using a computer and electronic theodolite in monitoring is the elimination of calculations which would also have taken a considerable time to complete. Another significant benefit with the use of the computer is that the precision of the results can also be assessed as it is possible to take extra readings to do this. Clearly, intersections or observations that produce unacceptable results can be rejected and remeasured while on site rather than have to return to the site at a later time. Following on from this, if unexpected results are obtained and confirmed, these can be reported much more quickly than by the traditional post-processing method.

Apart from those systems that have been developed 'in-house' for observing and computing intersections with electronic theodolites, a number of commercially made systems are available for measuring intersections, including the Leica ECDS system. This is described in section 15.9.

One of the disadvantages of 'real-time' computer intersection is the cost of suitable equipment and software (the Leica ECDS system currently (1994) costs about £50 000). In addition, two or more observers are required to measure simultaneously and more control points are needed to enable targets to be observed from a number of different directions as the method relies on the measurement of angles only and needs well-conditioned triangles for each intersection (see section 7.18).

15.7 Bearing and Distance

This technique relies on the measurement of angles and distances to monitor movement and overcomes some of the problems of intersection as follows. Firstly, the method is not dependent on control station and target geometry and the control network supporting a bearing and distance monitoring scheme does not need as many stations as that for an intersection scheme. Secondly, only one observer is required to take measurements although a number of observers could be used on a project to speed up the data capture process.

As with intersection, if a large number of points is to be monitored by bearing and distance, the method is only viable if electronic surveying instruments are used together with electronic data recording. It is also usual to link the field instruments directly to a computer, the computer processing 'real-time' coordinates. Most deformation surveys of this type are carried out using total stations or theodolite-mounted EDM systems with the best possible specifications and these are now available with precisions approaching 0.3″ for angles and (1 mm ± 1 ppm) for distance.

A disadvantage of bearing and distance is the limitation of the accuracy possible from the distance measuring component of a combined angle/distance measuring instrument. It is possible to use a first-order EDM instrument to measure distances with precisions approaching 0.1 mm but, at the moment, these have to be used separately from the theodolite and this greatly increases the observation times in the field and creates difficulties with data capture and data processing. Another disadvantage with the bearing and distance method is that targets have to be used for distance measurement as well as angle measurement. This means that targets have to incorporate some form of reflector which can be as simple as reflecting tape at short distances but requires a prism of some sort at longer ranges, both of which are more difficult to install. When monitoring a large number of points, the cost of such targets can become considerable remembering that each target and reflector has to be left in place for the duration of a survey. In addition to the cost problem with reflectors, they also tend to be more obtrusive and are often not capable of being made as vandal-proof as angular targets.

Instrumentation for bearing and distance can include any total station or theodolite-mounted EDM system with appropriate software and computer. Specialist equipment designed to carry out deformation surveys by bearing and distance include the MONMOS system developed by Sokkia (this is described in section 15.10) and the Wild APS System from Leica. The APS (Automated Polar System) is designed as an automatic measuring system for non-contact measurement and monitoring. The two main components of the system are the Wild TM3000 motorised precision theodolite fitted with any Wild Distomat EDM unit as shown in figure 15.5. At the start of an APS survey, all the points to be monitored are intersected using a joystick to drive the theodolite and horizontal angle, vertical angle, distance and point number are recorded for each target in the system computer. When the monitoring is to be repeated, the APS software drives the TM3000 to each target in turn and measurements are taken automatically to the new target positions without any operator assistance other than to initialise the measuring sequence through the computer. The automatic pointings are obtained by detecting the strength and centre of reflected EDM signals and if a target has moved considerably and no return signal is detected, APS invokes a search routine for that target. The precision quoted for the TM3000D version of APS is 15 ppm in target location. This can be improved to 3 ppm using the TM3000V/D.

Figure 15.5 *Wild APS (courtesy Leica UK Ltd)*

15.8 Coordinate Measuring Systems

In recent years, the use of coordinate measuring systems in deformation monitoring has increased. Such systems provide a portable non-contact method of acquiring 'real-time' three-dimensional coordinates of points on structures, often with a high degree of precision. The original idea for using coordinates as an accurate method for non-contact measurement came from the mechanical engineering industry where the control of dimensions in manufacturing processes is necessary. For small objects, on-line dimensional control is provided through the use of coordinate measuring machines but for large components this is not possible. In such situations, coordinate measuring systems are ideal since they can measure the dimensions of large objects and can be taken wherever they are needed. To be of use in an industrial environment, a coordinate measuring system must be capable of determining spatial coordinates with sub-millimetre accuracies and it must also be capable of producing 'real-time' coordinates. Both of these were made possible with the introduction of the electronic theodolite and with parallel developments in EDM and computing.

A number of coordinate measuring systems have been marketed including the Leica ECDS system which is based on two electronic theodolites and uses intersection and the MONMOS system from Sokkia which is based on a total station and uses bearing and distance.

Figure 15.6   *Leica ECDS (courtesy University of Brighton)*

15.9   Leica ECDS Coordinate Measuring System

The Leica ECDS (Electronic Coordinate Determination System) shown in figure 15.6 consists of two or more electronic theodolites linked to a computer for data processing. The theodolites should be of the 1″ category or better since the precision to which angles are recorded is directly related to the accuracy that the system can achieve. Dual-axis compensation is also essential so that single face pointings may be taken.

The technique of theodolite intersection on which ECDS is based is shown in figure 15.7 where two theodolites are set up at convenient positions in front of a structure at A and B. These define an arbitrary three-dimensional coordinate system with a local origin set at the centre of the left hand theodolite A and with the X-axis running horizontally from this so that it passes through the vertical at the other theodolite B. For those surveys where the theodolites occupy previously coordinated control stations, the coordinate system defined by these can be used instead of a local system. If, however, local coordinates have been measured and are required on another grid, these can be transformed from the 'theodolite system' to the 'object system' using transformation software.

Whatever coordinate system is used, the horizontal and vertical circles of the theodolites must be orientated to this. For a local system, this can be

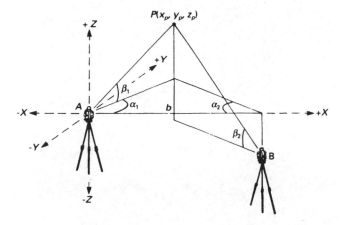

Figure 15.7   *Principle of theodolite intersection*

done by pointing the theodolites at each other (collimating the theodolites) or a *bundle adjustment* procedure can be used. The latter is always better and involves sighting a number of clearly defined points around the structure to be monitored. It is not necessary to know the coordinates of these points and as soon as they have been observed and angles recorded to them, the system computer calculates the relative positions between the theodolites and automatically sets the horizontal and vertical circles to a local orientation. If the coordinates of the bundle adjustment points are known, these can be included in the adjustment procedure and the theodolites will be orientated to the coordinate systems defined by these.

In order to introduce scale into the system, the baseline length b between the theodolites (see figure 15.7) must be known very accurately. Although b could be measured directly, the precision of such a measurement would seldom be acceptable and a different method is used that involves a *scale bar* (figure 15.8). The length of the scale bar is known very precisely and it is made from a material with a low coefficient of thermal expansion such as carbon fibre so that it maintains its length under normal temperature variations. The length of the baseline is determined indirectly from the scale bar by observing angles from the theodolites to targets fixed to both ends of it. Using an estimate for the length of the baseline b', the angles enable the length of the bar s' to be computed. By comparing this with the known length of the bar s, b can be deduced using $b = sb'/s'$. In this way, the precision of the scale bar is propagated into the baseline.

After the system has been initialised, the coordinates of points such as P in figure 15.7 are determined by taking horizontal and vertical circle readings from each theodolite to P to give the horizontal angles α_1 and α_2 together with the vertical angles β_1 and β_2. These are transmitted to the system

Figure 15.8 *Scale bar (courtesy University of Brighton)*

computer which continually calculates the X, Y and Z coordinates of points intersected. Some idea of the precision of each pointing is also given by ECDS and this enables readings that are not within a specified tolerance to be reobserved.

One of the more difficult problems with ECDS is the choice and installation of targets. As far as possible, a target should be omnidirectional, that is, whatever the viewing angle from each theodolite, it should be easy to sight and should always define the same point in three dimensions. Spherical targets are the best form of target and a number of different designs are available from Leica to suit a wide variety of applications. The circular 'stick on' version shown in figure 15.9 is quite popular but this tends to become elliptical if viewed from an acute angle.

The overall accuracy of ECDS depends on a number of factors. Assuming suitable theodolites are being used with angular precisions of 0.5″ or better, the accuracy is more dependent on the geometry of the observations (well-conditioned intersections are required for the best results) and the length of the baseline which should not be longer than 5–10 m. In addition, the position of the scale bar, the stability of the theodolite stands, the type of targets being used, their illumination and the ability of the observers to point accurately to them are important. Taking account of all the possible

Figure 15.9 *Leica 'stick on' ECDS target (courtesy Leica UK Ltd)*

errors, structures of up to tens of metres in size can be measured with accuracies of 0.1–1 mm.

15.10 Sokkia MONMOS Coordinate Measuring System

This system derives its name from MONo MObile 3-D Station and is designed specifically for high-precision three-dimensional measurement. It has three components: the Sokkia NET2 total station, an SDR33 data recorder and a range of special reflective targets.

The *NET2* (figure 15.10) is a high-precision total station capable of measuring angles with a precision of 2″ and distances with a precision of (1 mm ± 2 ppm). At present, the NET2 can be used either as part of the MONMOS system or as a stand-alone instrument.

The *SDR33* (see section 5.15 and figure 5.27) is connected to the NET2 and controls the MONMOS system. When first using the system, the software needed to run it is loaded from the SDR33 where it is retained for subsequent use even if the SDR33 is switched off. The software is menu-driven with the operator being guided through the observing and calculating procedures by a prompting technique. The SDR33 stores data generated by MONMOS which can be transferred to a computer for further analysis.

Some *reflective targets* have been developed by Sokkia that offer an alternative to corner-cube prisms for distance measurement. They are made in

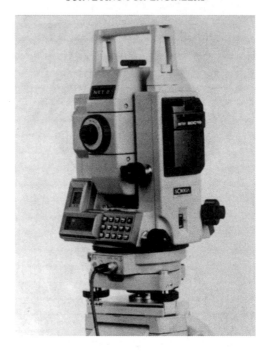

Figure 15.10 *Sokkia NET2 total station (courtesy Sokkia Ltd)*

a variety of sizes from 10 mm square to 90 mm square and are peeled from a sheet and stuck to the structure to be monitored using their adhesive backing. Measurement to them is possible up to angles of incidence as large as ± 45° to the normal and since they are low cost, they can be discarded after use.

The principle on which MONMOS is based is shown in figure 15.11. Two targets on or near the structure to be monitored are chosen to define a coordinate system and, to maintain accuracy, these should be as far apart as possible. In figure 15.11, points 1 and 2 are used to define a coordinate system and points 10 to 15 are to be measured on the structure. The NET2 is set up at any convenient location from which the targets can be clearly viewed (at T in figure 15.11) and there is no need to centre over a ground mark. To establish the coordinate system, the operator first defines the point which is to be used as the origin (point 1 in figure 15.11) and takes readings to it using the NET2. Then the X axis is defined by sighting and measuring point 2. The targeted points on the structure are now sighted in turn and each point is automatically stored in a file by the SDR33 under its own file number with coordinates and a description.

As with the ECDS, the accuracy of MONMOS depends on a number of factors but structures of up to 50 m in size can be monitored to about 1 mm.

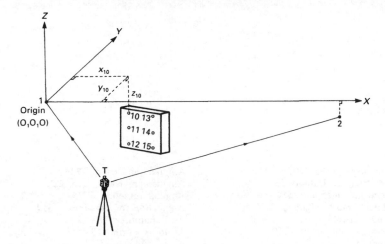

Figure 15.11 *MONMOS principle*

Further Reading

J.F.A. Moore (ed), *Monitoring Building Structures* (Blackie, Glasgow and London, 1992).

Publications of the Building Research Establishment, Garston, Watford. These include reports, current papers, digests and information papers.

Index